THE DISTRIBUTED FIBRE OPTIC SENSING HANDBOOK

THE DISTRIBUTED FIBRE OPTIC SENSING HANDBOOK

John P Dakin
Editor

IFS Publications, UK
Springer-Verlag
Berlin – Heidelberg · New York · Tokyo
1990

British Library Cataloguing in Publication Data

Dakin, John, *1947–*
The distributed fibre optic sensing handbook.
1. Distributed fibre optic sensors.
I. Title
620'.0044

ISBN 1-85423-064-6 IFS Publications
ISBN 3-540-19603-X Springer-Verlag Berlin Heidelberg New York Tokyo

Phototypeset by GCS, Leighton Buzzard
Printed and bound by Short Run Press Ltd, Exeter

Contents

Preface

THIS compendium of published papers covers the field of distributed optical fibre sensing, which involves measuring the variation of physical or chemical parameters along the length of an optical fibre. Clearly, this technology is far more attractive, in both logistic and economic terms, than using a large number of separate point sensors to perform the same task and also allows comparative measurements to be more readily taken as a continuous function along the length of the fibre.

The optical fibre has the main advantages of small size, low loss, electrical isolation and high resistance to chemical corrosion. Not surprisingly, therefore, distributed fibre sensors are receiving increased research attention. In addition, a number of distributed sensors are now becoming commercially available (e.g. the York distributed temperature sensor).

The 27 papers chosen include four general review papers and numerous papers covering more specialised techniques and application areas. The ability to measure such parameters as temperature or pressure, along the length of a fibre cable, has obvious engineering attractions, particularly when the relatively inert and unobtrusive nature of the fibre cable is considered. For example, a distributed temperature sensor can be incorporated into electrical cables, or high-voltage machines, such as generators or transformers, without the electrical compatibility problems associated with conventional metallic-wired types of sensor.

The basic problem to be solved when producing a distributed optical fibre sensor, is to simultaneously provide information, on both the position of measurement and the value of the quantity to be measured. Virtually all of the truly distributed sensors make use of the finite propagation time of optical signals, in order to determine the distance to the measurement point. Many of the methods are optical-guided-wave equivalents of the well-known radar (or, more correctly, 'LIDAR') principle, where a short pulse of light is launched into the fibre and the back-scattered signal is recorded as a function of time, which is equivalent to distance along the fibre. As a result of the high velocity of light, propagation times are usually very short (approximately 5μs to traverse 1km of fibre) and, therefore, it is often necessary to use an optical transmitted signal in the form of a short pulse, combined with very fast optical receivers, in order to achieve the desired distance resolution for the measurement, which may often be of a few metres or less.

The earliest distributed fibre sensor was, historically, the optical time domain reflectometer (OTDR) which used Rayleigh back-scattered light in order to monitor the parameters of the optical fibre itself (See M. K. Barnoski and S. M. Jensen, 'Fiber waveguides: a novel technique for investigating attenuation characteristics', *Applied Optics* Vol. 15, 1976, pp. 2112–2115). As the OTDR is such a well-known instrument, and because it was not originally intended for the monitoring of parameters external to the fibre itself, this early reference of Barnoski has not been included in this text. The reader interested in the concept of OTDR, however, can derive an introductory description from the review by J. P. Dakin, in chapter 1, and, in more detail, from the paper by A. Hartog in chapter 3.

There are also several alternative methods to that of back-scatter measurement. One method, by A. Rogers in chapter 2, uses counter-propagating beams, where a pulse of light, travelling in one direction, produces an external–parameter–dependent gain, in a signals are much stronger than with low-level light-scattering techniques, but care must be taken because use of a non-linear process can make calibration more difficult.

An alternative approach to two-way optical propagation methods, such as OTDR, or the use of counter-propagating interacting beams, is to use fibres capable of supporting two guided modes of different velocity, and arranging for mode cross-coupling to occur, in response to the external influence. The methods of Franks *et al,* and Kurosawa *et al,* chapter 2, both use a frequency-swept source and, in each case, the beat frequency produced when signals are cross-coupled between these modes, and subsequently detected at the far end of the fibre, provides information regarding the position of cross-coupling.

A further method is to use two-way propagation in a fibre-loop and perform a phase comparison of light emerging from each counter propagating path. The phase change for a given (time-varying) path length disturbance in the fibre is found, by Dakin *et al*, chapter 2, to be dependent on the position at which the path length disturbance occurs, and the rate of change of the disturbance.

This volume considers many more individual methods of distributed sensing than covered by the short descriptions above, and also considers, for comparison purposes, a number of methods that might, more aptly be described as 'mutliplexed' or 'pseudo-distributed' forms of sensor.

J.P. Dakin
Senior Consultant
York Ltd.

Distributed sensor systems fundamentals

THIS first chapter consists of review papers, selected for the purpose of providing a general introduction to the subject of distributed sensing. The first, by the consulting editor, covers the aspects of both multiplexed and distributed sensors and, hence, reviews the subject of this volume. However, the reader who is new to the subject of optical fibre sensors might care to start with the second paper by R. S. Medlock, which covers more basic aspects of the methods used for optical sensors. The third paper, by A. J. Rogers, gives a more detailed introduction to the mathematical concepts of several types of distributed sensor. The final paper in chapter 1 provides a discussion of several aspects (such as optical spatial domain reflectometry) not covered in the first three papers.

Multiplexed and distributed optical fibre sensor systems

J.P. Dakin
York Ltd, UK

THE ability to poll a large number of passive optical sensors or sensing regions from a single optoelectronics terminal will greatly enhance the commercial viability of any optical sensing technology. Particularly attractive are systems that permit the monitoring of not only the magnitude of a physical parameter or measurand, but also its variation along the length of a continuous uninterrupted optical fibre. This paper reviews the relatively new, yet rapidly developing, technologies which are leading to multiplexed and truly distributed optical fibre sensors, thereby extending the capability of a single measurement terminal and, hence, making the use of a more complex monitoring station more economically realistic. As with multiplexing in communications, the ultimate information gathering capacity is limited fundamentally by the bandwidth and signal/noise ratio of the detected optical signals, and similar care must be taken to avoid undesirable crosstalk between signals from apparently independent information sources, represented in this case by the individual sensor elements.

For the purpose of the review, it is convenient to define multiplexed sensors as those designed to collect data from a number of discrete sensing points, or regions, and distributed sensor types as those that operate on a continuous length of fibre, yet are capable of determining the variations of a parameter along the length of the fibre as a continuous function of distance.

On an historical note, one of the earliest schemes for sensor multiplexing in optical fibres was reported by Nelson *et al* (1980), where a fibre optic branched network was used in conjunction with an optical time domain reflectometer (OTDR) system capable of receiving and independently monitoring the separate returns from a series of reflective sensors. The earliest form of distributed sensor was the now well-known OTDR instrument. This system utilised an optical radar (or light detection and ranging (LIDAR)) concept to examine the continuity and attenuation of optical fibres, from a measurement of the backscattering against time characteristics when a short pulse of light is launched into a fibre waveguide (Barnoski *et al* 1976, Personick 1977). In spite of the relatively early development of the ODTR concept, it was several years before its use for the measurement of parameters external to the optical fibre was envisaged (Rogers 1980). Rogers suggested, for the first time, a method for the measurement of the spatial distribution of magnetic and electric fields, pressure and temperature using ODTR concepts in conjunction with polarised light sources.

In the following sections, the most significant developments in the technology will be described, with a classification according to the method used to address the various sensing elements. The relative advantages and disadvantages of each of the various approaches are summarised in Tables 1 and 2.

Sensor multiplexing methods

Spatial multiplexing (separate fibre paths)

Although a relatively low technology approach, the technique of using separate fibres to communicate with each separate sensing element is virtually

Table 1 A comparison of sensor multiplexing methods

Multiplexing method	*Advantages*	*Disadvantages*	*Comments*
Spatial multiplexing (use of separate fibre/channel)	*Simple concept. Excellent power budget. Single-channel failure may in some cases be tolerated.*	*Multiple detectors required. High cable and connector cost. Extra weight, volume and less-flexible cabling. Inter-channel comparison of analogue signals may be affected by variations in cable or connectors.*	*Technically less elegant, but simple to engineer. May be combined with TDM or WDM in source/ monitor terminal unit.*
Time-division multiplexing	*Simple electronic decoding. Single source and detector. Can have large numbers of channels/fibre.*	*Requires fast electronics, unless long optical path differences in sensor.*	*Attractive for large number of channels. May be used with single-fibre, dual-fibre or multiple-fibre links. Has been used in both coherent and incoherent systems.*
Wavelength-division multiplexing (optical-domain FDM)	*Low speed electronics. Good optical loss budget. (Same as separate fibre/ sensor except for losses in wavelength MUX/ DEMUX components).*	*Multiple sources and detectors necessary. Difficulties become severe for more than 10–20 sensor channels/fibre.*	*May be used with single fibre, dual-fibre or multiple-fibre links.*
Frequency-division multiplexing of sub-carrier. (Including FMCW techniques)	*Fairly simple decoding for small number of channels. Single source and detector possible, FMCW is an attractive method for high resolution yet DC response.*	*Difficulties increase with large number of channels. FMCW troubled by multiple paths. Also has coherence length limitations.*	*May be used with single-fibre, dual-fibre or multiple-fibre links.*
Coherence multiplexing	*Elegant method when used with a small number of channels. Use of broad-band sources reduces spurious fringes often associated with narrow-band interferometric sensors.*	*Difficulties increase very rapidly as number of sensors increases. Length matching of fibres may present practical difficulties.*	*Of technical interest but likely to have limited application outside laboratory environment.*
Conventional sensors with 'fibredyne' data collection highway	*Uses established sensing technology. Number of sensors mainly limited by frequency management (in case of FDM coding).*	*Sensors not passive – therefore no intrinsic safety advantages.*	*Not strictly an optical sensor, but a communication system well suited for data collection from multi-sensor installations.*

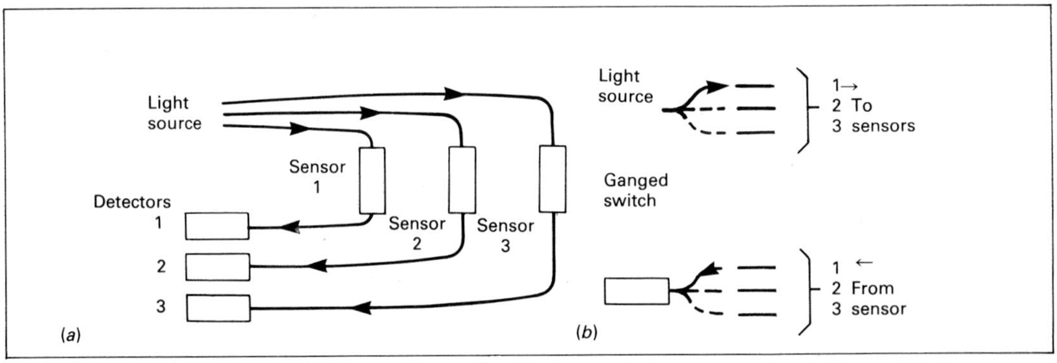

Fig. 1 Spatial multiplexing using separate fibre paths. (a) shows a basic arrangement and (b) a switched arrangement.

Table 2 A comparison of distributed sensing methods

Multiplexing method	*Advantages*	*Disadvantages*	*Comments*
Conventional OTDR methods			
Measurand-dependent attenuation	*Very simple; can use commercially-available OTDR equipment.*	*Relies on uniformity of scattering. It has high losses, unless localised change in measurand causes loss increase over short section only.*	*Very practical method for disturbance location and for other effects (such as ionising radiation) which may cause a high localised loss.*
Measurand-dependent elastic scattering	*Very simple, can use commercially-available OTDR equipment.*	*It has not yet been possible to use method with conventional commercially-available fibres. A large number of factors may limit use or give errors.*	*It has so far only been used with liquid-filled fibres, not convenient for most engineering applications.*
Raman OTDR	*Reasonably simple to engineer. Anti-Stokes/Stokes ratioing gives signal independent of fibre type or composition. Very low sensitivity to undesirable parameters.*	*Provides a much weaker signal than Rayleigh scattering, and therefore has slower response and requires sophisticated electronics.*	*A practical engineering method of providing distributed temperature measurements. Can be used over several km of fibre.*
Fluorescent OTDR	*Could potentially give a much larger signal than Raman or Rayleigh scattering.*	*No suitable low-loss fibres, with required short-fluorescent, life-time and high quantum efficiency, are yet available. likely to have slightly higher fibre losses than for the Raman method.*	*A method with future promise, but hampered at present by lack of suitable fibres.*
Optical frequency domain reflectometry (OFDR)	*A potentially-attractive method for high-resolution measurements.*	*Suffers from a number of practical difficulties.*	*At present, a good method only for qualitative measurements over short lengths.*
Transmissive FMCW	*An elegant and simple technique for disturbance location.*	*Can provide only qualitative measurement of disturbance and has nulls in the response under certain conditions.*	*A potentially-attractive technique for location of lateral strain points in fibres.*
Optical amplification by counter-propagating pulse	*A technically-interesting method for strain monitoring.*	*Will suffer errors if intrinsic Polarisation changes occur in fibre. Will also be highly sensitive to the pump pulse intensity. Uses rather sophisticated equipment.*	*Essentially a laboratory technique, less likely to be used in engineering systems.*

guaranteed to avoid crosstalk (see Fig. 1). As it is the easiest method to implement, it has already been used for several practical applications. Perhaps the most unusual use so far has been the application of 152 separate graded index fibres, each 0.6km long, for nuclear weapon diagnostics. The multiple fibre channels were used to guide light from each individual 'pixel' of an image on a fluorescent phosphor imaging screen to a monitoring area, presumably located at a safe distance! (Thayer *et al* 1981).

For many, more cost-sensitive, applications it is convenient to have a switched or scanned opto-electronic processor unit, which will permit a single light source and detector to be connected to each of the sensing heads on a time-shared basis. A multiplexed system for monitoring fluorescent dye concentration in water, via separate 600m lengths of optical fibre to each sensing head, has been reported (Dakin and King 1983). In this system, a mechanically scanned mirror system was used to inject light from an exciting laser sequentially into each separate sensor head via separate fibres, the returning fluorescent light travelling through separate fibres to a common photo-multiplier detector. This system was capable of monitoring fluorescent dye concentration down to levels of 10^{-10} by weight in water.

Although usually used to address independent sensors, multiple fibres have been used to permit precision digital sensing in systems where each 'bit' of binary coded information is carried by a separate fibre. This technique has been used in simple position-encoding schemes, using binary-coded discs of the type used in conventional optical shaft-angle encoders. (Miller and Lindsay 1978). A more complex system has been described for the 'digital' measurement of temperature, using a series of birefringent crystals, each of length dependent upon the significance of the 'bit' of information (James *et al* 1979). The temperature-dependent birefringence of each is separately monitored via polarising and analysing elements, which convert these changes in birefringence into changes in the amplitude of the transmitted light. The separate outputs combine to produce a binary word, which defines the temperature of the crystals.

Time-division multiplexing (TDM)

The concept of TDM, in combination with high-speed digital technology, has revolutionised many aspects of modern-day communications. Not surprisingly, therefore, it is an attractive choice for multiplexing sensors. The time differences between sensor returns, necessary for a TDM system, must be achieved by arranging differing total optical propagation delays for the signals from each sensing element. In addition, in order to distinguish the signal from each sensor, it is necessary to modulate the light source with a temporally varying signal.

The simplest form of coding, for ease of both modulation and demodulation, is to use a repetitive short pulse, of duration such that returning differentially delayed pulses from each sensor do not coincide at the detector, and with a repetition rate that is low enough to allow a pulse to return from the most distant element of the sensing array before the following pulse returns from the nearest sensing element.

One of the first multiplexing methods suggested for use with hydrophone sensor arrays was of this type (Nelson *et al* 1981). One of their proposed methods is shown in figure 2. This used two fibre optic access-coupled highways, one to distribute a transmitted optical pulse to each transmissive (loss-modulation type) sensor and a second to collect the signals from each individual sensor and guide the returning pulses to the detector.

The second of their proposed arrangements was reflective, using a single tapped highway with reflection-modulation sensors connected to the highway by means of access-type optical directional couplers (see Fig. 2. This was interrogated using a conventional OTDR arrangement, with a semiconductor laser source and an avalanche photo-diode detector. The latter system was tried in practice in a simple laboratory simulation.

More recently, Desforges *et al* (1986) have reported experimental results with reflective sensors, located at regions where the optical fibres are bent, to facilitate non-invasive coupling of light into and out of a continuous fibre.

The conventional technique of OTDR (see later section: Back-scattered sensors using the OTDR concept (general)), used to monitor the loss in optical fibres from the normal Rayleigh back-scatter signature, may be used to monitor discrete loss-modulation sensors of the 'microbend' type, situated along the length of a continuous optical fibre. This technique has been proposed to monitor strain or deformation in civil engineering structures (Asawa *et al* 1982, Bruinsma *et al* 1984) and for non-destructive testing of engineering composite materials (Claus *et al* 1985). However, care must be taken with such arrangements if quantitative results are required, because such microbend sensors (which achieve their loss modulation by periodic bending of the fibre) result in a high degree of mode conversion in multimode fibres, and the loss of each is mode dependent. Thus, there will, in general, be a certain degree of crosstalk in the loss measurements when multimode fibre is used with closely located sensors.

The main disadvantage of a simple pulsed source for multiplexed arrays is the poor optical efficiency due to the low duty cycle, a problem exacerbated by peak-power limitations of semiconductor light sources. A similar problem has, however, already been previously met, and dealt with, in radar systems, where an analogous approach to distinguishing differently delayed signal returns is necessary. (The radar field is one from which numerous techniques, applicable to

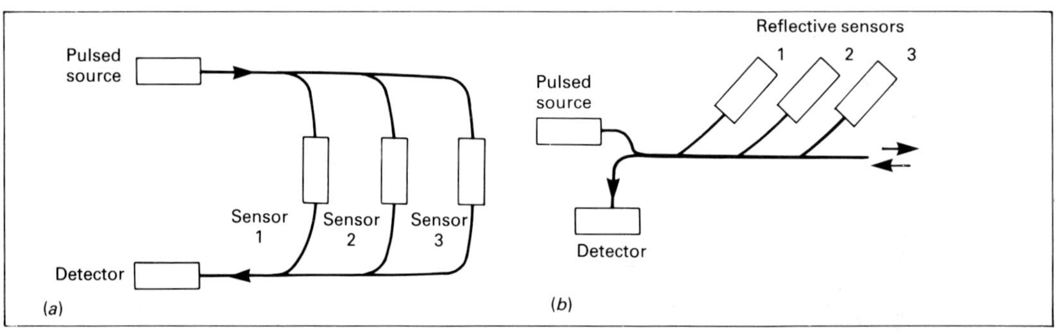

Fig. 2 Time division multiplexing of sensor: (a) shows a transmissive arrangement and (b) a reflective arrangement.

multiplexed and distributed optical sensors, have been drawn.) The radar technique relevant to time-division optical sensor multiplexing is that where a pulse code is transmitted with a 50% duty cycle and with an auto-correlation function corresponding to a periodic pulse train (Skolnik 1970). Simple cross-correlation of the detected signal, with differently phased versions of the code originally transmitted, can then be used to separate the pulses corresponding to the individual sensors.

An alternative means of reducing the peak power requirements for the optical source (also commonly used in radar systems) is to produce a temporally extended 'chirp' signal from an electrical pulse, using a dispersive surface acoustic wave (SAW) filter, and to use this signal to modulate the amplitude of the light source (Nelson *et al* 1981). The detected return signal is then subjected to an inverse transformation by a second SAW filter, in order to reconstitute a replica of the originally transmitted pulse from each reflected signal; the net result from a sensor array, with delays of different lengths, being a time-division multiplexed stream of pulses, similar to those that would be obtained without the SAW filters.

The TDM technique is also attractive when used with interferometric sensors. If an optical heterodyne method is used, where the returning signal is mixed with a reference signal of shifted frequency, the amplitude of the returning signal will, provided it remains constant, have no effect on the phase of the resulting beat signal. Such sensors, therefore, will require no amplitude referencing, as the output of the sensor is represented by the phase of the beat signal. A multiplexing scheme of this nature was first reported for hydrophone applications by Dakin *et al* (1984). This particular implementation (see Fig. 3(a)) involved

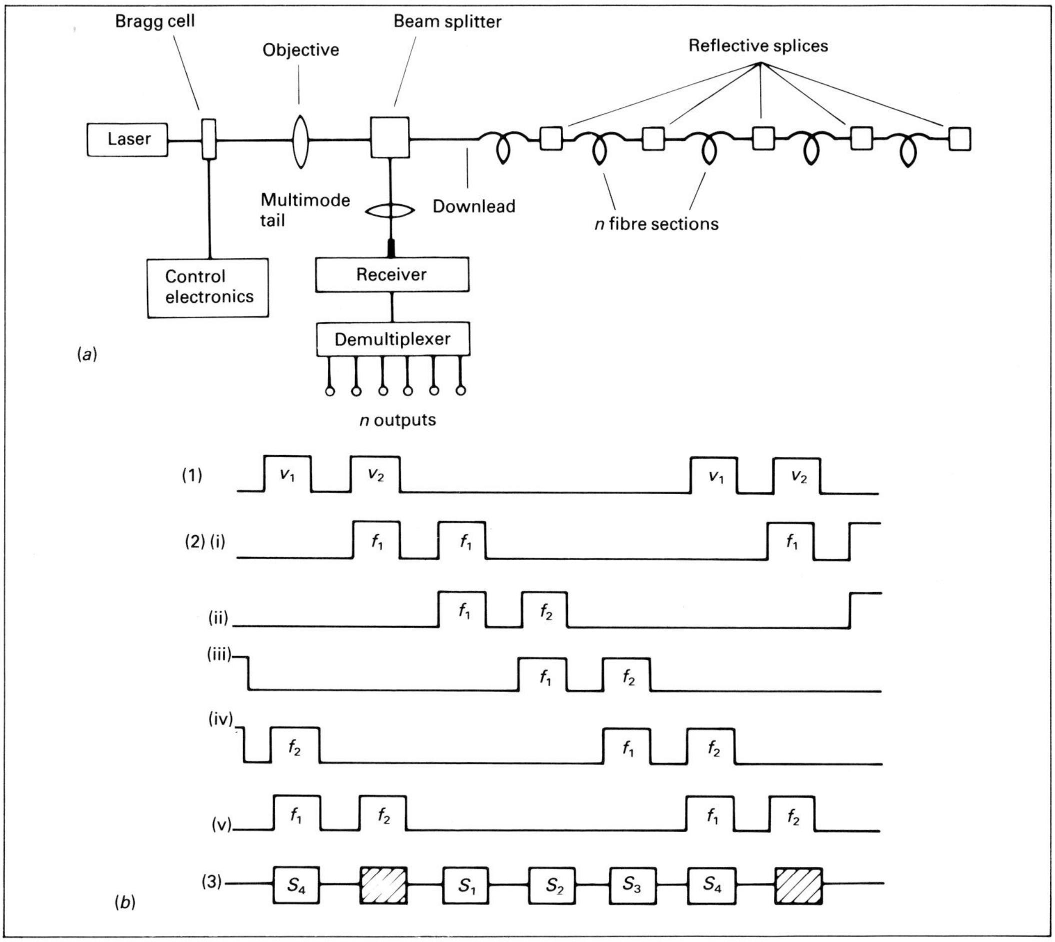

Fig. 3 (a) Diagram of optical fibre hydrophone array. (b) Timing diagram for a four-element array with: (1) the input to the Bragg cell; (2) the reflection from the end of (i) the downlead, (ii) the first sensor, (iii) the second sensor, (iv) the third sensor and (v) the fourth sensor; and (3) the output from the photo-diode

the transmission of a consecutive pair of optical pulses of slightly different frequency into a linear array of interferometric fibre strain sensors, each joined by partially reflective splices. The initial pulses transmitted did not overlap in time, but the returning pulses from adjacent splices gave rise to coincidence of the second pulse, from the nearer splice of a sensor element, with the first pulse, from the subsequent splice (see Fig. 3(b)). Thus, with suitable pulse separation and duration, a time-division multiplexed stream of heterodyne beat signals was obtained from the receiving detector, each heterodyne pulse corresponding to an element of the array, and carrying phase modulation proportional to the changes in optical path length, the latter resulting from insonification of the corresponding fibre sensing element. Phase modulation of each time-demultiplexed channel yielded the acoustic signals, free from any dependence on the amplitudes of the reflected light signals. Using a patented modification (Wade 1982) it is possible to modify the unbalanced interferometer arrangement with an optical pre-delay loop of fibre, thus reducing the laser phase noise contribution.

The attraction of coherent TDM approaches is the improvement in signal/noise capability, resulting first from the coherent detection process and second from the dynamic range improvement, which arises from phase demodulation of the intermediate frequency signal. The improvement in the detection process is greatest when a returning reflected signal is mixed with a strong local oscillator signal, derived at the monitoring station (Dakin and Wade 1982). The advantage of 'differential' sensing of the distance between adjacent reflective splices, achieved by the method shown in figure 3, may theoretically be retained if a recently developed three-wave mixing process is performed (Dakin *et al* 1986). Using this method, the beat signal between two weak received signals should be recoverable, after mixing with a strong local oscillator signal on the detector.

Wavelength division multiplexing (WDM)

The use of WDM, unlike the alternatives of TDM and sub-carrier frequency-division multiplexing techniques, has the advantage that there is no 'theoretical' loss penalty when compared with the single-fibre-per-sensor approach. The method involves guiding optical power to each sensor, and back to a corresponding sensor, via a route dependent on the wavelength designated for the interrogation of that particular channel (see Fig. 4). The path of light to each sensor is directed using WDM coupling components, similar to those used in communications systems. These components are, in theory, lossless, but will, in general, introduce a small loss in practice (typically 1–3dB for each pass), imposing, therefore, a small penalty in an otherwise lossless multiplexing method.

Although the spectral width of the fibre transmission 'window' is potentially enormous, when compared with the very much lower information rate theoretically required for telemetry, the potential of multichannel WDM is more usually limited by practical difficulties in achieving sufficient selectivity in the WDM filters, and by the problems caused by the broad linewidth of LED sources. Using full-width (i.e. unfiltered LED sources, the practical number of independent channels in the fibre window is generally less than ten. Far more channels are possible using 'spectral slicing' techniques with LED sources (Pendleton-Hughes *et al* 1985). This latter technique involves filtering out several separate channels from within the spectral linewidth of the LED a method that increases the losses by a factor at least in proportion to the number of channels required. Another method of achieving a large number of channels is to use narrow-linewidth laser sources. At present, the limit with lasers is set by practical considerations, such as the selectivity of channel filters, the availability, selection and stabilisation of laser sources, and the cost of a multilaser system of this nature.

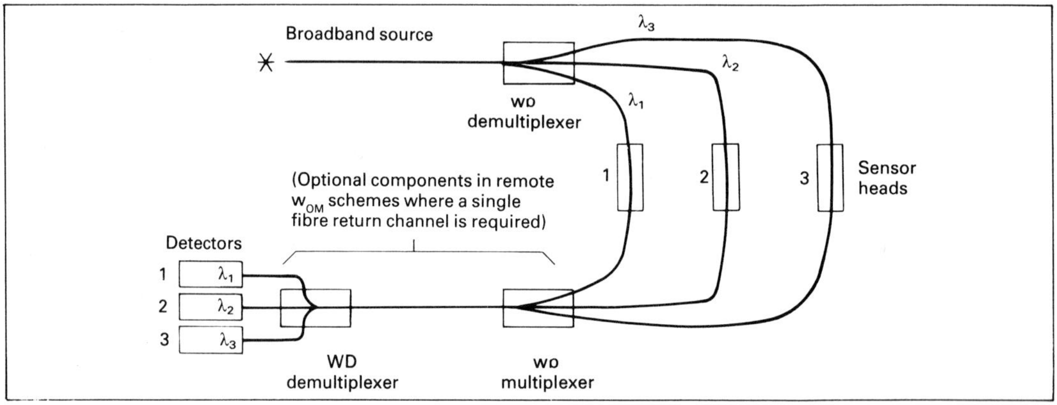

Fig. 4 Schematic diagram of a wavelength division multiplexed arrangement for remote detection of the state of three amplitude-modulation sensor heads.

In spite of the theoretically superior loss budget, the author is unaware of any practical sensor system that has used WDM methods to address a system of independent sensors. This is somewhat surprising, as the technique was suggested for sensing applications several years ago (Jones 1981) and was already a well-known option for communications. However, the WDM approach has been used by the Plessey Research Laboratory to permit the addressing of separate 'bits' in a ten-bit digital sensor designed to measure a 'single' physical parameter, in this case the angular position of a shaft in a fibre optic shaft angle encoder (Dakin and Liddicoat 1982). A diffraction grating was used, firstly in the encoder, to separate a white light source into ten wavelength channels, and again in the decoder to divide signals, according to wavelength, onto separate detectors. The digital encoding was performed by the well-known Gray-coded disc used in conventional optical encoders, but in the WDM version a different colour was used to interrogate each separate coded band on the encoder disc, yet allow transmission through a single optical fibre cable.

The multichannel WDM approach, for providing additional resolution in the monitoring of a single parameter, has also been proposed for remote image monitoring (Lear 1981). In this case, a diffraction grating system was again used in the sensor head, to enable the spatial position of each pixel of a one-dimensional white-light image to be uniquely coded, according to the optical wavelength transmitted to a remote monitor station. At this point, the image is recreated, using a second diffraction grating to perform the inverse transformation.

It is anticipated that the use of WDM systems to address arrays of independent sensors is likely to become more practically viable as WDM systems become more widely accepted for fibre optic communications systems. Until then, the cost and availability of carefully selected sources and filtering components are likely to be factors restricting their development.

Sub-carrier frequency-division multiplexing†

In point sensors, which are connected by separate fibre paths to a common detector, the technique of frequency-division multiplexing (FDM) may be used to separate the outputs from individual channels. The simplest way of using the technique is the relatively trivial method of transmitting signals from separate light sources, each modulated by electrical signals of different frequency, via separate fibres to each of the sensor heads. The outputs may then be combined (or added) into a common output fibre and detected on a common detector, yet still be separable by frequency-selective channel filtering. However, a more elegant FDM approach, using a single LED source, has been devised by workers at the University of Strathclyde in Glasgow (Mlodzianowski *et al* 1986). This method, using a simple transmissive system, is shown in figure 5(a). The three sub-carrier modulation signals add on the detector output, as shown in figure 5(b), with the resultant of their vector addition being dependent on their relative phase angles. These phase angles are a function first of the original phases of the transmitted envelope modulation and second of the different delays they have experienced in transmission through the optical network containing the point sensors.

These signals may be separated, by employing multichannel phase-sensitive detection (using the original modulation signals as reference signals), and subsequently by solving simple simultaneous equations on the resulting scalar quantities, in order to remove any crosstalk terms from light that has passed through the sensor heads of each other channel. Although not mentioned in the above reference, it should be possible, using appropriate lengths of fibre, in conjunction with careful selection of the operating frequencies, to derive independent outputs directly from the phase-sensitive detectors, although clearly this would require considerable care in the design and could be less generally applicable.

The above system has the advantage over TDM methods of using much simpler and slower electronics, and will, for low-frequency sensors, be easy to construct with a low-noise bandwidth in each of the phase-sensitive detection stages. However, although the electronics does not require a high-frequency response, care must be taken to ensure stability of electronic and optical delays, as serious crosstalk could otherwise result.

The frequency-modulated carrier wave (FMCW) multiplexing scheme

This method, like many of the preceding techniques, was adapted from the radar field and has much in common with the FDM method discussed in the previous section. However, in this case the RF 'sub-carriers' are generated at the optical detector by a heterodyne coherent mixing process. The transmitted signal in the frequency-modulated carrier wave (FMCW) method is an optical carrier wave, the frequency of which increases (or decreases) linearly for a period, *T*, after which time it flies back to its initial

†The term frequency-division multiplexing (FDM) is generally used in optical communications systems to imply systems in which an optical carrier is amplitude modulated by a composite electronic signal, consisting of a sum of modulated *sub-carrier* signals with different frequency channel allocations. In order to provide a more precise terminology, the prefix 'sub-carrier' has been inserted in the title of this section, but, for the remainder of this review, the abbreviation FDM will, for convenience, be used to describe such systems.

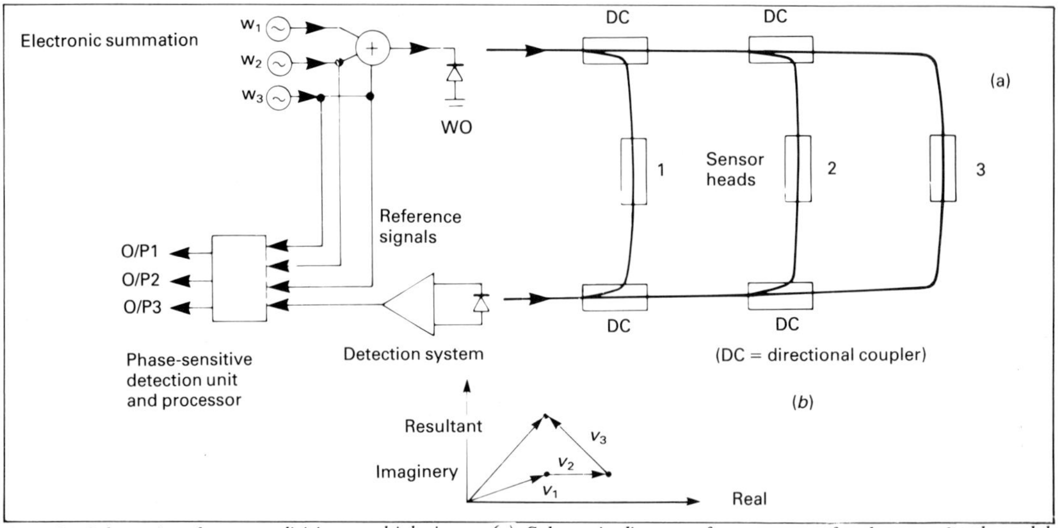

Fig. 5 Sub-carrier frequency-division multiplexing. (a) Schematic diagram of arrangement for three amplitude-modulation sensor heads. (b) Vector diagram showing summation of sub-carrier signals on an optical detector.

frequency, before repeating the process. If the source is connected to an interferometer with a differential path delay, the signals reaching the detector will differ by a frequency proportional to the optical path difference and the frequency slew rate of the source (see Fig. 6). Such a system was first proposed in single sensor form by Uttam and Culshaw (1982); see also Giles *et al* (1983).

If the source is connected via a series of interferometers with different delay paths (Al Chalabi *et al* 1985), a series of RF carrier signals is produced by the heterodyne process at the detector. The various sensor output signals may be isolated by electronically filtering out the corresponding frequencies. The sensor output signals may be represented by either the phase, frequency or amplitude of the recovered RF signals, depending on the scale of the path length change (small path length changes are more readily discernible as phase changes, large path length changes as frequency changes) or the transmission changes occurring in the sensor.

The technique of FMCW for sensor multiplexing suffers from several practical difficulties when used with a large number of sensors. The main problems arise from the large number of optical paths, which can produce a background of unwanted signals from both the expected sensor paths and additional stray paths due to multiple reflections (Sakai 1986). Also, when it is desired to extract precise phase information from the method, problems may arise from non-linearities in the laser frequency modulation ramp (due to the complex laser time constants and also from 'pulling' of the laser frequency by re-injection of retro-reflected light). Other problems arise because the technique requires an unbalanced interferometer for it to operate, and therefore has fringe contrast problems as the path differences increase (due to the finite coherence length of the source). It also necessarily suffers phase-noise degradation due to random frequency fluctuations of the laser source from the ideal sawtooth, as, unlike certain other methods (see previous section: Time-division multiplexing), the phase noise cannot be compensated for by using balanced optical paths. Limits on the permissible path imbalances greatly reduce the range of heterodyne frequencies that may be allocated for multiplexed sensors, a problem that may be exacerbated by non-linearities in the frequency ramp and undesirable amplitude modulation of the source.

However, in spite of some disadvantages, the FMCW technique is conceptually elegant and has the major advantage of being able to monitor steady state changes of the length of an interferometric sensor, simply from the absolute magnitude of the heterodyne frequency.

As a final note it should be pointed out that the FMCW method is not limited to situations in which the optical carrier is modulated. A recent paper by Mallalieu *et al* (1986) has shown that FMCW may also be operated using a frequency-modulated sub-carrier wave (perhaps this method will be termed FMSCW!, thereby removing the need for both coherent sources and monomode fibre, and simplifying the optical system for sensors that do not require the use of optical interferometry to achieve the ultimate in sensitivity performance.

Coherence multiplexing

The technique of coherence multiplexing is similar in

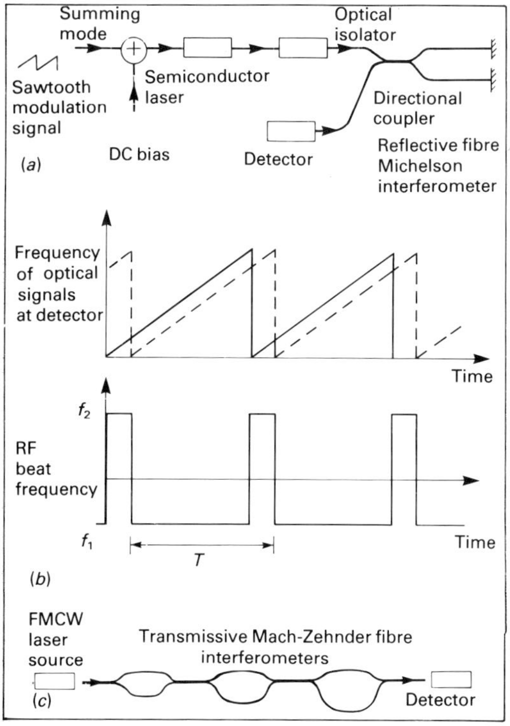

Fig. 6 The FMCW method of sensing and multiplexing. (a) Schematic diagram of the FMCW Michelson sensor. (b) Frequency against time for optical and electronic signals. (c) Multiplexed FMCW arrangement (Al-Chalabi et al *1985)*

concept to a technique used in spread-spectrum communications. A signal, with a superimposed random (or pseudo-random) modulation, may be demodulated to recover information by correlating the received signal with a similar random (or pseudo-random) reference signal. In sensing systems, a broadband source of short coherence length may be used as a transmitted signal as such a source will, in general, be subject to naturally occurring random phase or frequency excursions. If such a signal is guided via two equal, or near-equal, monomode fibre paths, then the signals suffer nearly equal delays, and have a strong correlation, provided the path difference is small compared with the coherence length of the source. Under these conditions, high-contrast interference fringes can be observed if the output signals are mixed on a square-law detector. If, however, the paths differ by very much more than the coherence length of the source the fringe contrast becomes close to zero.

The arrangement in figure 7, first proposed by Brooks *et al* (1985), shows how several remote Mach-Zehnder interferometers may be independently addressed using the coherence multiplexing technique, provided the optical path length differences $l_1 - l_0$, $l_2 - l_0$ and $l_1 - l_2$ are all much greater than the coherence length of the source. The sensing method is based on observing the fringe shifts, which occur as the path differences $l_1 - l_0$ and $l_2 - l_0$, in the sensors 1 and 2, are changed by Δ_1 and Δ_2, respectively, compared with the corresponding fixed lengths $l_1 - l_0$ and $l_2 - l_0$ in the receiver interferometers.

As a means of separately interrogating the outputs from a small number of remote sensors, the method has attractions. However, as the number of sensors increases, the number of possible optical paths increases dramatically and the use of a very short coherence length source, such as an LED, soon becomes necessary. This then presents extreme difficulty in the path equalisation required, in order to ensure adequate fringe contrast. (Use of a laser source with such a system may present potential difficulties due to frequency pulling or coherence enhancement if re-

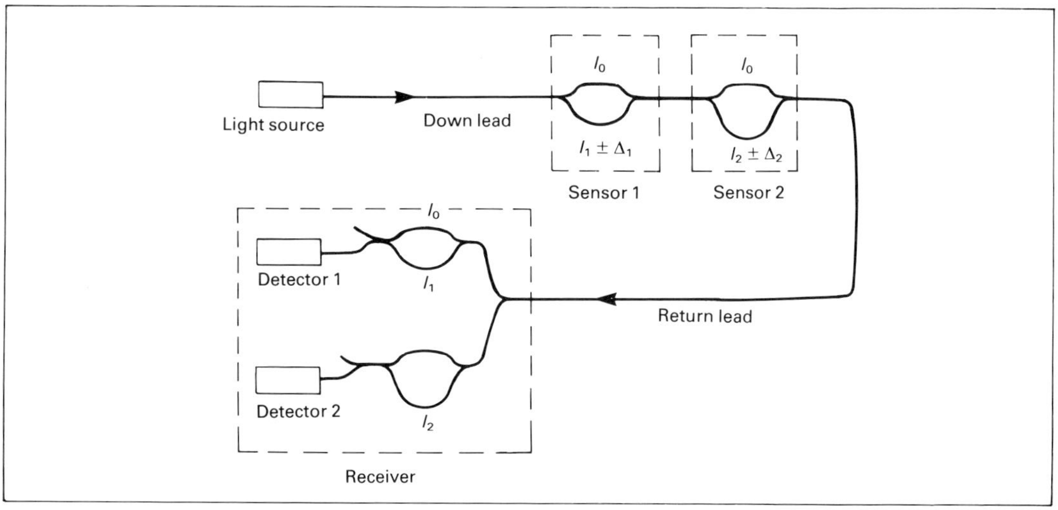

Fig. 7 Schematic arrangement for coherence multiplexing of two remote Mach-Zehnder interferometric sensors.

flections from the sensor chain are allowed to couple back to the lasing cavity). Further practical disadvantages of the system are the need for each sensor to have a unique path length difference (and hence, without careful design, a different sensitivity) and the need for adequate stabilisation of the receiver interferometers, to avoid undesirable errors due to drift in these reference interferometers.

In conclusion, therefore, it appears likely that, although elegant in conception, the method is less likely to find practical application in systems with more than two or three sensors.

Conventional sensors with an acoustic-sensing fibredyne highway

As a final sensor multiplexing method, a hybrid technology will be described, which uses conventional sensors, having electrical outputs, in conjunction with a continuous optical fibre sensing highway. This highway is essentially a long, continuous, acoustic sensor, based on interferometric techniques, to which signals are coupled using piezo-electric transduction elements, in contact with the fibre or cable. In order to distinguish the separate signals applied to the common highway, it is possible to use electrical sub-carrier signals, with a different frequency allocated to each sensor, and to allow the sensor output signals to modulate these carriers, before they are applied to the piezo-electric transducers.

The first version of the method, referred to as a 'Fibredyne' system, was built by workers at University College in London (Kingsley *et al* 1978). The originally proposed version used a Mach-Zehnder twin-fibre interferometer arrangement, in which one fibre acted as a passive reference arm and the second was used as the strain-sensitive highway, which was subjected to an acoustic influence from the piezo-electric transducers

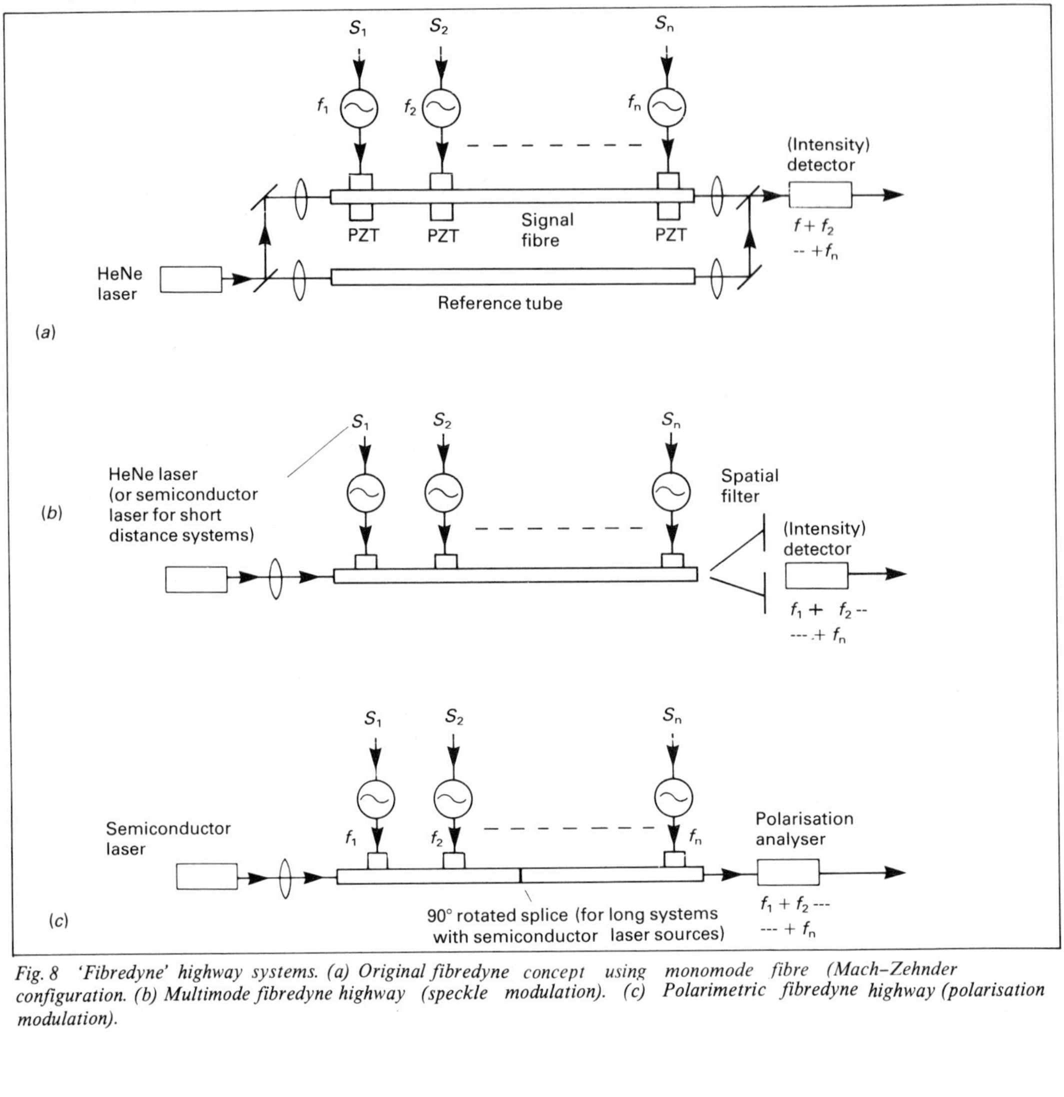

Fig. 8 'Fibredyne' highway systems. (a) Original fibredyne concept using monomode fibre (Mach–Zehnder configuration. (b) Multimode fibredyne highway (speckle modulation). (c) Polarimetric fibredyne highway (polarisation modulation).

(see Fig. 8(a)). The method was initially proposed as a wideband fibre highway and had a high sensitivity, but would, without use of a polarisation controller, be subject to polarisation fading when used with a normal monomode fibre. A later version, from the same research team, used a single multimode fibre and relied on variations of interference 'speckle' patterns as the transducer caused mode conversion in the fibre (see Fig. 8(b)). This method had the advantage of not requiring a separate reference fibre but, as a result of the multiple intermode beating processes involved, exhibited a somewhat lower sensitivity, a much reduced efficiency in the use of available optical energy (only a fraction of the speckle pattern was incident on the detector) and the possibility of multimode fading in addition to the polarisation fading (unless some form of spatial or phase diversity is introduced to ensure there is no fading). The method, however (Culshaw *et al* 1981, Crossley *et al* 1983), was engineered for a sensor telemetry highway, in which, using relatively low-frequency carrier signals ($\sim$1MHz) and an even lower baseband-channel frequency, ($\sim$10kHz; low, yet more than sufficient for most sensor requirements), a number of stations could be fed into the highway. This was carried out acoustically, as before, but in these later systems the acoustic waves were transmitted via the plastic protective sheathing of a fibre optic highway *cable*, the reduction in coupling efficiency being compensated for by acceptance of a reduced information bandwidth.

More recently, a method has been devised using a polarisation-maintaining fibre with lateral acoustic excitation (see Fig. 8(c)), which has the effect of modulating the phase of one polarisation mode, relative to that of the orthogonally polarised mode, resulting in the modulation of the polarisation of the transmitted light (Dakin and Pratt 1986). This method is more efficient in its use of the received optical energy and cannot fade, provided the correct phase bias is maintained between the two received polarisation modes (in order to set the quiescent bias condition to the quadrature point of the sinusoidal response of the polarimetric interferometer).

Distributed sensors

In the second half of this review, attention will now be paid to distributed methods of sensing in continuous lengths of fibre.

Back-scattered sensors using the OTDR concept (general)

All the sensors discussed in this section make use of radar-type back-scattering, or other similar inelastic light generation processes, to make truly continuous

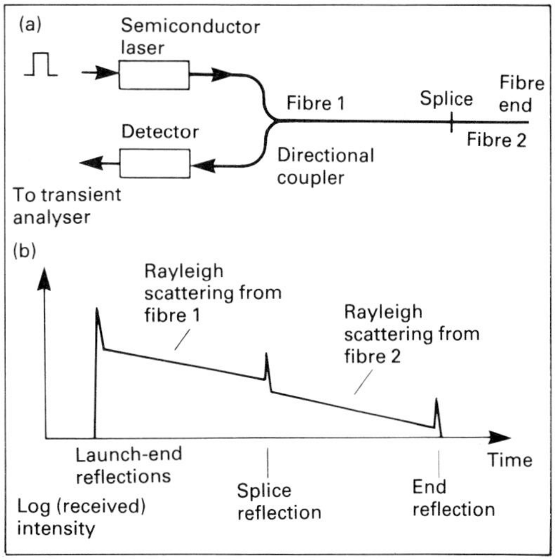

Fig. 9 Basic concept of OTDR: (a) optical arrangement and (b) OTDR returns

measurements on unbroken optical fibres. The basic method of optical time domain reflectometry, devised by Barnoski *et al* (1976) was, as mentioned in the introduction, the first type of distributed optical fibre sensor. The ODTR method is depicted in figure 9, in its most usual current implementation. A pulsed semiconductor laser is coupled into a section of fibre via a directional coupler, which serves also to couple the back-scattered light fraction, captured and returned via the fibre on test, to the avalanche photo-diode detector (APD).

The Rayleigh component of the scattered light represents, for good quality fibres, well over 98% of the returning signal, except during the intervals when more intense specularly reflected pulses return from discrete discontinuities, such as connectors and splices. For uniform fibres, the detected Rayleigh component varies as the product of the launched energy and the bi-directional attenuation between the source and detector, since the scattering coefficient does not vary significantly along the length. As a result, the method has proved extremely useful for measuring spatial variations of fibre attenuation. However, if the geometry or numerical aperture of the fibre varies significantly, changes in the guidance properties of the fibre (primarily in the modal V number) will cause additional variations in the back-scattered signature (Di Vita and Rossi 1979, Conduit *et al* 1981). These workers showed that variations may be compensated for if the OTDR signatures are taken from opposite ends of the same fibre.

In the following sub-sections, several sensing methods using OTDR, and variations based on various inelastic scattering and fluorescent methods, will be described.

Monitoring of variations in attenuation using OTDR

The original objective of the OTDR method was to examine attenuation variations in manufactured and installed lengths of optical fibre.

One of the first suggestions that it may be possible to construct distributed sensors using the attenuation characteristics observed was made by Theocharous (1983), who proposed a method for the measurement of temperature. If an OTDR return signal is differentiated with respect to time and normalised by division by the instantaneous value of the signal, a measure of the fibre attenuation is obtained. If, therefore, a fibre with temperature-dependent attenuation is used, variations of temperature along the length may be monitored.

The technique appears at first sight to be attractive, but a short semi-qualitative analysis suggests that the number of distance-resolution elements is likely to be low. To measure the average temperature within each distance-resolution element, an accurate loss measurement is necessary, in order to determine the smaller temperature-dependent changes in loss. A temperature-dependent loss of at least 1dB in the resolution element (whatever its length) will probably be necessary, in order to accurately detect changes in this loss in a relatively noisy OTDR return. Thus the round-trip attenuation in this example will be 2dB per resolution element, giving a likely limit to the number of measured resolution elements of perhaps ten, particularly bearing in mind the poorer signal/noise ratio from the later sections of fibre.

A means of using OTDR monitored attenuation variations for distributed radiation dosimetry was presented by (Gaebler and Braunig 1983). In this application, a short section of fibre, which was exposed to ionising radiation, suffered excess attenuation, enabling simultaneous detection and location of the radiation exposure. If used as a sensor for the monitoring of exposure over a short section of a more extensive overall length, the quiescent attenuation before radiation is merely that of a normal low-loss communication fibre, enabling the increased attenuation of the short exposed section to be measured with reasonable accuracy.

Several more, loss-modulation effects in fibres are also amenable to OTDR interrogation. These include plastic-clad silica fibre systems, which show increased attenuation when cooled (Pinchbeck and Kitchen 1985), and the commercially available 'Herga' Fibre (available from Herga Electrics, Bury St. Edmunds, UK), which exhibits high micro-bending loss when subjected to lateral pressure, as a result of its novel spiral plastic sheathing arrangement.

Variations in Rayleigh back-scatter characteristics

As already discussed, the use of OTDR to monitor fibre attenuation depends on the constancy of the Rayleigh back-scattering coefficient along the length of the fibre. However, this may vary significantly in two basic ways, even in fibres of uniform geometry and composition. The first form of variability occurs in monomode fibres, using polarised illumination and polarisation-sensitive detection. This diagnostic method, known as polarisation optical time domain reflectometry (POTDR), relies on the high degree of preservation of polarisation exhibited by Rayleigh and Rayleigh–Gans scattered light in silica fibres.

The POTDR method was first suggested by Rogers (1980), who pointed out its potential for distributed measurements of magnetic field (via Faraday rotation), electric field (via the Kerr quadratic electro-optic effect), lateral pressure (via the elasto-optic effect) and temperature (via the temperature dependence of the elasto-optic effect). The first experimental measurements were reported by Hartog *et al* (1980), who used the technique for a distributed measurement of the intrinsic birefringence of a monomode fibre, and Kim and Choi (1981) who measured the birefringence induced by the bending of a wound fibre. Ross (1981) carried out the first measurement of a variable external field, using the Faraday rotation of polarisation as an indication of the magnetic field environment of the fibre. A comprehensive theoretical treatment of the POTDR method has been presented in the specialist paper of Rogers (1981).

The POTDR technique appears to be attractive for the measurement of a large number of parameters. However, its main drawback is, as with many potentially useful sensing methods, the variety of parameters to which it may respond; the sensitivity to strain and vibration being particularly troublesome. In addition, POTDR requires the use of monomode fibres, which can, when used with narrow linewidth laser sources, have particular problems due to coherent addition from multiple Rayleigh back-scattering centres (Healey 1984).

In addition to the variation in the polarisation of back-scattered light, the magnitude of the Rayleigh scattering coefficient may vary with temperature, enabling a simple OTDR arrangement, without polarisation sensitivity, to be used for distributed temperature measurement.

In vitreous fibres, however, the majority of back-scattered energy arises from Rayleigh or Rayleigh–Gans scattering from small refractive index variations. The latter arise from high-temperature thermally-induced density changes, which are 'frozen' into the glass structure as it is cooled from the melt. As a result, the scattered light intensity varies very little with the temperature of the glass. In liquids, however, the scattering arises from real-time thermodynamic fluctuations in the refractive index and, therefore, the scattering shows a significant temperature coefficient.

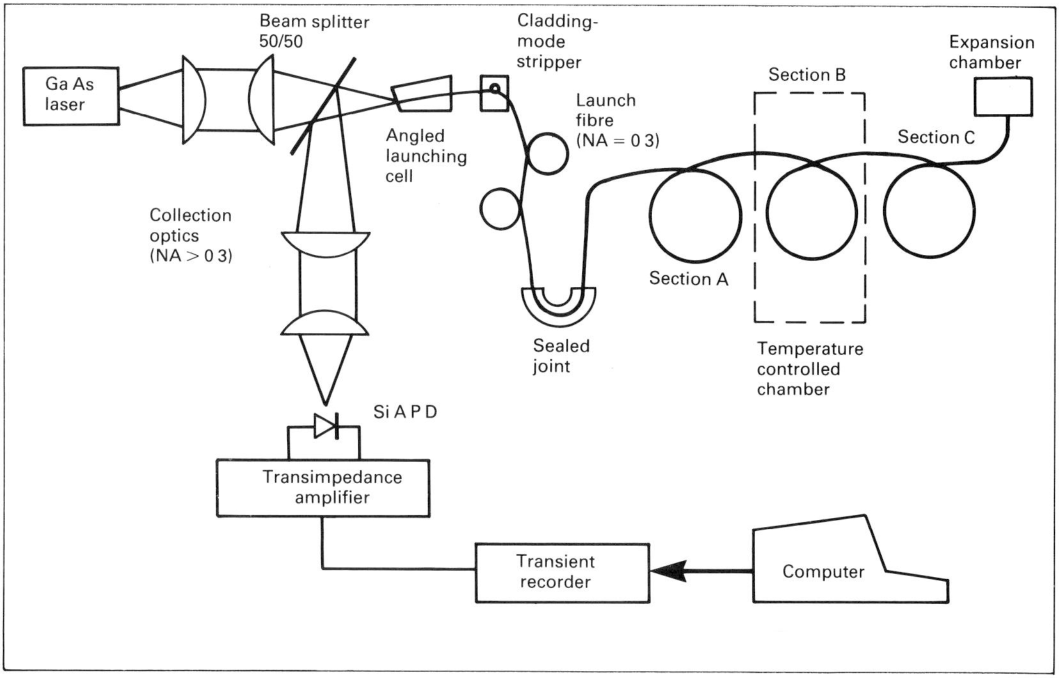

Fig. 10 Rayleigh-scattering temperature profiler using liquid-filled fibre.

This effect has been exploited in the system described by Hartog and Payne (1982), using the apparatus shown in figure 10. The OTDR signals from the liquid-filled fibre are shown in figure 11. The temperature profile may be clearly seen from the measurements, which may be made with a somewhat greater signal/noise ratio than for conventional OTDR measurements on a solid-cored fibre. However, the use of step-index liquid-filled fibres presents difficulties, which, at best, complicate the method and, under some circumstances, restrict its use:

- The use of liquid-filled fibres is inconvenient because of the need for expansion reservoirs.
- There are temperature range restrictions imposed by the freezing and boiling points of the liquid.
- If the central region of a long length of such a fibre is rapidly cooled, it is possible to create voids by rapid thermal contraction, before pressure differentials can refill the tube against the resistance of viscous forces.
- Impurities or dust particles may increase the scattering cross-section or increase the localised loss of the fibre, giving the appearance of false temperature variations in the OTDR return.
- The numerical aperture of the liquid-filled fibre shows a significant temperature dependence, generally reducing as the temperature is raised, allowing hot zones in particular to affect the calibration of results further along the fibre. (In the absence of mode-conversion, this effect may be reduced by the use of mode filters.)

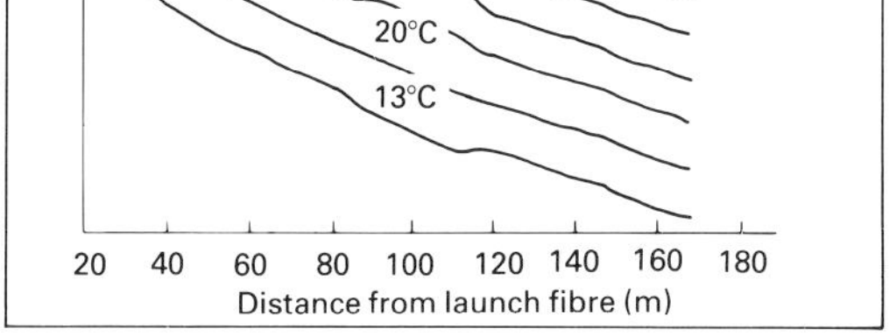

Fig. 11 Results from the system shown in figure 10

However, in spite of the above potential drawbacks, the system worked well over moderate temperature ranges and was capable of monitoring the temperature distribution to ± 0.2° C resolution, over several hundred metres of fibre, with a distance resolution of the order of 2m. This was a remarkable advance, considering that this was the first experimentally demonstrated temperature-profile measuring method for use with optical fibres.

Distributed anti-Stokes Raman thermometry (DART)

If the 'spectral' variation of back-scattering from a germania-doped silica fibre is examined (see Fig. 12), it may be seen that there is a strong central line, primarily due to Rayleigh (or Rayleigh–Gans) scattering, but also containing an unresolved contribution from Brillouin scattering. At each side of the central line, however, there are side-lobes due to Raman scattering. These may be used to detect temperature profiles in conventional vitreous communications fibres using modified OTDR techniques (Dakin *et al* 1985a, b). From standard texts on Raman scattering, the ratio, R, of anti-Stokes (higher-frequency band) intensity at wavelengths λ_a, and λ_s, respectively, with equal frequency separation from the central line, is given by the relationship:

$$R(T) = (\lambda_s/\lambda_a)^4 \exp(-hcv/kT)$$

where h is Planck's constant, c is the velocity of light in vacuo, k is Boltzmann's constant, T is the absolute temperature and v is the frequency of the incident light.

Therefore, a measurement of the ratio of Stokes and anti-Stokes back-scattered light in a fibre should provide an absolute indication of the temperature of the medium, irrespective of the light intensity, the launch conditions, the fibre geometry and even the composition of the fibre. In practice, however, a small correction has to be made for the difference in fibre attenuation between the Stokes and anti-Stokes wavelengths.

The Raman technique appears to have only one significant practical drawback: that of a relatively weak return signal, the anti-Stokes Raman-scattered signal being between 20 and 30dB weaker than the Rayleigh signal. In order to avoid an excessive signal averaging time, measurements have been taken using relatively high launched powers from pulsed lasers. In the first experimental demonstration of the method (Dakin *et al* 1985a), a pulsed argon-ion laser was used in conjunction with Corning telecommunications-grade 50/125μm GRIN fibre (see Fig. 13). The OTDR results obtained are shown in figure 14. The anti-Stokes/Stokes intensity ratio is clearly a strong function of temperature. Figure 15 shows a curve from Dakin *et al* (1985b), where, with a different thermal profile along the fibre, the ratio of anti-Stokes and Stokes intensity has been computed to avoid the exponential decays due to fibre attenuation which are apparent in figure 12. This paper also described preliminary results from a system based on a more realistic semiconductor laser source. A recent result of an anti-Stokes return from this solid-state system, normalised to remove the effects of fibre attenuation, clearly shows the hot and cold regions of fibre (see Fig. 16). The method is believed, with further development, to be capable of measuring to 1K and 1m resolution of temperature and distance, respectively, in a measurement time of approximately 20 seconds.

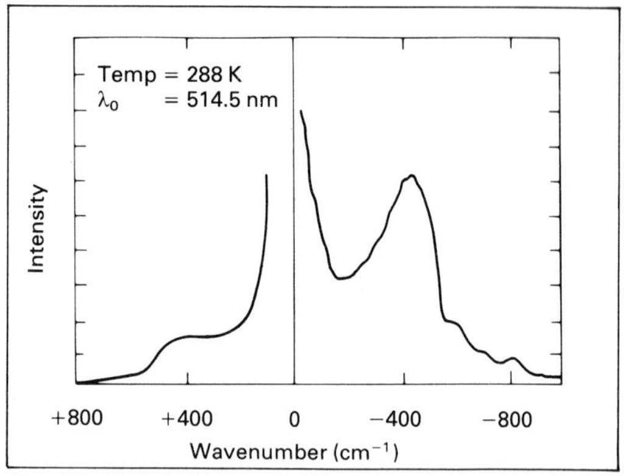

Fig. 12 Raman back-scatter spectrum of germania-doped silica fibre

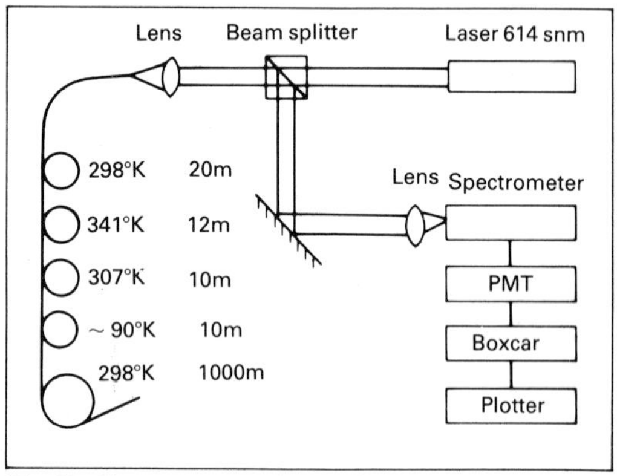

Fig. 13 Raman temperature profile equipment first experimental layout.

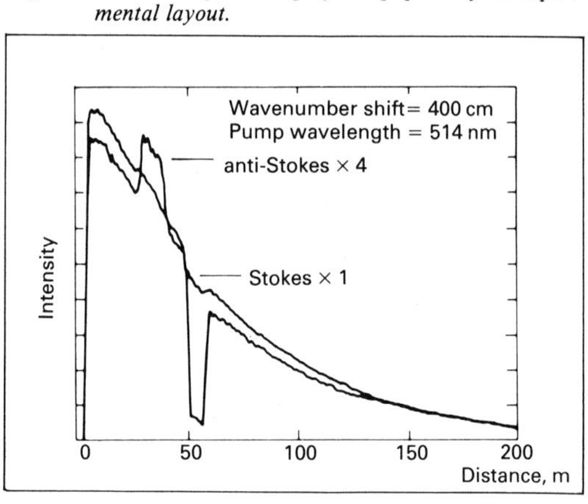

Fig. 14 Raman back-scatter signal (argon-ion laser source).

Time-domain fluorescence monitoring

The re-emission spectrum of most fluorescent materials

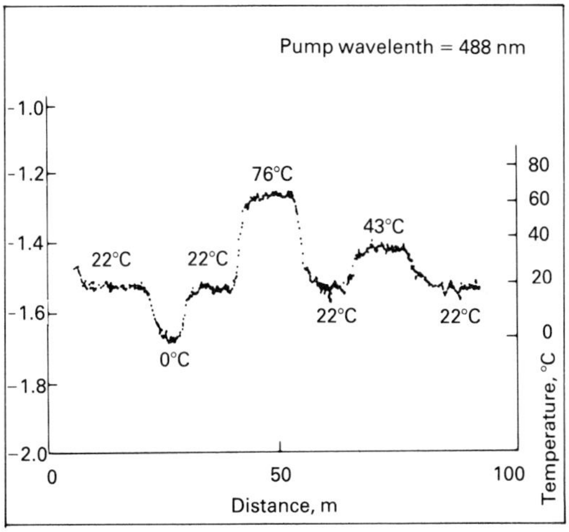

Fig. 15 Logarithm of anti-Stokes/Stokes scattered intensities.

generally exhibits a significant temperature variation. Thus, if an optical arrangement similar to that used for Raman OTDR is constructed, with a laser exciting source as before, with the detector filters now selected to examine regions of the fluorescent decay spectrum having the maximum possible differential temperature variation, a distributed temperature sensor should be possible. The potential attraction of the method, first proposed by Dakin (1984), is that the fluorescent quantum efficiency may be many orders of magnitude higher than that for Raman scattering and higher doping levels may greatly enhance the signals in short distributed sensor systems. However, there remains a problem with the availability of suitable fibres.

Silica-based optical fibres, with appropriate doping for high fluorescent efficiency, have been prepared using rare-earth doping (Poole *et al* 1985), but these are likely to have limited distance resolution due to the long fluorescent lifetimes. Reduction in lifetimes in glasses may generally be achieved by increased coupling to non-radiative processes, but this also reduces the fluorescent efficiency. Polymer fibres are believed, by this author, to offer more promise in short distance systems, as these may be doped with organic dye materials with an excellent combination of high quantum efficiency (~50% or better) and fluorescent lifetimes of the order of only a few nanoseconds.

A later paper (Dakin and Pratt 1986) made a more detailed comparison between distributed temperature sensors based on the techniques of temperature-dependent absorption, scattering, Raman scattering and fluorescence. Although fluorescent doping necessarily increases the loss in the optical fibres, it is found that this is, for short distance operation, more than offset by the much higher fluorescent light levels theoretically attainable.

The optical frequency domain reflectometry (OFDR) technique

This method is essentially similar to the FMCW technique discussed earlier. If an FMCW system is operated in back-scattering mode, in a continuous monomode fibre, the beat signal produced at the detector increases in frequency as a function of the distance from which the light is retro-scattered (Eickhoff and Ulrich 1981, Kingsley and Davies 1985). If the detected beat signal is displayed on a conventional electronic spectrum analyser, the power in each frequency increment represents the scattered light received from a section of fibre situated at a distance corresponding to the frequency offset observed. As the frequency-slew rate of current-ramp-driven semiconductor laser diodes may be very high (100GHz s^{-1} is easily achievable) and the frequency resolution of commercial electronic spectrum analysers is a few Hz or less, the technique has a far superior distance resolution capability than OTDR methods. The second reference mentioned above suggests the

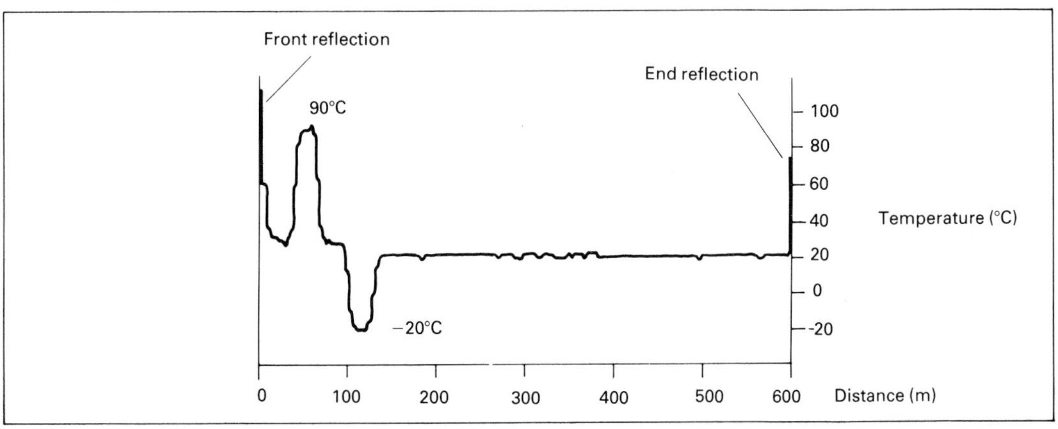

Fig. 16 Recent Raman temperature profile result.

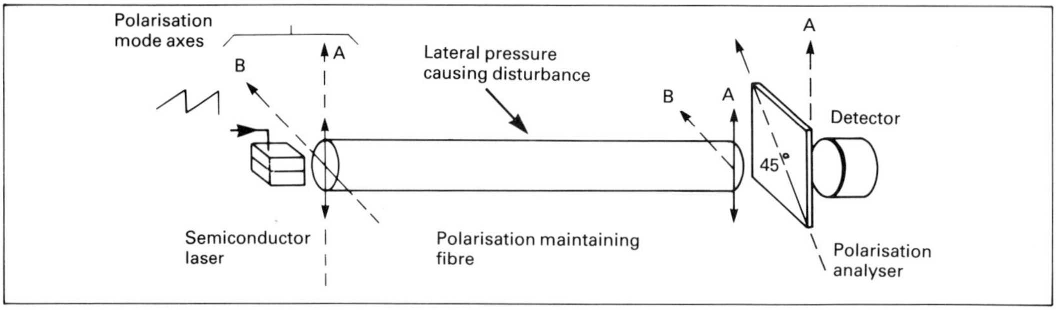

Fig. 17 Transmissive FMCW disturbance location sensor

technique for distributed measurements in integrated optical waveguide circuitry!

However, a major potential problem with optical frequency domain reflectometry (OFDR) is the coherence function of the source, which will modulate the received spectrum and therefore distort any spatial variation of scattering that it is desired to observe. Another approach to the problem uses, as with the FMCW method discussed earlier, a frequency-source (Ghafoori-Shiraz and Okoshi 1986). This removes the problems due to source coherence, but presents a reduced resolution due to the lower frequency-slew rate possible with electronic sub-carrier systems.

Although several publications have reported OFDR results for fibre attenuation monitoring, none have so far used the method for distributed sensing of parameters external to the fibre.

The transmissive FMCW method for disturbance location

The FMCW methods may be used to locate discrete points where mode coupling in a fibre has occurred, provided the fibre is capable of supporting two modes with significantly different phase velocities. External disturbances, causing cross-coupling from an initially excited single mode, will produce, on a suitable detection arrangement situated at the far end of the fibre, a beat signal, of frequency dependent on the distance from the source, at which mode coupling to the second mode has taken place. This approach, first suggested by Franks *et al* (1986), is depicted in figure 17. This particular implementation used a birefringent fibre, allows use of the FMCW techniques, over lengths launched into only one of the two principal polarisation modes of the fibre. The disturbance to be monitored was a lateral pressure, which caused coupling into the orthogonal mode.

A convenient attribute of the technique is that the relatively close velocity matching between the polarisation modes, of even so-called high-birefringence fibre, allows use of the FMCW techniques, over lengths much longer than the coherence length of the source.

Two potential difficulties exist with their approach, however. First, mechanical strains of certain critical magnitudes may cause coupling of power from one polarisation mode to the other and then completely back again, resulting in no net beat signal. Second, disturbances in the direction along a fibre polarisation axis will cause no mode coupling. Otherwise, except for these somewhat unlikely conditions, the technique appears a simple and elegant method of locating the position of lateral disturbances on a continuous fibre.

Distributed sensing using amplification as a result of a counter-propagating optical pump pulse

If an optical signal from a steady carrier wave source is transmitted through a fibre to a detection system, the power level received will be dependent on the total attenuation in the fibre. If, however, an intense optical pulse is transmitted in the optical fibre in the opposite direction, the detected signal is affected by any non-linear gain processes that may be created by the effects of the pump.

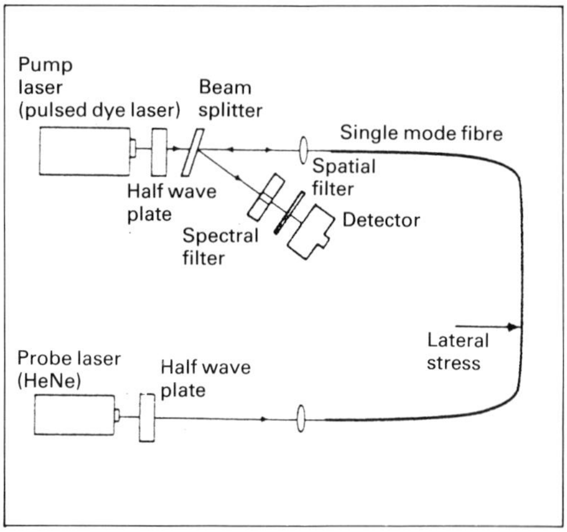

Fig. 18 Experimental arrangement of a system for stress location using Raman amplification by counter-propagating pump pulse.

The first report of such a system was presented by Farries and Rogers (1984). This monomode fibre system used a pulse from an NdYAG-pumped dye laser at 617nm to provide Raman gain in a continuous 633nm transmitted signal from a helium–neon laser monitor source. The Raman gain is very sensitive to polarisation and, therefore, the arrangement (see Fig. 18) in a fibre of low intrinsic birefringence is capable of detection and location of lateral stresses, as these cause polarisation mode conversion modifying the degree of Raman gain.

The technique has strong academic interest, as it is the first of a class of sensors. However, it suffers, in its present form, from a number of practical disadvantages. As it uses convenient optical sources, it will be critically dependent on the pump power level of the dye laser source, and is likely to suffer badly from undesired variations due to environmental factors such as bends in the fibre.

Concluding remarks

From the preceding review it may be clearly seen that, although a relatively new field, the number of techniques for achieving multiplexed and distributed sensors is increasing rapidly. Of course, not all of these techniques will find practical application, and it will be interesting to observe which endure the passage of time to be developed into system instrumentation.

It is already becoming difficult to generalise on preferred approaches, but the time-division multiplexing methods and OTDR methods (including Raman and fluorescent variants) appear to have major advantages for multiplexed and distributed systems, respectively.

The achievement of more sensors per monitoring station, and the correspondingly greater ease of making comparisons of the measurand value at each sensor head, is a factor likely to increase the practical use of optical sensor systems. The major area requiring evolution of technology is the extension of these methods to address (and separate effectively!) several different physical parameters simultaneously on a common optical highway.

Further reading

B. Culshaw, Distributed and multiplexed fibre optic sensor systems. *Proc. 11th Course of the Int. School of Quantum Electronics, Erice, Sicily*. Published by Martinus Nijhoff, 1986.

D.E.N. Davies, Signal processing for distributed optical fibre sensors. *Proc. Int. Conf. on Optical Fibre Sensors, Stuttgart*, 1984, pp. 285–95.

D.E.N. Davies, Optical fibre distributed sensors and sensor networks. *IEE Colloquium 1986/74 Distributed Optical Fibre Sensors*. SPIER, 586, 1986, pp. 52–57.

A.J. Rogers, Intrinsic and extrinsic distributed optical fibre sensors. *SPIE*, 566, 1985, pp. 234–42.

S.A. Al Chalabi, B. Culshaw, D.E.N. Davies, I.P. Giles, and D. Uttam, Multiplexed optical fibre interferometers: an analysis based on radar systems. *Proc. IEE*, 132, 1985, pp. 150–6.

References

C.K. Asawa, S.K. Yao, R.C. Stearns, N.L. Nota and J.W. Downs, High-sensitivity fibre-optic strain sensors for measuring structural distortion. *Electron. Lett.*, 18, 1982, pp. 362–4.

M.K. Barnoski and S.M. Jensen, Fiber waveguides: a novel technique for investigating attenuation characteristics. *Appl. Opt.*, 15, 1976, pp. 2112–5.

J.L. Brooks, R.H. Wentworth, R.C. Youngquist, M. Tur, B.Y. Kim and H.J. Shaw, Coherence multiplexing of fiber-optic interferometric sensors. *J. Lightwave Tech.*, LT3, 1983, pp. 1062–72.

A.J.A. Bruinsma, P. Van Zuylen, C.W. Lamberts and A.J.T. de Krijger, Fibre optic strain measurement for structural integrity monitoring. *Proc. 2nd Int. Conf. Optical Fibre Sensors, OFS '84, Stuttgart* (Berlin: VDE) 1984, pp. 399–401.

R.O. Claus, B.S. Jackson and K.D. Bennett, Nondestructive testing of composite materials by OTDR in imbedded optical fibres. *Proc. Fiber and Laser Sensors III*. SPIE, 566, 1985, pp. 243–8.

A.J. Conduit, D.N. Payne, A.H. Hartog and M.P. Gold, Optical fibre diameter variations and their effect on backscatter loss measurement. *Electron. Lett.*, 17, 1981, pp. 308–10.

S.D. Crossley, I.P. Giles, B. Culshaw, and A.A. Sadler, An optical fibre data telemetry system for use in remote or hazardous locations. *Proc. Int. Conf. on Optical Techniques in Process Control, The Hague* (Cranfield: BHRA) 1983, pp. 121–32.

B. Culshaw, P.R. Ball, J.C. Pond and A.A. Sadler, Optical fibre data collection. *Electronics and Power*, February, 1981, pp. 148–50.

J.P. Dakin, *UK Patent Application GB 2156513A* (priority date 28 March 1984, published 9 October 1985). 1984.

J.P. Dakin and A.J. King, Limitations of a single optical fibre fluorimeter system due to background fluorescence. *Proc. 1st Int. Conf. on Optical Fibre Sensors, London*, (IEE Conf. Pub. 221) 1983, pp. 195–9.

J.P. Dakin and T.J. Liddicoat, A wavelength multiplexed optical shaft encoder. *Measurement and Control*, 15, May, 1982, pp. 176–7.

J.P. Dakin and D.J. Pratt, Improved non-invasive fibredyne highway based on polarimetric detection of strain in polarisation-maintaining fibres. *Electron Lett.*, 21, 1985, pp. 1224–5.

J.P. Dakin and D.J. Pratt, Fibre-optic distributed temperature measurement – a comparative study of techniques. *Proc. IEE Colloq. on Distributed Optical Fibre Sensors* (London: IEE) IEE Digest No. 1986/74, 1986, pp. 10/1–10/4.

J.P. Dakin, D.J. Pratt, G.W. Bibby and J.N. Ross, Distributed anti-Stokes ratio thermometry. *Proc. 3rd Int. Conf. on Optical Fiber Sensors, San Diego* (post deadline paper) 1985a.

J.P. Dakin, D.J. Pratt, G.W. Bibby and J.N. Ross, Distributed optical fibre Raman temperature sensor using a semiconductor light source and detector. *Electron. Lett.*, 21, 1985b, pp. 569–70.

J.P. Dakin and C.W. Wade, *British Patent Application No 8220793* 1982.

J.P. Dakin, C.A. Wade and G. Ellis, A novel 3-wave mixing approach to coherent communications. *Proc. 12th ECOC, Barcelona* (post deadline paper) 1986.

J.P. Dakin, C.A. Wade and M. Henning, Novel optical fibre hydrophone array using a single laser source and detector, *Electron. Lett.*, 20, 1984, pp. 51–3.

F.X. Desforges, P. Graindorge, L.B. Jeunhomme and H.J. Arditty, Progress in OTDR optical fiber sensor networks. *SPIE*, 718, Paper 31 1986.

P. De Vita and U. Rossi, The backscattering technique: its field of applicability in fibre diagnostics and attenuation measurements. *Opt. Quantum Electron*, 11, 1980, pp. 17–22.

W. Eickhoff and R. Ulrich, Optical frequency domain reflectometry in single mode fiber. *Appl. Phys. Lett.*, 39, 1981, pp. 693–5.

M.C. Farries and A.J. Rogers, Temperature dependence of the Kerr effect in a silica optical fibre. *Electron Lett.*, 19, 1983, pp. 890–2.

M.C. Farries and A.J. Rogers, Distributed sensing using stimulated Raman interaction in a monomode optical fibre. *Proc. 2nd Int. Conf. Optical Fibre Sensors, OFS '84, Stuttgart* (Berlin: VDE), 1984, pp. 121–32.

R.B. Franks, W. Torruellas and R.C. Youngquist, Birefringent stress location sensor. *SPIE*, 586 1986.

W. Gaebler and D. Braunig, Application of optical fibre waveguides in radiation dosimetry. *Proc. 1st Int. Conf. on Optical Fibre Sensors, London* (London: IEE) 1983, pp. 185–9.

H. Ghafoori-Shiraz and T. Okoshi, Fault location in optical fibres using optical-frequency-domain reflectometry. *J. Lightwave Technol.*, LT-4, 1986, pp. 316–22.

I.D. Giles, D. Uttam, B. Culshaw and D.E.N. Davies, Coherent optical fibre sensors with modulate laser sources. *Electron. Lett.*, 19, 1983, pp. 14–15.

A.H. Hartog and D.N. Payne, Fibre optic temperature distribution sensor. *Proc. IEE. Colloq. Optical Fibre Sensors* (London: IEE) 1982.

A.H. Hartog, D.N. Payne and A.J. Conduit, POTDR: experimental results and application to loss and birefringent measurements in single mode fibres. *Proc. 6th ECOC, York* (London: IEE) (post deadline paper) 1980.

P. Healey, Fading in heterodyne OTDR. *Electron Lett.*, 20, 1984, pp. 30–2.

K.A. James, W.H. Quick and V.H. Strahan, Fiber optics: the way to true digital sensors. *Control Eng.*, February, 1979, pp. 30–3.

B.E. Jones, Simple optical sensors for the process industries using incoherent light. *Proc. Inst. of Measurement and Control Symp. on Optical Sensors and Optical Techniques in Instrumentation, London* (London: IMC). 1981.

B.Y. Kim and S.S. Choi, Backscattering measurements of bending-induced birefringence in single mode fibres. *Electron. Lett.*, 17, 1981, pp. 193–5.

S.A. Kingsley and D.E.N. Davies, OFDR diagnostics for fibre and integrated-optic systems. *Electron. Lett.*, 21, 1985, pp. 434–5.

S.A. Kingsley, D.E.N. Davies, B. Culshaw and D. Howard Fibredyne Systems. *Proc. Fibre Optic Communications Conf., Chicago* (Information Gatekeepers) 1978.

R.D. Lear, Time dependent recordings of images transmitted over optical fibers. *Proc. Conf. on Fiber Optics in Adverse Environments, San Diego.* SPIE, 296, 1981, pp. 228–33.

K.I. Mallalieu, R. Youngquist and D.E.N. Davies, RF-band and FMCW for passive multiplexing of multimode fibre optic sensors. *Proc. IEE. Colloq. on Distributed Optical Fibre Sensors*, (London: IEE) IEE Digest No 1986/74, 1986, pp. 4/1–4/3.

G.E. Miller and T.A. Lindsay. Feasibility demonstration of fibre optic digital status monitoring devices. *Boeing Aerospace Company (Seattle) Report No D296–10048–1* 1978

J. Mlodzianowski, D. Uttam and B. Culshaw, A multiplexed system for analogue point sensors. *Proc. IEE Colloq. on Distributed Optical Fibre Sensors.* (London: IEE) IEE Digest No 86/74, 1986, pp. 12/1–12/3.

A.R. Nelson, D.H. McMahon and R.L. Gravel, Passive multiplexing system for fiber optic sensors. *Appl. Opt.*, 19, 1980, pp. 2917–20.

A.R. Nelson, D.H. McMahon and H. Van de Vaart, Multiplexing system for fibre optic sensors using pulse compression techniques. *Electron. Lett.*, 17, 1981, pp. 263–4.

S. Pendleton-Hughes, N. Weston and A.C. Carter, Forty channel wavelength multiplexing for short-haul wideband communications networks. *Proc. 11th Eur. Conf. on Optical Communications (ECOC '85) Venice* (Genoa: IIdC) 1985, pp. 649–52.

S.D. Personick, Photon probe – an optical fibre time domain reflectometer. *Bell Syst. Tech. J.*, 56, 1977, pp. 355–66.

D. Pinchbeck and C. Kitchen, Optical fibres for cryogenic leak detection. *Proc. Conf. on Electronics in Oil and Gas, London* (London: Cahners Exhibition Ltd.), 1985, pp. 469–501.

S.B. Poole, D.N. Payne and M.E. Fermann, Fabrication of low-loss optical fibres containing rare-earth ions. *Electron. Lett.*, 21, 1985, pp. 737–8.

A.J. Rogers, Polarisation optical time domain reflectometry. *Electron. Lett.*, 16, 1980, pp. 489–90.

A.J. Rogers, POTDR a technique for the measurement of field distributions. *Appl. Opt.*, 20, 1981, pp. 1060–74.

J.N. Ross, Measurement of magnetic field by POTDR. *Electron. Lett.*, 7, 1981, pp. 596–7.

I. Sakai, Frequency-division multiplexing of optical fibre sensors using a frequency-modulated source. *Opt. Quantum Electron*, 18, 1986, pp. 279–89.

M.I. Skolnik, Radar Handbook (New York: McGraw-Hill), 1970, pp. 20.18–21.

D.R. Thayer, P.B. Lyons, L.D. Looney, J.P. Manning and R.M. Malone, Preparation, installation and calibration of a 152 fiber imaging experiment at the Nevada test site. *SPIE*, 296, 1981, pp. 191–4.

E. Theocharous, Differential absorption distributed thermometer. *Proc. 1st Int. Conf. on Optical Fibre Sensors, London* 1983, pp. 10–2.

D. Uttam and B. Culshaw, Optical FM applied to coherent interferometric sensors. *IEE Colloq. on Optical Fibre Sensors* (London: IEE) IEE Digest No 1982/60 1982.

C.A. Wade, *British Patent Application No 8207961* (priority date 18 March 1982) 1982.

Review of modulating techniques for fibre optic sensors

R.S. Medlock
ABB Kent plc, UK

APPENDIX 1 gives examples of the types of modulating signals used for measurement purposes and most of these have been applied to modulate light in fibre optic sensor systems. However, radiant signals embrace a wider field than that of optics so there are a few examples of certain forms of radiant energy being capable of modulating optical systems, e.g. the Cherenkov effect. In order to examine modulating techniques in a logical way it is usual to classify them as five separate effects resulting in a change of intensity, wavelength, phase, polarisation or time. This classification can be arbitrary depending on how one frames the rules. For example, an interferometric sensor can be regarded as a device measuring either a time delay or a phase lag. Again, a photo-elastic force sensor modulates polarised light but the measurement can be in the form of an intensity signal. In spite of the ambivalence of the classification system it is generally the simplest and the preferred method of analysing modulating techniques.

The characteristics of optical fibres are important parameters in the modulating process. A variety of characteristics and materials can be found in the various types of fibre, examples of which are monomode, multimode fibre with specific birefringent and specially coated fibre. Monomode fibres are used where the modulation requires the maintenance of light coherence, as in the case of time or phase modulation generated in interferometers over useful distances. Multimode fibres are useful for intensity and wavelength modulation. Birefringent fibres are used with polarising techniques. Coated fibres offer opportunities for compound modulation, which means that the coating can provide a primary sensory effect which in turn modulates an optical signal in the fibre by virtue of mechanical extension of strain or strain induced birefringence or both. In addition there are miscellaneous fibres offering specific characteristics, examples being liquid core and scintillation types, and no doubt as sensor technology advances there will be further introductions.

Modulation techniques can be applied intrinsically, i.e. some optical property of the fibre can be exploited for measurement purposes over a length of fibre which can then become a distributed sensor. Altenatively, the fibre can function simply as a conductor of light to an extrinsic sensor which modulates the incoming light and returns it through the same or separate fibre to a means of detection.

Intensity modulation

This was employed in the earliest optical sensor developments. It offers the virtues of simplicity, reliability and low cost to sensors employing one of its many configurations. The analogue applications of this technique have one main drawback: for good accuracy and stability it needs some form of referencing to avoid errors arising from source intensity variations, variable losses in fibres and connectors, and sensitivity changes in detectors. Some clever techniques have been developed to provide referencing but this adds to the complexity and cost. Figures 1–10 illustrate the principles of some members of this family of sensors which range from simple on/off switches through analogue modulating types to digital output devices.

Many sensors of the intensity type are modulated by a small linear or angular displacement. Sensors involving relative displacement of fibres are shown in figures 1 (a)–(d). Shutter types (see fig. 2(a)–(f)) have fixed fibres with collimating lenses and variable aperture devices operated by a primary displacement sensor such as a diaphragm, bourdon tube, thermal expansion element or a turbine type of flow meter. The shutter need not be made of solid material. Liquids in tubes can be used in a similar manner but if the liquid is transparent its function is more complex as its refractive index in relation to that of the fibre, determines the light coupling between the input and output fibres. Figure 2(d) shows a special type of shutter modulator which provides Moiré fringes. Such a system can measure displacement by the method of counting fringes. This avoids the referencing problem but introduces ambiguity in measurement.

One of the earliest fibre optic sensors was the Fotonic*[1] or reflective sensor which can take many forms—multicore, twincore, or single fibre with coupler (see Fig. 3(a)–(c)). The characteristic of these sensors is typically that shown in figure 3(d). Figure 3(e) shows a particular application of reflective modulation in the sensing of the float position of a variable area flow meter. Light reflection has also been used for level measurement. The principle is shown in figure 3(f). As the level changes, light angled on to the surface of the liquid is reflected and moves laterally. This movement can be measured with a multi-element charge coupled device.

A fourth type of intensity modulation relies on losses from the core or cladding. Such devices vary the degree of fibre bending which results in the loss of the higher modes (i.e. the light rays which are internally reflected near the critical angle). A typical example is the microbend sensor[2] which is illustrated in figure 4(a). It can be shown that the loss is a maximum for a given fibre and a small displacement when the bending is applied periodically with a bend pitch x, where:

$$x = C\pi r n (NA)^{-1}$$

C is a core factor = 1.42 for step index fibres (2 for graded index, r is the core radius and n is the core index of refraction. This type of sensor is applied to a commercial vortex shedding flow meter to measure the frequency of the shedding by detecting the small pressure change generated by each vortex.[3] Figure 4(b) shows another type of vortex detector consisting simply of a fibre stretched across a pipe.[4]

Figure 4 illustrates other means of modulating the loss from a fibre. The losses can be from a normal multimode fibre (as with the microbend sensor) or from its unclad core (see Fig. 4 (c)) or from its cladding. Fibres have been made in which the indices of the core

*Trademark of MTI Latham, NY, USA.

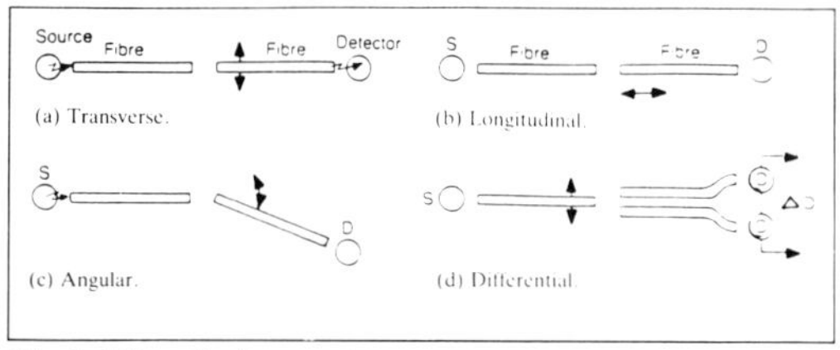

Fig. 1 Fibre displacement sensors

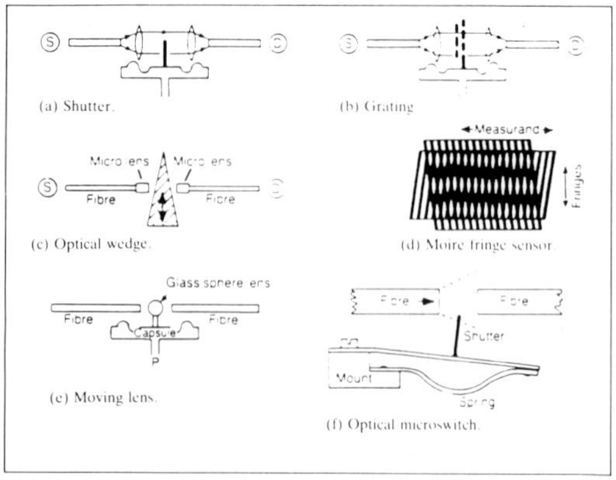

Fig. 2 Shutter modulated sensors

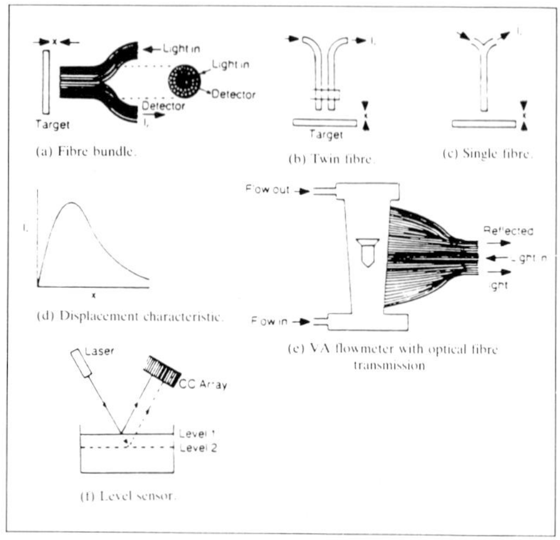

Fig. 3 Reflective sensors

and cladding are nearly equal but have a differential temperature coefficient. If for example the temperature rises, the indices could approach equality with increasing loss from the core.

Conversely, ordinary plastic coated silica fibre can act as a low temperature detector because the coefficient of refractive index with temperature is greater for the silicone cladding than it is for silica and at a temperature below −25°C the cladding increases its index to a value above that of the core with a resulting extraneous loss of light.[5,6]

For measuring temperatures above ambient the role of the core and cladding can be reversed and this can be

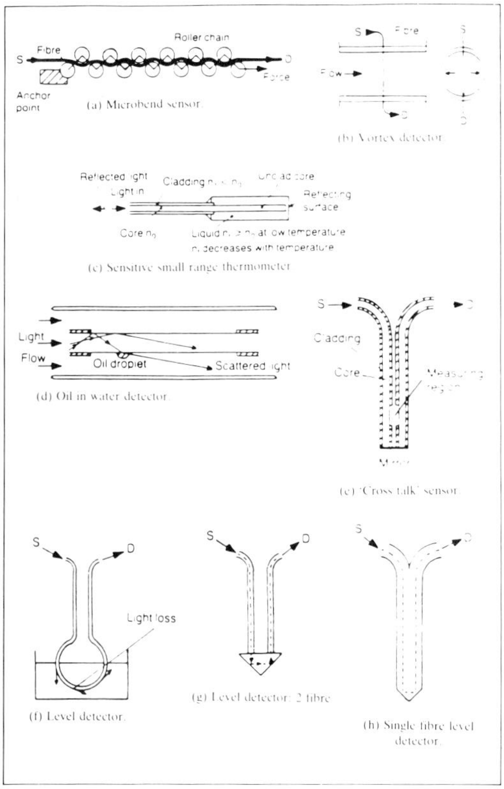

Fig. 4 Fibre loss sensors

achieved by having a liquid-in-glass light guide, in which the liquid has a higher index than glass at ambient conditions but decreases fairly rapidly with temperature. There are not many liquids with the required high index but hexachloro 1,3-butadiene is one that has been used. Other liquids include hexachloropropene and 1,4-difluoro octachlorobutane.[7] Influences other than temperature can similarly alter the index of the cladding. This is the principle of an oil in water detector which depends on the diffusion of oil into the cladding to change the index and allow the escape of light from the core.[8] Oil in water can also be detected by using a fibre core from which the cladding has been removed. By selecting a core with an index similar to that of the oil (but higher than water) the core loss increases as oil dispersions settle on it (see fig. 4(d)).[9]

A similar principle operates in the case of 'cross talk' sensors. These are made by removing short lengths of adjacent cladding from two parallel fibres in contact with each other and bridging the two cores with a transparent material (preferably liquid) which has a high temperature/index coefficient.[10] This device has three applications:

- As a temperature sensor.
- As an instrument for measuring the refractive index of a fluid similar to that of the core.
- As a level sensor in which the 'cross talk' increases as the liquid (which must have an index of more than 1.40) rises up the sensing gap.

The last is suitable as a brake fluid level sensor (see Fig. 4(e)). If the fluid has a low index it would be necessary to operate the device indirectly by enclosing the sensor in a small flexible closed end tube filled with a liquid having the required optimum index.

A simple level detector can be made from a loop of fibre as shown in figure 4(f).[11] When a liquid starts to submerge the loop the higher light modes are lost from the fibre into the cladding and the liquid. A larger attenuation can be achieved by having two fibres (an input and an output) cemented to a 90 degree glass prism (see fig. 4(g)). In air, the light ray is reversed through an angle of 180 degrees by internal reflection within the prism. In liquid, the light escapes from the prism. In another version (see Fig. 4(h)) the principle is the same but a single fibre is used with its end ground to form a 90 degree prism.[12] Light is fed into one arm of a Y coupler and the return light is measured as it emerges from the second arm.

The next class of intensity modulators involves the phenomenon of frustrated total internal reflection (FTIR) or attenuated total reflection (ATR), FTIR, (see fig. 5(a) involves the extraction of energy from the evanescent wave existing near the surface of a prism when light undergoes internal reflection at the prism/air interface.[13] The energy of the evanescent wave decreases exponentially from the prism's surface but for practical purposes can be considered to extend about 1μm. A light absorbing surface brought within this distance, extracts the energy and the reflected beam in the prism undergoes attenuation. Such a system can provide sensitive modulation for displacements of 0 to 1μm. ATR (see fig. 5(b)) operates on a different principle. The modulation is brought about by altering the distance of a reflecting surface such as silver from the prism face.[14] When 'p' polarised light in the prism is undergoing total internal reflection at a critical angle of incidence, reflectivity can vary from

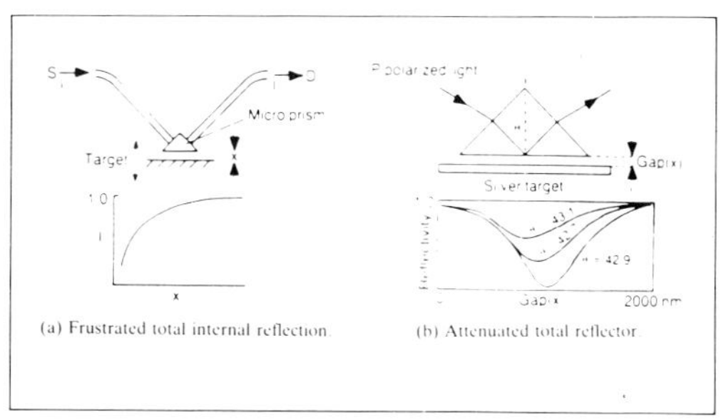

Fig. 5 Evanescent field sensors

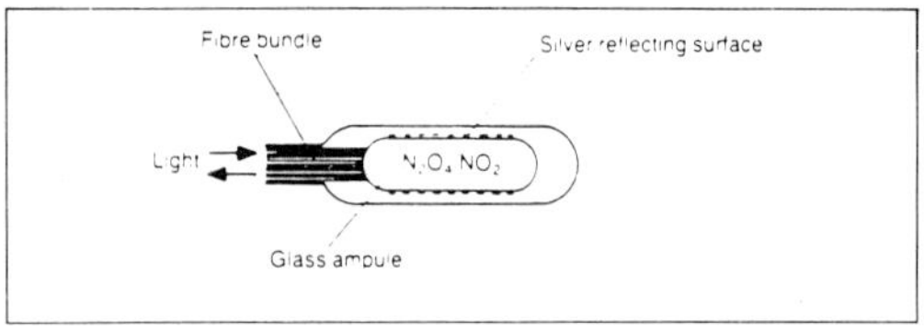

Fig. 6 Absorption sensor

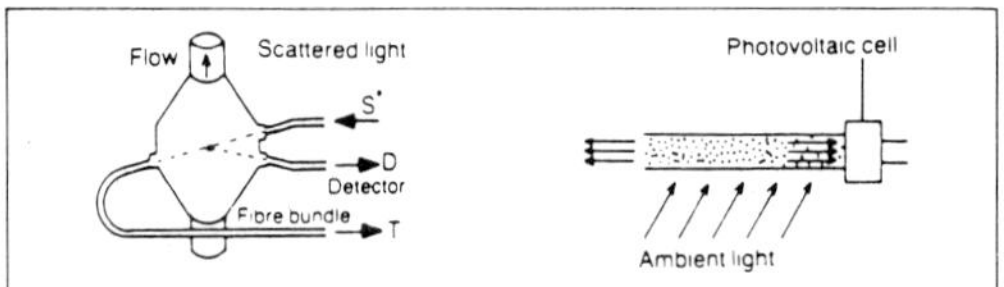

(a) Oil in water sensor. (b) Light intensity sensor.

Fig. 7 Light scattering sensors

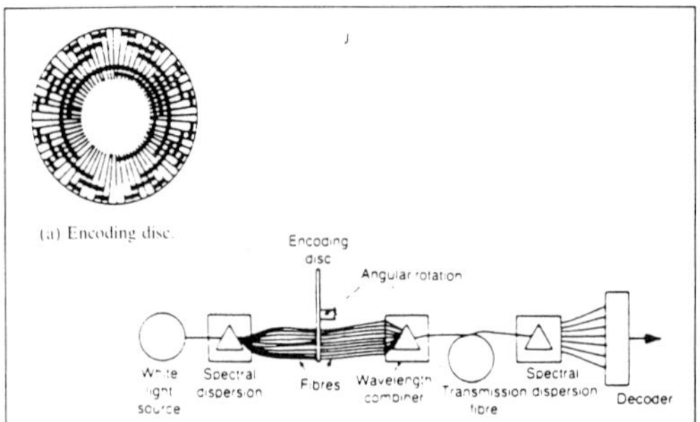

(a) Encoding disc.

(b) Transmission of 8-digit signal on one fibre.

Fig. 8 Digital encoding sensor

zero at a gap width of about 1μm to about 96% for a zero gap.

Intensity modulation by colour absorption has been used for some types of sensors for both physical and chemical parameters. One of the early developments measured the absorption in an equilibrium mixture of N_2O_4 and NO_2 sealed in a capsule at the end of a fibre bundle (see Fig. 6). Absorption of the light in the gas is temperature dependent. This device was intended for temperature measurement of high voltage transformer windings.[15] Turbidity and light scatter are two more parameters which lend themselves to fibre optic measurement by intensity modulation. The presence of oil in water can be measured by the method shown in figure 7(a) which depends on the measurement of light loss or light scatter from oil droplets in the sample.[16] Another application of light scattering operates in an inverse mode in which a special fibre incorporates reflecting, fluorescing or scattering particles from which light emission is propagated in both directions along the fibre to a detector. This technique can provide a distributed sensor for measuring light intensity (see fig. 7(b)). Light emission in a fibre can also be provided by scintillation and by Cherenkov radiation.

When measurement is required in digital form, free from drift and intensity errors, a linear or angular digital encoder can be used. In its simplest form 8-bit encoding will require 8 fibres each being modulated at two levels of light intensity representing 0 and 1 digits. A single transmission fibre system has been devised using eight spectral bands each of which illuminate a separate track on a Gray encoded digitiser. At the receiving end the eight spectral bands are demultiplexed by a diffraction grating and measured by a detector array (see fig. 8(a), (b)).[28]

To round up this section, two other intensity modulated sensors are illustrated in figures 9 and 10.

Figure 9 illustrates a refractive index sensor. Light emerging from a fibre into air produces a cone of light whose half angle is θ. If the emerging light then passes into a medium having an index greater than 1.0 the cone angle is diminished and the received flux in the other fibre increases. The ratio of received flux via the liquid to that via the air is a linear function of the refractive index of the medium.[18] The device illustrated in figure 4(f) can be used for refractive index measurement as well as for liquid level. It was designed for measuring acid level in batteries.[19]

Figure 10 illustrates an integrated optic switch. Coherent light entering a wave guide produced from a titanium diffusion in lithium niobate can be switched electrically by applying a voltage across the electrodes or by surface acoustic wave diffraction.

Wavelength modulation

The variety of wavelength modulating techniques reported in the literature is much smaller than that of the previously described intensity modulations. This is because the demodulation is more complex and often involves a form of spectrometer. However, a spectrometer with its angular dispersion is not always necessary as a simpler demodulation technique can sometimes be used based on optical filtering or two wavelength detection and processing (see fig. 11). Yet there is one big advantage of using wavelength; it is inherently less susceptible to instability of some system components which might cause attenuation in the fibres or connectors.

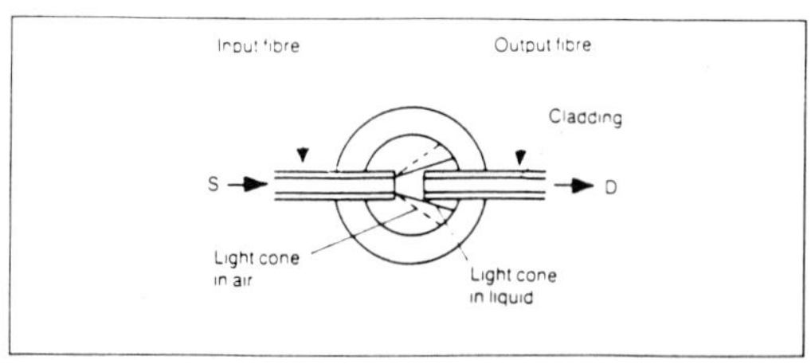

Fig. 9 Density sensor

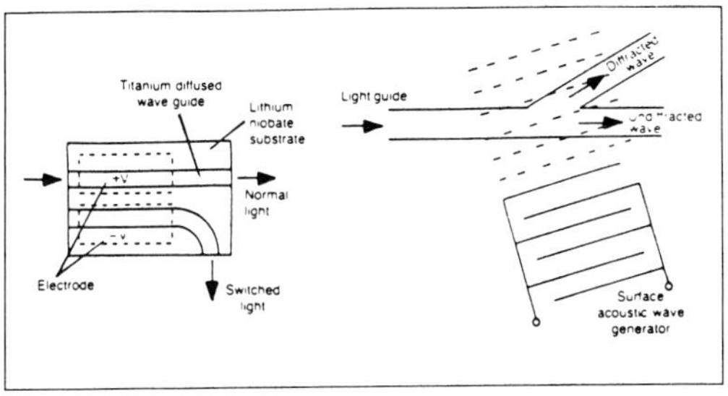

Fig. 10 Two methods of switching direction of light in waveguides

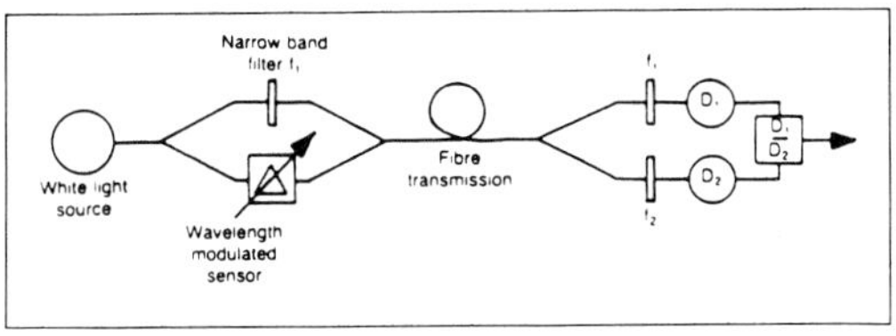

Fig. 11 Two wavelength detection

One of the most direct measurements involving wavelength or colour measurement is that of the remote reading optical radiation pyrometer. This is simply a fibre enclosed in a black body cavity whose function is to transmit the broad band radiation from the cavity to a spectrometric device which can compute the temperature of the cavity by examining two narrow bands in the received spectrum. This technique has been used in a commercial optical fibre thermometer (Accufibre) for which the following remarkable specification is claimed (see fig. 12).[20]

- Range 500°C–2000°C.
- Resolution $7.5 \times c10^{-6}$°C.
- Accuracy 0.0025%.
- Bandwidth 50,000Hz.

Two other modulating techniques have been employed in commercial fibre optic temperature sensors. ASEA has developed a system based on photoluminescence.[21] Light from an LED is transmitted by a fibre to impinge on a crystal of gallium aluminium arsenide in the sensor head. Light is re-emitted at another wavelength which is temperature dependent and analysed by a two wavelength detector system. The Luxtron 'Fluoroptic' sensor operates on similar principles but uses UV excitation of a rare earth phosphor (see fig. 13).[22]

The second technique is the one used by United Technologies who have described a sensor consisting of a neodymium doped fibre which provides a differential spectral absorption characteristic with temperature.[23] In this case the ratio of the responses at two wavelengths is a function of temperature (see Fig. 14).

Wavelength modulation can be obtained by a displacement technique. With the simple apparatus

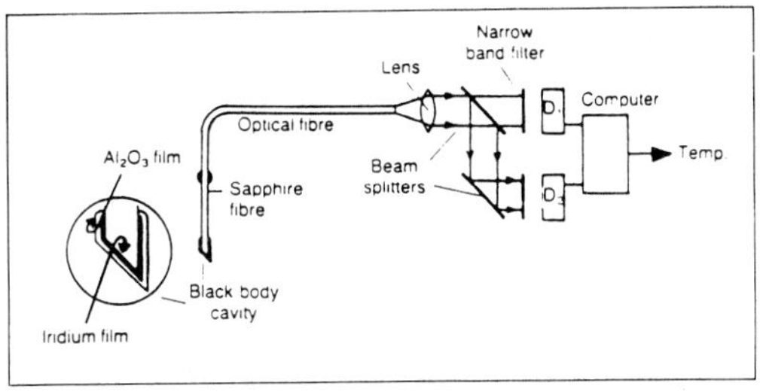

Fig. 12 'Accufibre' optical radiation pyrometer

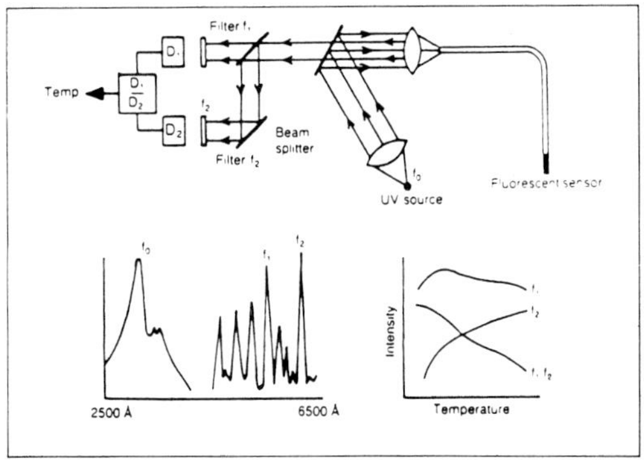

Fig. 13 'Luxtron' fluorescent temperature sensor

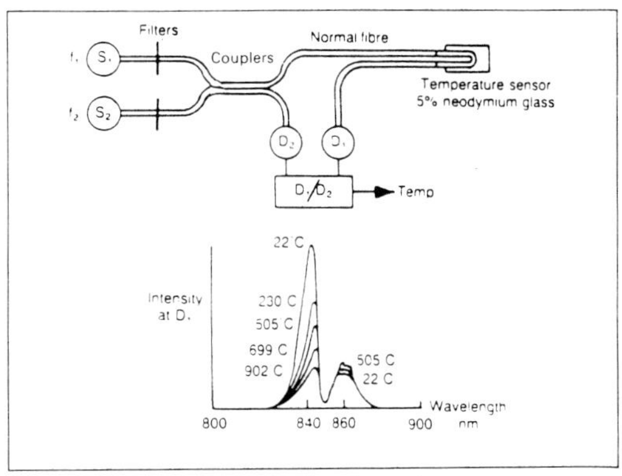

Fig. 14 Temperature sensing by absorption

shown in figures 15(a)–(c), modulation can take place by rotation of a prism or diffraction grating or by linear displacement of the receiving fibre. In an alternative arrangement a zone plate can undergo linear displacement which will alter the wavelength of the light reflected into a receiving fibre.[24,25]

Other wavelength modulating systems do not depend on optical components to produce a spectral pattern. There is a number of materials, liquid, gaseous and solid, which exhibit a colour change under specific physical conditions. Thermo-chromic substances such as cobalt chloride in an iso propyl alcohol/water solution shows a marked colour change between 25°C and 75°C and this can be spectrographically measured and interpreted by a two wavelength detector system (see fig. 16(a), (b)).[26]

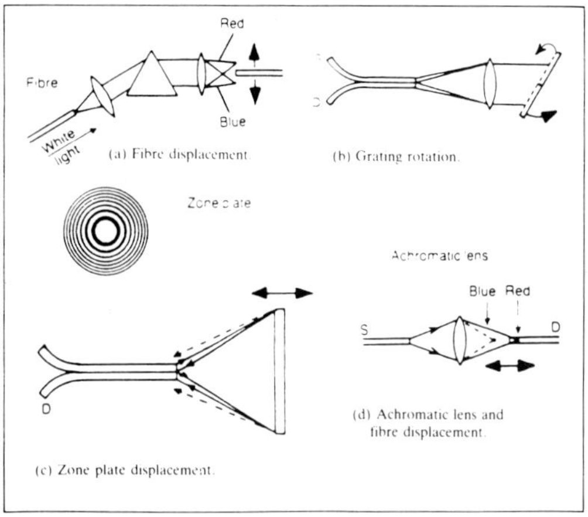

Fig. 15 Wavelength displacement sensors

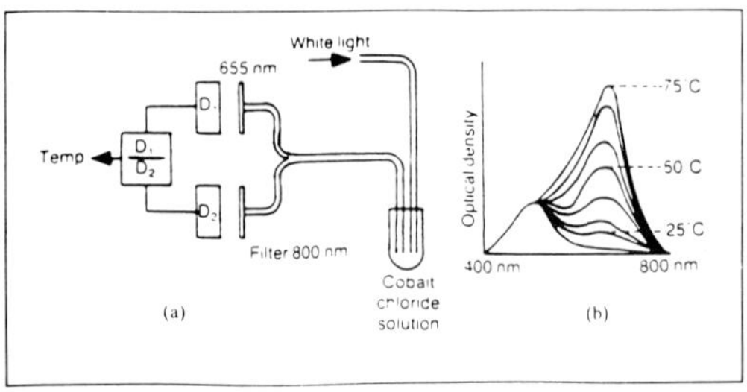

Fig. 16 Thermochromic sensor

Another wavelength modulating technique utilises the parameter sensitivity of filters. Many filters of the birefringent, interference and semiconductor types are temperature dependent and these can be used to modulate light in a fibre optic system. An interesting example of a temperature modulating wavelength filter is that demonstrated by Christiansen.[27] It consists of a fine dispersion of a transparent material in a liquid having a refractive index which varies with temperature and crosses over that of the dispersed material. The indices of the two components equalise at a unique temperature and wavelength and the filter transmits strongly at that wavelength. At another temperature there is another wavelength at which the indices equalise and allow transmission to take place. So, temperature can be determined spectroscopically by measuring the wavelength of the transmitted light when the filter is illuminated by a broad band source.

For chemical analysis involving colour change, the cladding of a fibre can be replaced by one which is wavelength-selective.[28]

Modulation by temperature can be achieved through the wavelength dependence of birefringence in lithium niobate.[29] Essentially a sensor of this type consists of a white light source, a connecting fibre, a lithium niobate crystal sandwiched between crossed polarisers and a grating monochromator with detector (see Fig. 17).

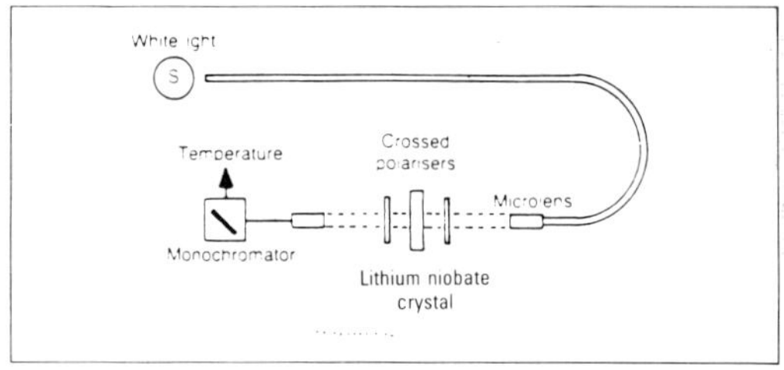

Fig. 17 Birefringent temperature sensor

The chromatic transmission function of the system is given by:

$$I = I_0 \sin^2 [\pi(n_o - n_e)d\lambda^{-1}]$$

where I_o = source intensity, I = transmitted intensity, n_o = refractive index (ordinary ray), n_e = refractive index (extraordinary ray), d = crystal thickness and λ = wavelength.

The modulation occurs because $n_o - n_e$ is dependent upon temperature. The sinusoidal function in the formula means the output modulated signal varies from a minimum to a maximum over a wavelength change of approximately 0.15nm/°C.

Finally in this section frequency modulation can be achieved by the technique known as laser Doppler velocimetry (LDV). This technique uses the Doppler frequency shift which occurs when light is scattered from moving particles in a flow stream and can provide a very large dynamic range (approximately 10^{-6} to 10^{5} ms^{-1}) at a point location. The wavelength of the reflected light is decreased or increased respectively according to whether the direction of incident light has a vector component which is in the direction of flow or in the reverse. The reflected wavelength λ' is then equal to $\lambda_0(1 - u/c)$ where λ_0 is the incident light wavelength, c is the velocity of light and u is the velocity of the particle in the same direction. Laser Doppler technology has become very sophisticated and many designs have been reported using optical fibres which can indicate directional velocity in two and three dimensions. Figure 18 shows the basic system.[30]

Phase modulation

Phase modulation is a technique which generally requires coherent light sources and monomode fibres with one notable exception — the fibredyne system which uses coherent light in a multimode fibre.[31] This produces a speckle pattern image at the output end of the fibre due to multiple phase interactions. Movement or strain in the fibre modulates the multimode light paths and creates pattern changes which can be detected as an amplitude effect. This system is useful

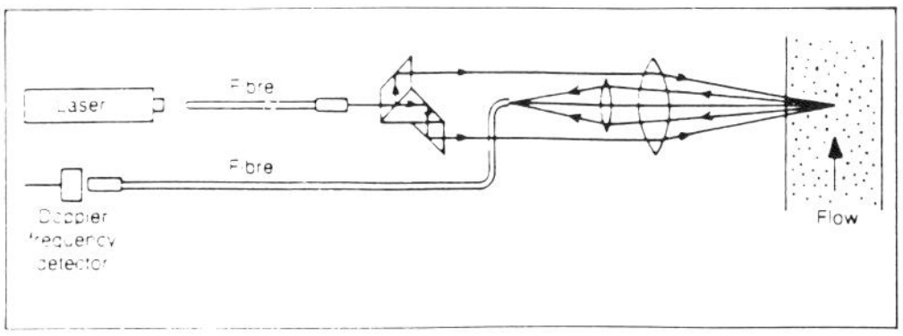

Fig. 18 Laser Doppler anemometer

for dynamic measurements arising from vibrations, accelerations or acoustic effects.

Phase modulation has both good and bad features. Its main advantage is its extremely high degree of sensitivity. According to Culshaw[32] the modulation sensitivities can be 100 radians/meter/°C or 10 radians/meter/microstrain or 10 radians/meter/bar pressure strain. As phase changes of 10^{-4} radians can easily be detected by interferometers (and sometimes change as little as 10^{-8} radians) phase modulation offers one of the most sensitive measuring techniques available to the sensor designer. However, the main advantage of sensitivity is offset by the following problems:

- It is sensitive to several parameters particularly temperature and strain which can swamp the desired measurement of small displacements. Also temperature and strain are not independent variables as temperature changes can cause strain changes.
- As a phase detection system, its output is sinusoidal and for maximum sensitivity and linearity a quadrature relationship is required between the measuring and reference phases. This is technically achievable but adds to the complexity of the system.
- It is difficult to ensure that coherent sources (particularly laser diodes) maintain adequate wavelength stability. Again it is possible with feedback paths to stabilise the wavelength but the penalty is added complexity. A wavelength change can cause a phase change.
- In order to maintain coherency, monomode fibres are required and these by their nature are difficult to couple to light sources.
- Because of the repetitive sinusoidal nature of interferometer outputs and of the extreme sensitivity of the technique, the output signal can be ambiguous unless there is constant referencing to a standard input condition.
- The reference arm stability is crucial for high accuracy and it is difficult to protect this arm from all the variables which can produce phase changes. Temperature and strain are particularly troublesome in this respect.
- Passive interferometers cannot provide a remote reading facility without considerable additional complexity.

Most of these difficulties prevail with a homodyne system (i.e. one without frequency changing) but heterodyne systems solve some of them but not that of remote reading unless it is acceptable to modulate the reference arm by an electrically operated device, in which case the sensor system is no longer a passive one.

There are many forms of interferometer but the main ones are shown in figures 19(a)–(d) and include the Fabry Perot, the Mach Zehnder, the Michelson and the Sagnac. They are shown both in discrete optical component form and as their equivalent optical fibre systems.

Modulation of the Fabry Perot device is by alteration of the distance or travel time between the reflecting elements which can be accomplished by temperature, pressure or strain. Using coated fibres (magnetostrictive or piezo-electric) magnetic and electrical fields can be measured.

The Mach Zehnder and Michelson interferometers are configured with two paths and the parameter to be measured alters the fibre length or the transit time of the light, thus altering the phase relationship with respect to the reference path. This phase shift produces a sinusoidal amplitude function which is not always desirable for measurement purposes; particularly when the signal fades at 180 degrees displacement. Also, as already noted, intensity signals need referencing as they are dependent on the light losses in the system and the drift of optical components. This disadvantage can be mitigated by ratioing the interference signal to the sum of the individual signals in the two paths.

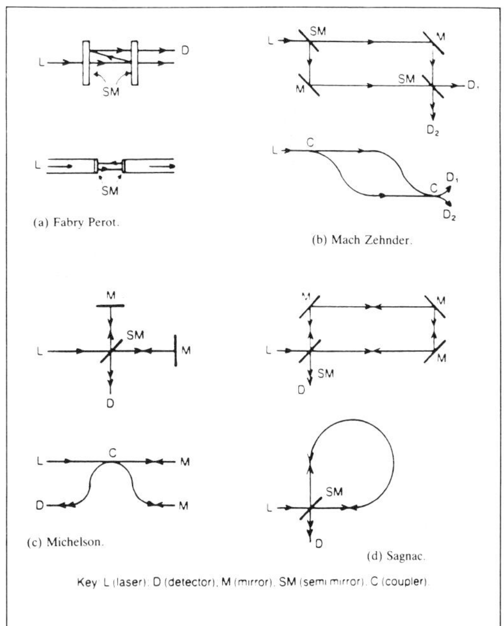

Fig. 19 Interferometers: top line, optical components; bottom line, fibre equivalents

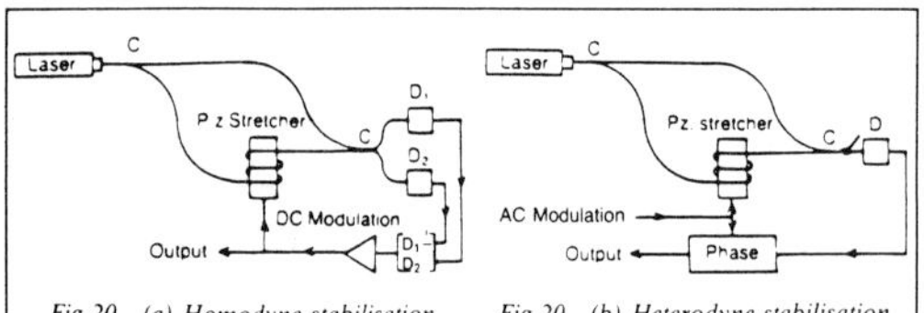

Fig 20 (a) Homodyne stabilisation Fig 20 (b) Heterodyne stabilisation

Fig. 20(a) Homodyne stabilisation
Fig. 20(b) Heterodyne stabilisation

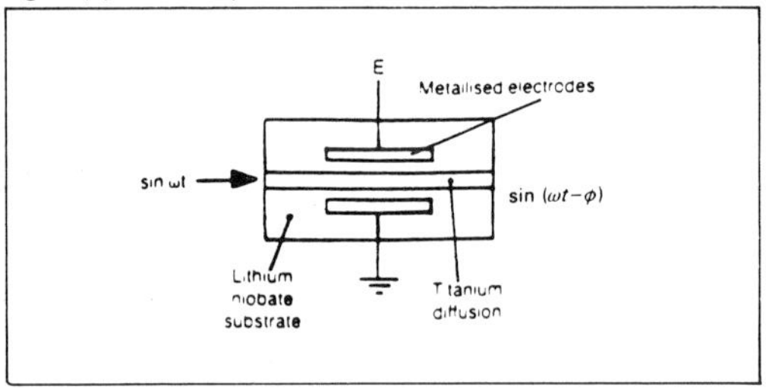

Fig. 21 Electro-optic phase shifter for interferometry

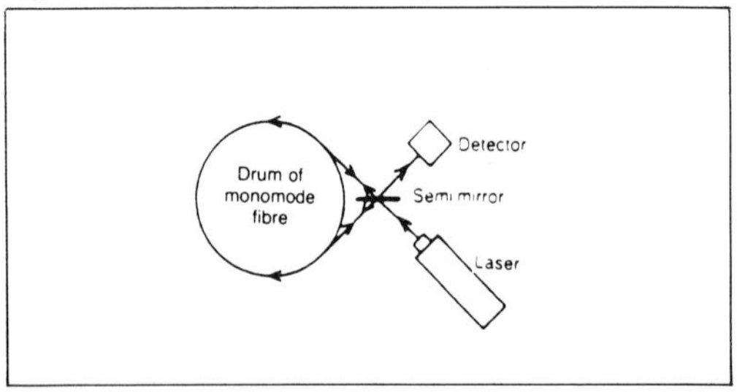

Fig. 22 Basic fibre optic gyro system

The problem of maintaining a quadrature bias (i.e. so that the signals from the two paths are kept at 90 degrees phase difference) is more difficult. It can be done by a feedback system which adjusts the reference path to maintain this phase difference. One such homodyne system operates by having a DC modulated piezo cylinder in the reference arm which can stretch the reference fibre.[33] The other is to modify the frequency of a laser diode source by altering its injection current.[34]

Stabilisation of interferometers can also be carried out by heterodyne techniques. One method is inject a stable high frequency optical modulating frequency (e.g. 10–100MHz) into the measuring arm of the interferometer (see fig. 20(b)). This produces an intermediate frequency whose phase shift can be measured electronically. The optical modulating frequency can be provided by a Bragg cell (which can be expensive) or more simply by a high frequency modulated piezo-electric fibre stretcher. Ramping the frequency output of a laser diode by a sawtooth injection current is also used as a non-ambiguous stabilising method.[35] The measurement can then be made by cross correlation of the detector output and the ramp signal. If a frequency swept laser source is mixed with its own time delayed signal, the frequence difference is a measure of the time difference between two interference beams. This enables an interferometer to be operated passively and remotely without ambiguity.

Integrated optics based on titanium diffusion in lithium niobate can provide a waveguide whose index of refraction can be changed by applying an electrical field E in accordance with the formula (see Fig. 21):

$$\Delta n = 0.5 n^3 r E$$

where r is the electro-optic coefficient.

Such a device could be incorporated into the reference arm of an interferometer and connected into a feedback circuit which could maintain a stable quadrature phase relationship in the same way as the DC operated piezo fibre stretcher.

Most applications of interferometers involve either temperature or strain measurement. Sometimes the strain signal is applied directly, in other cases indirectly, as for example by the effect of differential expansion with temperature or by the effect of static pressure on a compliant coating attached to the measuring fibre. Other parameters such as magnetic and electric fields can be measured interferometrically by using appropriate coated fibres.

The sensor which has received most attention in the fibre optic field is the fibre gyro (see fig. 22). Light from a coherent source is fed into a coil of monomode fibre (preferably a single polarisation fibre) from both ends in two counter-propagating beams. The length of fibre is typically 0.5 to 1.0km. The arrangement is basically a Sagnac ring interferometer. When the coil is rotated one beam stays longer in the coil than the other which leads to a phase difference between the two emergent beams. Various techniques, as described earlier, are used to measure the phase modulation due to rotation. The gyro can be extremely sensitive — a detection figure of 0.022 degrees/hour has been reported — but the design and stability problems are much greater than might be expected from this simple explanation of its operating principle.

Another interferometric sensor is the hydrophone in which acoustic energy creates displacement or strain in the measurement arm of an interferometer.

Polarisation modulation

Polarisation modulation offers great scope for variety and sensitivity, partly because there is wealth of polarisation phenomena to draw upon, and partly because of the recent introduction of monomode fibres with special polarisation characteristics namely the 'polarisation-maintaining' fibre, the 'single polarisation' fibre and the 'ultra-low birefringence' fibre. The latter fibre has an overall retardance of less than 1 degree at 633nm for any length.

The first of these is a highly birefringent fibre. The

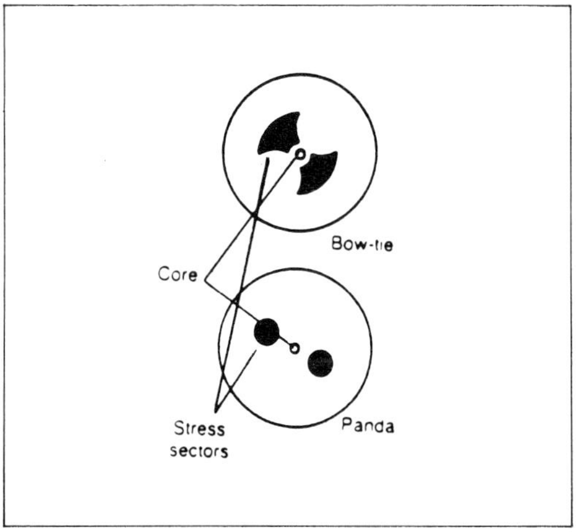

Fig. 23 Polarising fibres

birefringence is the result of anisotropic thermal stress between the core and the stress-producing bow-tie side-pits, due to differential thermal expansion coefficients.[36] Two types are shown in figures 23(a) and (b).[37] Such fibres can maintain linearly polarised light because of the high value or birefringence which masks the effect of external bends and twists.

Single polarisation fibres are similar to the polarisation-maintaining fibres except that one polarisation state suffers a high attenuation (50dB/km) while the other has a low loss (5dB/km). The difference in the loss is due to the fact that one state is strongly affected by bends or external forces.

The availability of high birefringent fibre has made possible the development of remote polarisation sensors, i.e. the sensor and the signal processing electronics can be remote from each other whereas with conventional fibres additional birefringent effects could take place in the fibre connecting the optical and electronic components. These effects could be variable depending on strain and temperature changes in the connecting paths.

Single polarisation fibres can be used as intrinsic sensors as the birefringence changes with both longitudinal strain and temperature. Such fibres are also invaluable for altering the state of polarisation when subjected to electrical and magnetic fields. However, polarised light passing through conventional fibres can also have its state of polarisation modulated by external electrical and magnetic fields. Alternatively, the modulation can take place in an electro-optic crystal such as lithium niobate connected to conventional fibres as light guides.

There are several polarisation effects, viz:

- The Faraday magneto-optic effect. A magnetic flux in the same direction as a beam of linearly polarised light in a fibre cases a rotation equal to:

$$V \int_0^L H dl$$

where V is the Verdet constant (for a given material, wavelength and temperature), H is the magnetic field and L is the optical path length.
- Electrogyration effect. This is the electrical counterpart of the magneto optical effect. It occurs when an electric field acts upon certain transparent crystals, such as quartz, carrying linearly polarised light. The rotation is proportional to the electrical field E multiplied by the optical path length L.
- Kerr effect. This occurs when linear polarised light passes through an isotropic medium (of which there is a great variety) which is subject to an electrical field, thereby creating a linear birefringent property in the medium. If the light is polarised at 45 degrees to the field direction it will emerge elliptically polarised. The phase retardation S is given by the expression KE^2L. K is a constant for a given wavelength, material and temperature (E and L as before).
- Pockels effect. This is similar to the Kerr effect but the phase retardation is a linear function of the electrical field and the effect is only exhibited in a few crystalline materials.
- The photo-elastic effect. Certain transparent media become birefringent when one axis is strained by a mechanical force. Media such as araldite and polyurethane exhibit this property strongly but there are many other materials having similar properties. When these materials are placed between crossed polarisers and stressed the birefringence varies the intensity of the light from the analyser in a $(\cos)^2$ relationship. This effect has been used in the sensor shown in figure 24.[38,39]
The limited measurement range which occurs with monochromatic light due to the cyclic $(\cos)^2$ relationship and the disadvantages of intensity measurement can be avoided by using a white light source as described in reference 39.
- Optical activity in solutions. The other form of optical activity which has a potential application for fibre sensors is that exhibited by certain solutions particularly some sugars. In these applications the sensors measure concentration in terms of optical rotation.

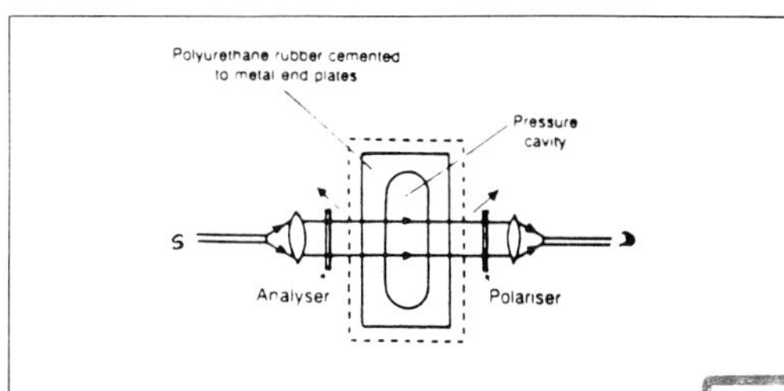

Fig. 24 Photo-elastic pressure sensor

An application which has been intensively investigated by Rogers and others is the measurement of current and voltage of high voltage transmission lines (see fig. 25). Current can be measured by the Faraday magneto-optic effect by propagating linear polarised light through a low birefringent monomode fibre (or better still through polarisation maintaining fibre) which is wound several times round the high voltage conductor.[40] The magnetic field due to the current is codirectional with the fibre and produces a rotation of the plane of polarisation which can be measured by a polarisation analyser. This detection can be made with a polarisation-prism beam-splitter of the Wollaston type. The prism analyses the incoming light into two orthogonal linearly polarised beams which can be individually measured by two photo-detectors A and B. By computing the ratio $(A - B)/(A + B)$, a measurement of the rotation can be made which is independent of light intensity.

The voltage can be measured using the electrogyration effect (qv). In one system[41] which measures both voltage and current simultaneously, quartz crystals are used as modulators since they exhibit both magneto-optic and electrogyration effects (see Fig. 26). This can be exploited if two crystals are fixed at opposite sides of a high voltage conductor. Separation of the two variables can be achieved because the electrogyration effect is the same for both crystals but the magneto-optic effect is reversed. Thus the addition of the photo-detector signals measuring the polarisation analyser outputs can provide a voltage measurement and a subtraction can provide a current measurement.

An alternative method of modulating light to measure an electrical field is to use the Pockels effect (qv).[42] A crystal of $Bi_{12}SiO_{12}$ is used because it exhibits a linear light retardation effect with only a small temperature dependence. A similar arrangement is described by Shibata using a crystal of $Bi_4Ge_3O_{12}$.[43] This gives a phase shift of linear polarised light of:

$$(2\pi n^3 \gamma E\ (2L))/\lambda$$

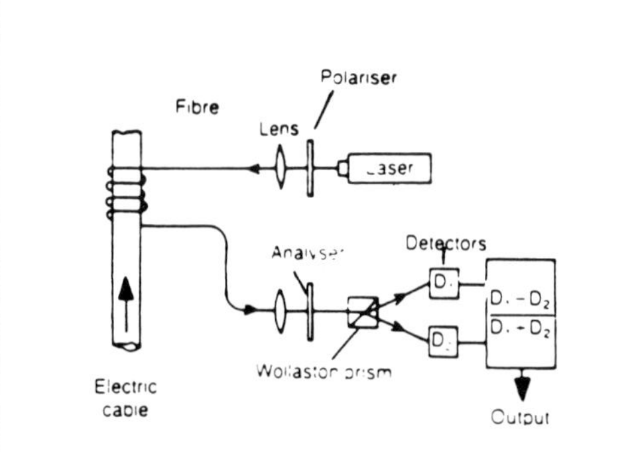

Fig. 25 Electric current transducer

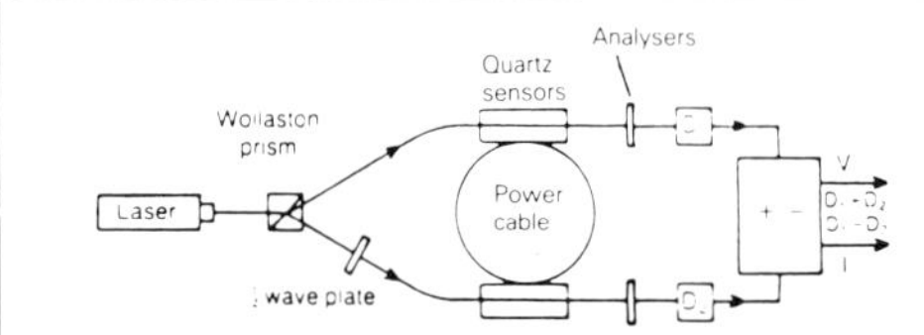

Fig 26 Combined voltage and current sensor

Fig. 26 Combined voltage and current sensor

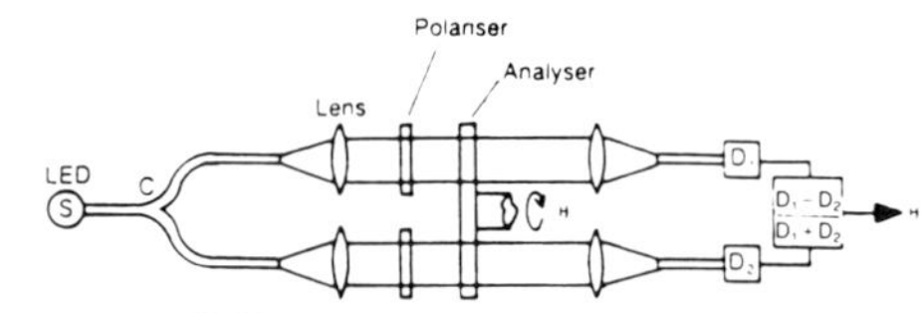

Fig 27 Angular displacement polarimetric sensor

Fig. 27 Angular displacement polarimetric sensor

where λ is the wavelength, n the refractive index, γ is a coefficient, E is the electric field and L is the length of the crystal. $(2L)$ is the optical path as the beam is reflected.

The optical activity of crystalline bodies can be strain and temperature dependent and thus these variables can be measured by the extent to which they can rotate the plane of polarisation. The general form of the equation for polarisation sensors of this type is:

$$I = I_0 \sin^2[\lambda(n_0 - n_e)d\lambda^{-1}]$$

where I is the transmitted intensity, I_0 the source intensity, n_0 and n_e are the indices of refraction for the ordinary and extraordinary rays, d the path length, and λ the wavelength. As $(n_0 - n_e)$ is a function of temperature and strain, these crystalline bodies can form the basic components of fibre sensors for strain and temperature measurement.

A simple form of polarisation sensor for the measurement of angular displacement has been suggested by Jackson[44] and is illustrated in figure 27. An LED source is coupled to a multimode fibre and the beam is split 50:50 to pass after collimation through two polarisers with their optical axes at 90 degrees to each other. A single rotatable polariser is mounted with its optical axes at 45 degrees to the other two polarisers. The two beams of light are then conducted to two linear photo-detectors. In these circumstances the two outputs from the detectors and A and B should be the same, but if the rotatable polariser should move through an angle θ then the value of θ can be computed from the function $(A - B)/(A + B)$ and the accuracy of the measurement is not influenced by changes in the source intensity.

Polarisation techniques can suffer from ambiguity problems in the same way as those of interferometers but the severity of the problem is far less, and in theory, at least, a solution can be found in heterodyne detection systems.

Time duration modulation (TDM)

Time duration modulation (TDM) has a lot to commend it for sensor development. Apart from the fact that time intervals can be measured cheaply and with a very high degree of accuracy, systems operating in the time domain can offer the following advantages in many cases.

- Drift in measurement systems can be eliminated or reduced.
- Elimination of the problem of establishing a zero reference.
- Transmission of data by pulse position, duration or frequency can be made without serious degradation of signal.
- Ambiguities such as those which handicap interferometers can be avoided.
- Linear functional relationships prevail.

For the purposes of this section the inverse function of time, i.e. frequency, comes within the scope of TDM and sensors working in the frequency domain will be considered first. The range of frequencies can be quite large from a few hertz to the gigahertz band. Earlier reference was made to the microbend sensor and its commercial exploitation as a sensor for a vortex shedding meter. Such a sensor operates in the low frequency domain (10–100Hz) and needs no intensity referencing. Similarly, the multi-mode interforerometer, the fibredyne, used as a vortex flow meter detector, requires no referencing. These examples indicate the attraction of sensors in general which generate a frequency output signal. Currently, there is a whole class of commercial sensors with a time dependent electrical output signal which is finding increasing favour with users partly for the five reasons given above and partly because a frequency signal is more compatible with digital processing than the time honoured analogue variety.[45] It is reasonable to suppose that many electrical frequency based sensors currently available could be converted to their fibre optic equivalents.

As an example, a commercial electronic pressure sensor has used the resonant wire technique successfully for many years. More recently, a fibre optic equivalent has been developed in which the force provided by pressure on a diaphragm alters the tension of a taut vibrating wire and, thus, alters its resonant, frequency (see Fig. 28).[46] The energy for maintaining the wire in the resonant mode is so small that it can be provided by an LED coupled to a multimode fibre over a distance of a kilometre. In order to maintain the wire in resonance a fibre optic reflective sensor detects movement of the wire and transmits its signal to a distant photo-detector. This feeds into an electronic amplifier which in turn drives an LED to supply the energy through another fibre to maintain the wire in resonance in an amplitude controlled feedback loop. The frequency of resonance (f) is available as an output signal from the amplifier and is related to the tension in the wire T by the equation:

$$f = (1/2L)\sqrt{(T/m)}$$

where m is the mass/unit length of the wire, but $T = T_O + T_P$ where T_O is the initial tension in the wire when the gauge pressure is at a reference level and T_P is the algebraic change in tension for a change in pressure on the diaphragm.

Other resonant elements, like piezo-electric quartz, are possible alternatives to the wire, particularly as their Q value is very high and the frequency maintaining energy is correspondingly low. In Britain Patent application GB 2146-120-A it is proposed to maintain the quartz crystal in resonance by direct action of light from a fibre by photoacoustic action. The crystal vibration is arranged to modulate light in a fibre loop which is used as a feedback signal to maintain resonance. The frequency of resonance is then a function of the force applied to the crystal (see Fig. 29).

By coating a quartz crystal with an absorption layer which can selectively and reversibly absorb or liberate a gas or vapour, the resonant frequency can be altered.

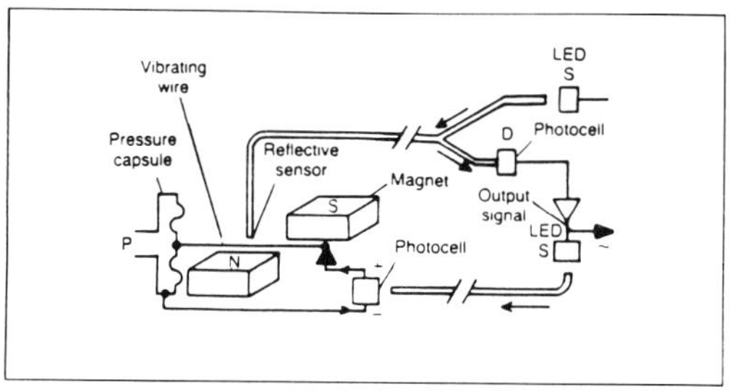

Fig. 28 Resonant wire sensor

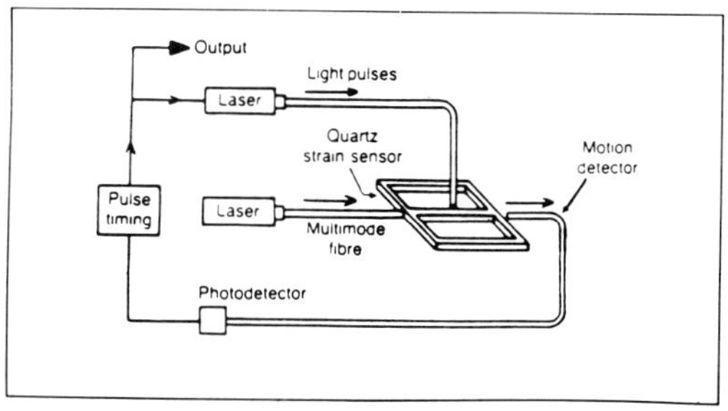

Fig. 29 Resonant quartz sensor

This is due to a change in the vibrating mass which is related to the ambient concentration of the gas or vapour.[47] This technique could provide a useful range of passive analytical fibre optic sensors. As a further future development the technique could be used with micromachined silicon devices as described in reference.[48] Another possibility could be photo-electrical excitation of surface acoustic wave devices as oscillating sensors for pressure, temperature and chemical measurements.

The next class of time modulated sensors is dependent upon transit times of light in a medium to measure position or level. These require the measurement of the time taken by a light beam to travel the distance to be measured. A typical example is that of liquid level measurement which has some similarity to the echo sounding technique used in the better known ultrasonic liquid level gauges. However, a modification is required when light is used instead of sound. This is because sound has a velocity of approximately $300 ms^{-1}$ while light has a velocity of 300×10^6 m/s^{-1} or one million times as great. For long distances, e.g. for measuring the distance of the earth from the reflectors left on the moon during Apollo space missions, it was possible to measure the round trip transit time of a very short light pulse from a laser. For short distances, like measuring the level of oil in a tank, a modified technique is required using high frequency amplitude modulation of a laser beam and the measurement of phase difference.[49] This is achieved in the same way as that described in the section dealing with interferometry (see 'Phase Modulation')—a laser diode is amplitude modulated at a frequency of about 40MHz which gives a wavelength of about 7.5 metres. In order to avoid ambiguity from a single measurement of phase shift, the total path length should not exceed the wavelength, i.e. the distance from the source to a reflecting surface should not exceed 3.75 metres (see Fig. 30). If distances in excess of this need to be measured either the modulating frequency needs to be lowered, which will give a low resolution, or two modulating frequencies should be used.

Previous reference has been made, in the section 'Phase Modulation', to the technique of ramping the frequency output of a laser to provide a non-ambiguous method of measurement. The application of frequency ramping (the chirp technique) has been used extensively in other fields, particularly that of radar. The advantage of such a technique is that the measurement is made in the time domain instead of by phase shift and the ambiguities no longer exist.[50,51] Figures 31 (a) and (b) shows the principle of measuring time delay as an alternative to phase measurement in an interferometer. Provided the ramping is linear, a mixing of the source frequency signal with the delayed signal gives rise to a difference frequency which is a measure of the delay.

Another method which has been suggested for measuring the time delay resulting from the effects of strain and temperature, is based on an optical resonator (see Fig. 32). This is a feedback arrangement in which a light signal is fed through a loop of fibre and the emerging signal is detected, amplified, converted back to light and fed back into the loop.[52] Such an arrangement operates as a feedback oscillator with an operating pulse frequency inversely proportional to the delay. Such a system requires an intrinsic fibre sensor loop of at least 100 metres to provide a resolution of say 1%. The time delay in 100 metres of fibre is approximately 0.5 microseconds corresponding to a pulse frequency of 2MHz. For 500 microstrain in the fibre the frequency change would be 1000Hz and a resolution of 1% would require the measurement of 10Hz in 2MHz. Unfortunately, it would be difficult to apply a uniform strain on the fibre over 100 metres without introducing a temperature error.

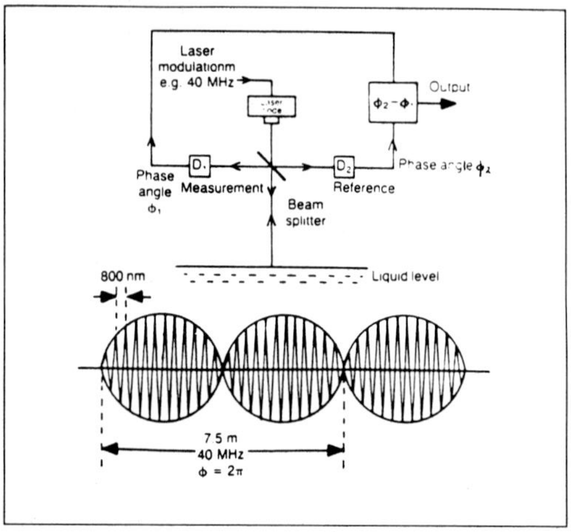

Fig. 30 Distance from the source to a reflecting surface

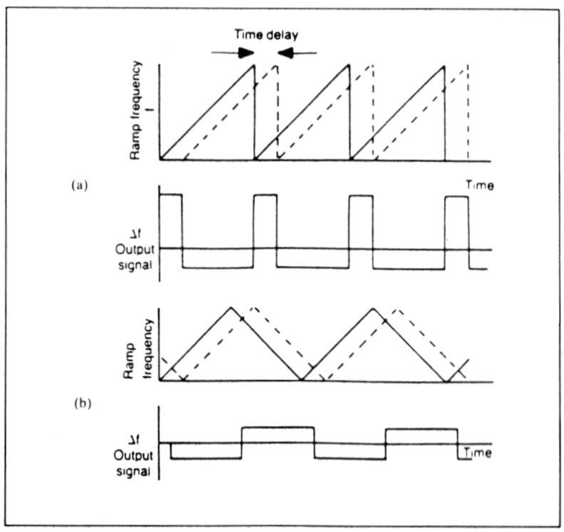

Fig. 31 Time delay measurement by frequency ramping

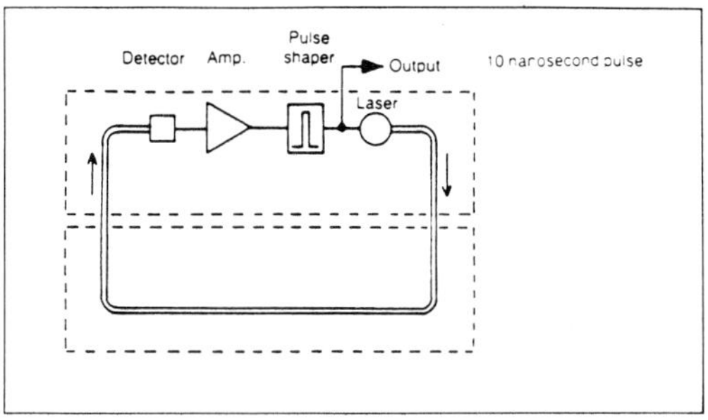

Fig. 32 Optical ring resonator

Hartog and Payne[53] have described an intrinsic fibre optic temperature distribution sensor which is based on the change in the Rayleigh scattering coefficient with temperature and employs optical time domain reflectometry (OTDR). The method is to launch a short pulse of light through a liquid core fibre (200 micron bore silica tube filled with hexachloro-1,3-butadiene) and to measure the power ratio of the returned scattered light which is a linear function of temperature. Other modulating influences such as pressure, strain or vibration can be measured by the OTDR technique. This produces a time delay function from which the spatial distribution of the variable can be determined.

Another time delay system of a hybrid nature is to convert the measured parameter into an electrical pulse time modulated signal which can be converted into light pulses and propagated through a fibre to a receiver for demodulation.[54] Such a sensor would not be passive as it would require a low power electrical source to carry out the first stage of modulation (e.g. photo-voltaic conversion of continuous light transmitted in a separate fibre or a long life battery incorporated in the sensor).

Finally, there is one other time modulated technique which depends on the decay time of the excited state of a luminescent body. The rate of decay is increased with increasing temperature.[55,56]

Concluding remarks

An attempt has been made to review a representative sample of the many types of modulation which have been investigated in the past few years in connection with the development of fibre optic sensors. A few of the techniques described have already been introduced into commercially available devices. Some still require further development before they become commercially acceptable. Others may remain in limbo until they can be resurrected and rejuvenated with new technology. It is hoped that this review will provide a background of the subject for those hoping to commence studies in this fascinating field.

References

1. Mechanical Technology Inc., Latham, New York.
2. J.N. Fields *et al J. Acoust, Soc. Am.* 67, 1980, p. 816.
3. Bailey Meter Company, USA Tech literature.
4. J.H. Lyle and C.W. Pitt, *Electron. Letts* 17, 1981, p. 244.
5. Pilkington Cryogenic Monitoring System: Tech. literature.
6. R.J. Murphy and D.M. Turner, *US Patent Application 2,062,877*, 1981.
7. *British Patent 1,390,426.*
8. Ishikawajima. Harima Heavy Industries Tech. literature, 1983.
9. G.B. Pitt *et al Trans I. Mar. E.* 92, Paper 8, 1980, pp. 66–75.
10. S. Ramakrishnan and R. Th. Kersten, *OFS Conf. Proc.*, 1984, pp. 105–110. Stuttgart.
11. M. Sharm and R.E. Brooks, *SPIE* 224, 1980, pp. 46–52.
12. J.E. Geake *J. Sci Instrum.* 31, 1954, pp. 260–261.
13. I. Fromm *Europatent 0,025,565*, 1980.
14. G.T. Sincerbox and S.G. Gordon, *Laser Focus*, Nov. 1981, pp. 55–58.
15. J.D. Montgomery and F.W. Dixon, *ISA/IMC Proc. Conf. "Promecon 81"*, London, 1981, pp. 81–90.
16. D. Snell and G.D. Pitt, *BHRA Proc. Int. Conf. Opt. Tech. in Process Control*, 1983, pp. 27–41.
17. J.P. Dakin Plessey Technical Journal *Systems Technology*, No. 38, May 1984.
18. D. Bell *et al.* West Glamorgan Institute of Higher Education.
19. K. Spencer *et. al. IEE Proc. 1st Int. Conf. on Opt. Fibre Sensors*, 1983, pp. 96–99.
20. Accufibre Corporation. Technical Literature, Vancouver, Washington, 1983.
21. Ovren *el al. BHRA Proc. Int. Conf. Opt. Tech in Process Control*, 1983, pp. 67–81.
22. K.A. Wickersheim *US Patent 4,075,493*, 1978.
23. E. Snitzer *et al. IEE Proc. 1st Int. Conf. on Opt. Fibre Sensors*, 1983, pp. 79–82.
24. M.C. Hutley *OFS Conf. Proc.*, 1984, pp. 111–116. Stuttgart.
25. M.C. Hutley *NPL News*, No. 363, Summer 1985, pp. 3–6.
26. M. Brenci *et al. OFS Conf. Proc.*, 1984, pp. 155–160. Stuttgart.
27. C. Christiansen *Ann. Phys. Chem.* 23, 1984, p. 298.
28. J.H.W. Cramp and R.F. Reid, *Europatent 0,061,884*, 1982.
29. J.M. Knox *et al. IEE Proc. 1st Int. Conf. on Opt.Fibre Sensors*, 1983, pp. 1–5.
30. A.A. Boiarski and S.A. Kingsley, *ibid*, pp. 214–216.
31. B. Culshaw and M.J. Hutchings, *Electron. Letts.* 15, 1979, pp. 569–571.
32. B. Culshaw *The Radio and Electronic Engineer* 52, No. 6, 1982, pp. 283–290.
33. D.A. Jackson *et al. Appl. Opt.* 19, 1980, p. 2926.
34. A. Dandridge and A.B. Tveten, *Opt. Lett.* 7, 1982, p. 279.
35. A.D. Kersey *et al. IEE Proc. 1st Int. Conf. on Opt. Fibre Sensors*, 1983, pp. 43–47.
36. R.D. Birch *et al. ibid*, pp. 83–90.
37. T. Hosaka *et al. OFS Conf. Proc.*, 1984, pp. 395–398. Stuttgart.
38. B.E. Jones and R.E. Spooncer, *ISA/IMC Proc. Conf. "Promecon 81"*, London, 1981, pp. 91–96.
39. B.E. Jones and R.E. Spooncer, *IEE Proc. 1st Int. Conf. on Opt. Fibre Sensors*, 1981, pp. 173–177.
40. A.J. Rogers *Electric Power Applications* 2, No. 4, 1979, pp. 120–124.

41. A.J. Rogers *CEGB Research*, June 1979, pp. 3–11.
42. T. Yoshino *et al. OFS Conf. Proc.*, 1984, pp. 55–58. Stuttgart.
43. K. Shibata *IEE Proc. 1st Int. Conf. on Opt. Fibre Sensors*, 1983, pp. 164–168.
44. D.A. Jackson Optical Sensors Association. Internal report No. 84/11F, 1984.
45. R.M. Langdon *J. Phys. E. Sci. Instrum.* 18, 1985, pp. 103–115.
46. B.E. Jones and G.S. Philip, *Proc. Europ. Conf. Sensors and their Applications*, Inst. of Phys., London, 1983, pp. 86–88.
47. W.H. King, Jr. *Res. Dev.* 20, 1969, pp. 28–34.
48. C.J. Greenwood *J. Phys. E. Sci. Instrum.* 17, 1984, pp. 650–652.
49. D.A. Jackson *IEE 1st Int. Conf. on Opt. Fibre Sensors*, 1983, pp. 100–103.
50. D. Uttam and B. Culshaw, *IEE Colloquium*, May 26th Digest No. 1982/60, 1983, pp. 10/1–10/4.
51. D. Uttam *et al. IEE Proc. 1st Int. Conf. on Opt. Fibre Sensors*, 1983, pp. 182–184.
52. M. Johnson and R. Ulrich, *Electron. Letts.* 14, 1978, pp. 432–433.
54. R.T. Murray *Fibre optic sensors, Development and Design for Industrial Applications*, Oct. 18th 1984. Oyez Scientific and Technical Services.
55. K.A. James *et al. Control Engineering*, Feb. 1979, pp. 30–33.
56. J.S. McCormack *IMC Proc. Symposium 'Optical Sensors and Optical Techniques in Instrumentation'*, Nov. 12th 1981.

Appendix 1

Examples of modulating signals used for measurement purposes:

- Physico/mechanical: force, weight, displacement (linear or angular), pressure, differential pressure, vibration, sonic energy, velocity, acceleration, flow, level, hardness, thickness, viscosity, density, opacity, turbidity, time, moisture content, particle concentration.
- Electrical: voltage, current, frequency, phase angle, inductance, capacitance, resistance.
- Magnetic: field strength, hysteresis, coercivity, induction.
- Thermal: temperature, thermal conductivity, heat flow, emissivity.
- Radiation: radiofrequency, infrared, visible, ultra violet, X-rays, nuclear.
- Chemical: concentration, composition (gas, liquid, solid), biological activity.

Point and distributed polarimetric optical fibre sensors

A.J. Rogers
King's College London, UK

THERE is a large range of demands which are made on optical fibre sensor performance, depending on the application area. The range extends from the simple but robust and reliable two-state 'alarm' intensity sensor, to the extremely sensitive Mach-Zehnder type of interferometric sensor, with its sophisticated control and processing electronics. Between these lies the polarimetric optical fibre sensor, which possesses many of the advantages of the interferometric sensor, in particular its high sensitivity, but also benefits from the simplicity which is consequent upon the fact that its action is basically that of a self-referencing Mach-Zehnder device.

Figure 1 makes this clear. Figure 1(a) illustrates the basic Mach-Zehnder arrangement. This is essentially an optical phase 'bridge' which allows detection of a phase disturbance, produced by an external measurand in one of the two arms, by observing the interference pattern which results when the light propagations in the two arms are recombined.

If the Mach-Zehnder arrangement is to be implemented in optical fibres, it is clear that these must be monomode fibres, since it is only with monomode propagation that a single, definable phase and polarisation state can be assigned to the light at each point along the propagation path, thus allowing the elementary conditions for optical interference to be satisfied. The optical fibre gyroscope[1] and the hydrophone[2] are good examples of interferometric Mach-Zehnder devices. In figure 1(b) we see the essence of the polarimetric device. The two arms of the bridge now lie within the same fibre, the distinction between them lying in their differing polarisation states. The action of the measurand is, again, to influence the phase difference between the two polarisation states, which, in turn, results in a change in the final, composite state formed from the superposition of the other two, and which emerges from the fibre. This composite state is then detected, to indicate measure and amplitude. In addition to the self-referencing brought about by this action, this type of device is especially valuable for application to those measurands which act directly on the polarisation properties of the fibre, such as magnetic field, electric field and pressure/strain.

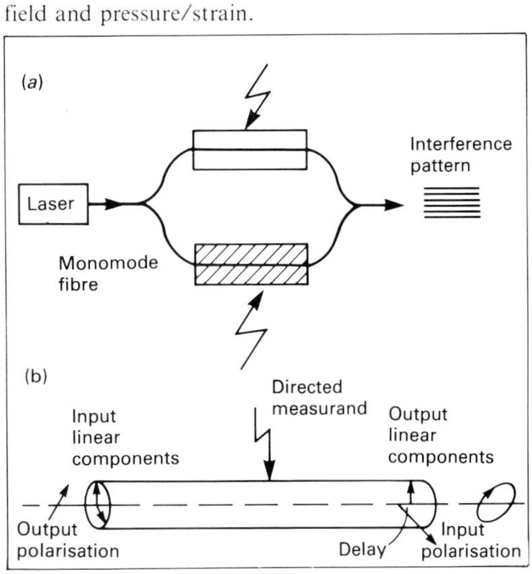

Fig. 1 The Mach-Zehnder principle

Point and distributed sensors

The one-dimensional nature of the optical fibre endows it with a number of advantages for measurement sensing applications.

First, it provides a long path for interaction between the measurand and the propagating light. For 'point' measurements within the measurand field the spatial extension within the field may be limited by coiling the fibre, if necessary; but the long interaction path also allows the measurand to be 'line-integrated' along the fibre to give, for example, electric current via integration of magnetic field[3], voltage via integration of electric field[3] or just in order to provide an average value over a chosen path, if that should be the desired quantity.

Correspondingly, and perhaps more importantly, line differentiations of the measurand may also be performed along the fibre over any chosen path. The implication of this is that the measurand may be spatially resolved along the chosen path, with a resolution depending on the speed of the detection and processing electronics (of order 0.1m with sub-nanosecond response times). With these, so-called, distributed measurement systems[4] one has available, simultaneously, both spatial and temporal information, thus giving rise to an extra level of knowledge about the behaviour of large structures such as buildings, bridges, boilers, airframes, power lines, power transformers, etc.

Polarimetric methods have a part to play in each of these types of sensors, and some device and system examples will shortly be described in order to illustrate their role.

Elements of monomode fibre polarisation optics

For most measurement sensing purposes the lengths of optical fibre in use are sufficiently small for the optical loss to be negligible: at most we are normally considering only about 500m of fibre. We shall thus, for our present purposes, consider the fibre to be a lossless polarisation element.

Any such polarisation element will possess two polarisation eigenmodes: these are the two polarisation states which will pass through the element without change of form. For the generalised element, these two states are elliptical in form, the two ellipses being orthogonal but with the same ellipticity[5]. The two states lie on opposite ends of a diameter of the Poincaré sphere (see Fig. 2), and the action of the element on any other polarisation state is equivalent to a rotation of the sphere about that diameter, through an angle equal to the phase delay inserted by the element between its two eigenmodes. The full properties of the Poincaré sphere are quite well known.[6,7]

The fact that the two elliptical eigenmodes travel at different velocities within any given element is, of course, the origin of the polarisation action of the element and it is known as elliptical birefringence. Any elliptical birefringence may be resolved uniquely (within the necessary ambiguity of 2π) into linear and circular components. This resolution is prescribed as follows. For ellipticities $\pm e$ for the two eigenmodes (the sign refers to the direction in which the tip of the electric vector circumscribes the ellipse) and a phase delay Δ between the two eigenmodes we have:[5]

$$e = \pm \tan \eta; \qquad \tan 2\eta = 2\rho/\delta$$
$$\Delta^2 = (4\rho^2 + \delta^2)$$

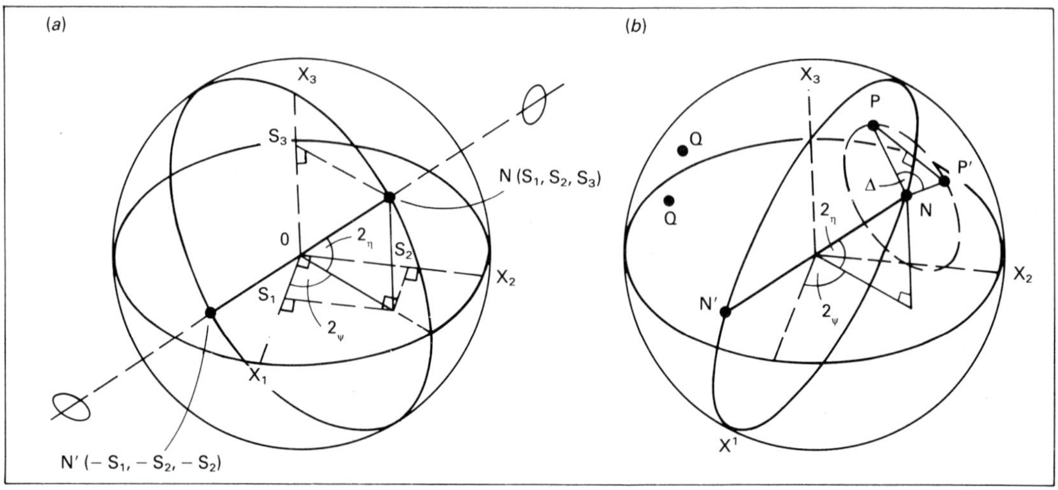

Fig. 2 The Poincare sphere

where 2ρ and δ are, respectively, the circular and linear birefringence components. (The factor of two in the case of the circular birefringence is a convenience resulting from the fact that the angle ρ, through which a linear-polarisation direction is rotated by the action of circular birefringence, is half that of the birefringence itself; the rotation is the measured quantity). Thus, the components 2ρ and δ are fixed, except in regard to absolute sign which is determined from a knowledge of the faster of the two elliptical eigenmodes.[5] Thus, to specify 2ρ and δ it is necessary only to specify the eigenmodes for a given element. This may be done by measuring any two distinct pairs of input and output polarisation states via the Stokes parameters.[8]

In many cases of practical importance the two eigenmodes are either the two orthogonal linear states or the two oppositely rotating circular states, with the action of the measurand being to alter their relative phase displacement. For these conditions, clearly, the eigenvector remains fixed in direction, while the measurand controls the angle through which the sphere is rotated about it.

Polarimetric point sensors

A description will now be given of some optical fibre polarimetric sensors which provide a single measurand value averaged over a defined region, and which are usually referred to as 'point' sensors (even though the spatial extent over which the averaging occurs may, in some cases, be quite large).

Alternating current measurement

A polarimetric device for the measurement of alternating electric current at high voltage has been extensively reported[3,9,10], and the principle is illustrated in figure 3. Linearly polarised light from a semiconductor laser is launched into a monomode fibre which loops a number of times around a current carrying conductor. The action of any magnetic field component lying along the fibre axis is to produce a rotation of the linear-polarisation direction, via the Faraday magneto-optic effect which, thus, constitutes a magnetic-field-induced circular birefringence. The total birefringence throughout the fibre length is proportional to the line integral of the magnetic field around the conductor and is thus a direct measure of the current through the looped fibre (and independent of any currents which do not thread the loop). Thus, the device constitutes a method for measurement of current at high voltage using a passive insulating element which avoids any physical contact with the conductor. The primary problems are associated with intrinsic and induced linear birefringence.

The method of measurement constructs the function:

$$P = (I_1 - I_2)/(I_1 + I_2)$$

where I_1 and I_2 are the intensities of the two normal linearly polarised components of the emergent light (see Fig. 3(a)). We find[9] with linear and circular birefringence components δ and 2ρ, respectively, that to first order:

$$P \cong 2\rho\frac{\sin\delta}{\delta} \qquad \delta \gg 2\rho \qquad (1a)$$

$$P \cong \sin 2\rho \qquad \delta \ll 2\rho \qquad (1b)$$

It is clear that a large value of δ will reduce the sensitivity of the measurement by a factor $\sin\delta/\delta$, but the most troublesome effects are caused by variations in δ as these will lead to output noise.

Conventionally-fabricated monomode optical fibre is found to possess intrinsic linear birefringence resulting from small departures from true circularity in the core and from residual asymmetric stresses in the core and cladding structures.

This birefringence is temperature dependent and thus leads to drift in the calibration of the device. The problem has been effectively overcome by the development of 'spun' monomode fibre,[11] which has reduced the intrinsic linear birefringence to very low levels ($<1°\ m^{-1}$) by spinning the material preform as the fibre is pulled, thus averaging all asymmetries to zero over each spin period.

With the use of 'spun' fibre there remains the asymmetric strain introduced by bending the fibre to form the necessary looped geometry for the device. It has the total bend birefringence is given, approximately, been shown [12] that for a circular loop of radius R, by:

$$\delta_b \cong \frac{\pi}{2R}\ \text{degrees}$$

where R is in metres. For example, for a bend radius of 0.15m, leading to a bend birefringence of 10°, the effect would be to reduce the current sensitivity (value of P) by only ~ 0.5%, which is quite acceptable. Moreover, the temperature dependence of δ (which results from the temperature dependence of the stress-optic coefficient) will cause a sensitivity variation of only ~0.03% over the ambient range, which is also acceptable for most power system applications. Thus, for R in excess of about 0.15m, there should be few problems in practice.

There remains the problem of vibrationally induced birefringence, both circular and flinear. Any twist effect on the fibre can be minimised by firmly clamping each end, thus ensuring that the overall twist (and thus the overall circular birefringence) is zero. However, any twist waves induced by vibration will introduce local circular birefringence via the stress-

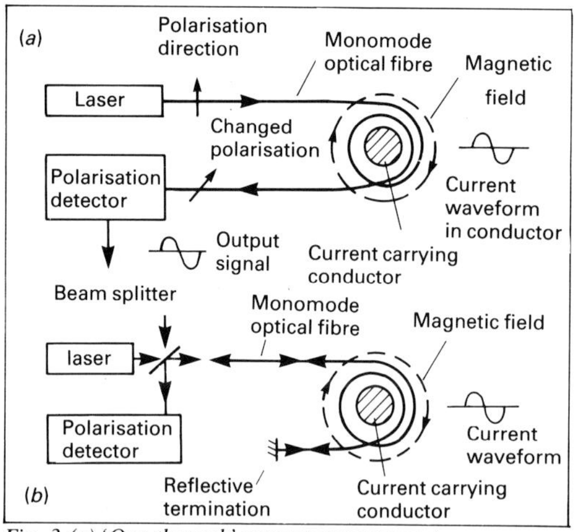

Fig. 3 (a) 'One-through' current measurement
(b) 'Double-pass' current measurement

optic coefficient,[13] and these may not average to zero in the fibre propagation time. Care must be taken in the device design to isolate the fibre from twist stress in order to reduce the possibility of this. For vibrationally induced linear birefringence, again, care should be taken in the design to decouple the fibre from all sources of linear vibration. By placing the fibre in a gel-filled tube, very considerable isolation may be achieved for both types of source.

An alternative approach to the solution of the problem of linear-birefringence interference in this device is to use a fibre with a very large intrinsic circular birefringence and some structures have been studied in the laboratory.[31] Such a fibre cannot yet (as far as the author is aware) be manufactured, but it is clear from equation 1(b) that, if 2ρ can always be maintained at a much greater value than δ, then δ can be ignored. (Some circular birefringence may be introduced by twisting the fibre, but the amount is obviously strictly limited by the necessity to avoid physical damage to the fibre.) An obvious difficulty in this case is the possibility that the drift inherent in a large 'bias' value of ρ will itself give rise to a significant noise level. This can be effectively solved by using a 'back-reflection' arrangement. In this (see Fig. 3(b)), the far end of the fibre is silvered so that the light performs a go-and-return passage through the fibre. The important point now is that, whereas the intrinsic circular birefringence is reciprocal (i.e. the polarisation properties are independent of propagation direction and will, thus, cancel on the double passage), the magnetic-field-induced birefringence is non-reciprocal and will have twice its single-passage value. It is perhaps not obvious that, under these conditions, the (cancelled) intrinsic circular birefringence can still be effective in swamping the linear birefringence. That this is indeed the case can best be appreciated by reference again to the Poincaré sphere (see Fig. 4).

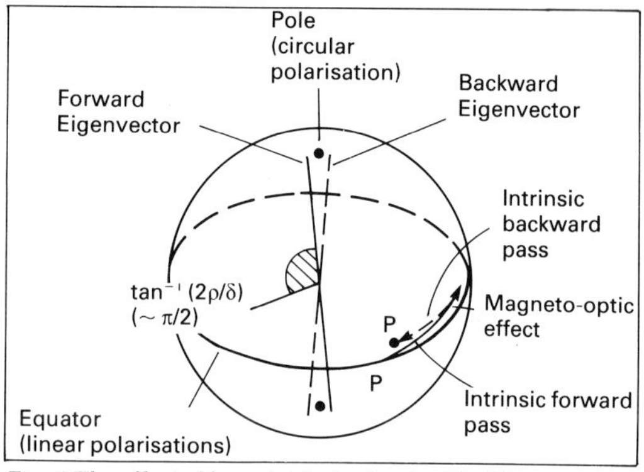

Fig. 4 The effect of large intrinsic circular birefringence

The eigenmodes for both forward and backward propagations are close to the circularly polarised poles. Hence, the action in each direction of propagation (equivalent to rotations about the eigenmode diameters) maintains the state of polarisation close to the equator; i.e. close to the linear state. In other words, it is the ratio $2\rho\delta$ which controls the action in each direction, rather than the resultant value of either quantity after the double passage.

Direct-current measurement

The fact that the Faraday magneto-optic effect allows fibre measurement of direct current, in addition to alternating current, offers a potentially important advantage over conventional current-measurement transformers.

An investigation into the possibility of engineering a DC device has yielded some important additional features of the polarisation behaviour in monomode fibres.

The most notable of these was the apparently marked increase in the vibrationally induced polarisation noise. Investigation showed that this was true only when the fibre was bent in more than one plane, and analysis soon revealed that this was a purely geometrical effect. In essence, it is due to the fact that bending the fibre in one plane will rotate the direction of polarisation of a propagating component linearly polarised in the plane of the bend, due to the necessity for the electric vector to remain normal to the fibre axis (see Fig. 5). If the fibre, additionally, is bent into another plane, the polarisation direction of the light will depend upon the bend radius in the first plane (see Fig. 5). Hence, the emergent polarisation state will be dependent on the fibre geometry, and small vibrational effects can lead to relatively large variations in the output polarisation direction.

As these geometrical effects will be optically reciprocal, the resulting noise may be removed by using the go-and-return reflection system described in the last section. However, this may introduce 'etalon-

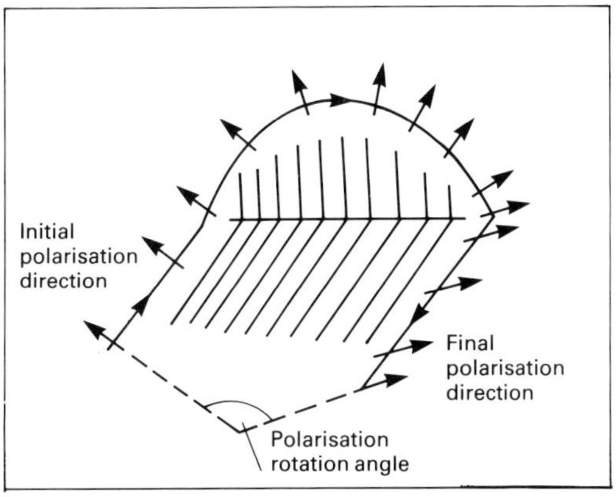

Fig. 5 The geometrical polarisation rotation effect

ing' effects which lead to drift in the DC device and, thus, it now becomes necessary to use a low-coherence light source such as a multimode semiconductor laser.

Finally, it was also found that the polarisation properties of such components as microscope objectives, polarising prisms, optical filters, etc., varied significantly with aperture position, so that small drifts in the launch geometry or source polar diagram (due, for example, to temperature-dependent stresses) caused corresponding drifts in the output polarisation state of the light from the fibre. These drifts were large enough to be intolerable for a practical DC device. Thus, a successful DC device will require either a much better quality of discrete components, or a carefully engineered integrated optical launch package.

It follows, also, that alternating-current measurement devices would benefit from a proper consideration of these new features in the engineering design, as the same disturbances will be present, even though they are less significant.

Voltage measurement

Although the preferred method for optical fibre measurement of current is the line integral of magnetic field using the magneto-optic effect, the preferred method for measurement of voltage is the line integral of electric field using the electro-optic effect.[1]

The electro-optic effect (Pockels, Kerr) comprises an electric-field-induced linear birefringence. The Pockels effect provides birefringence which varies linearly with field, and can exist only in crystalline media, whereas the Kerr effect is a quadratic effect and can occur in both crystalline and non-crystalline media. In both cases, the effect is reciprocal and, thus, contrasts with the magneto-optic effect. The quadratic nature of the Kerr effect precludes its use directly for line integration, but a convenient method for line integration using the Pockels effect in a fibre is shown in figure 6(a).

A fibre which provides a linear birefringence proportional to the axial electric field is strung from

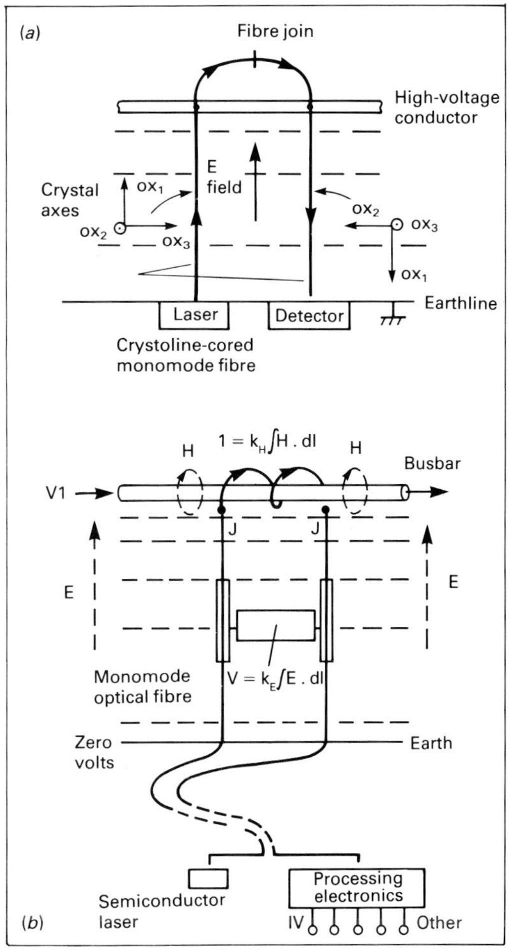

Fig. 6 Optical current and voltage measurement
(a) Voltage measurement alone
(b) Combined current and voltage measurement

earth up to a high-voltage inductor.

Light which propagates along the fibre will thus experience two components of linear birefringence: an intrinsic component, due to the (necessary) crystal structure of the core; and the electric-field-induced component allowing the required line integration of the electric field. Another similar fibre, with crystal axes rotated through 90°, is now joined to the first, and brings the light back down to earth. Apart from the fact of bringing the light back down to a convenient earthy measurement point, the reason for the second fibre is two-fold: first, the intrinsic birefringence component of the fibre cancels that of the first fibre, whereas the electric-field-induced component adds to that of the first fibre, as the electric field is now of the opposite sign, second, provided that the fibres lie physically close together, vibrationally induced birefringence will also cancel in the two fibres. (Temperature effects should also cancel if the fibres lie close, as the intrinsic birefringence will then remain the same.)

Suitable fibres for the realisation of this device do not yet exist: it is very difficult to fabricate crystalline-cored fibre. However, the method has been investigated experimentally, using crystalline quartz rods, and has been shown to behave as expected.[15]

A natural extension of these ideas is to combine electro-optic voltage measurement with magneto-optic current measurement (see Fig. 6(b)). As the linear and circular birefringence components are fundamentally separable, both voltage and current indications are separately available, allowing, also, real- and reactive-power indications to be derived.

Temperature measurement

One method by which high birefringence monomode fibre may be fabricated relies on the asymmetric strain introduced into the core by including, in the cladding, rods of material which differ in their thermal expansion coefficient from the silica in which they are embedded.[16,17] As the birefringence is strain-induced it is also temperature dependent, via the temperature dependence of the strain-optic coefficients.

An arrangement which uses this dependence in a polarimetric temperature sensor is shown in figure 7.[18] Light is launched into one of the eigenmodes of a section of 'hi-bi' fibre and this conveys linearly polarised light up to a similar section of 'sensor' fibre which has birefringence axes rotated by 45° with respect to those of the 'lead' fibre. Equal light powers are thus launched from the 'lead' fibre into each of the two eigenmodes of the sensor fibre. The light is reflected from the far end of the sensor fibre and returns to the front end with a polarisation state which depends on the phase difference which has been inserted between the modes by the fibre, this phase difference being, of course, temperature dependent. Light is now launched back into both of the eigenmodes of the lead fibre and the power level in the originally unaddressed mode of the lead fibre is used as an indication of the temperature of the sensor fibre.

The light power in this mode will vary sinusoidally with temperature (about a DC level) and hence there exists an ambiguity in the temperature measurement which has birefringence axes rotated by 45° with count of power maxima from a calibrated point, or by using, simultaneously, sensors of different lengths and comparing the various counts. Temperature sensitivities of order 0.005°C have been reported using this method.[19]

One of the difficulties (apart from the stated ambiguity) with the use of this method is that, with 'hi-bi' fibre, the 'arms' of the bridge are unbalanced. The consequence of this is that considerable phase noise is introduced into the detection process unless the source coherence length is greater than the difference in optical path length, and this necessitates the use of high-coherence laser sources.

One way around this problem is to use a compensated-sensor arrangement.[20] In this, the second section of fibre is divided into two equal lengths with the birefringence axes of one length rotated through 90° with respect to the other. The static birefringence is now cancelled for propagating light, giving a balanced Mach-Zehnder once more. Only one of the two sections is now exposed to the temperature field. The coherence requirements are, consequently, relaxed a great deal and an LED source has been used successfully with this arrangement.[21]

Polarimetric distributed sensors

Polarimetric methods are as directly applicable to spatially-distributed sensing as to point sensing. Several distributed polarimetric methods have already been studied and these will now be described. This is also an area which is rich with unexplored possibilities.

A particular further advantage which polarimetric methods hold for distributed sensors is that they do not depend on any measurand-induced signal attenuation, so that the dynamic range does not vary with the measurand distribution. This makes them much better suited for distributed measurement over distances of, say, a few hundred metres.

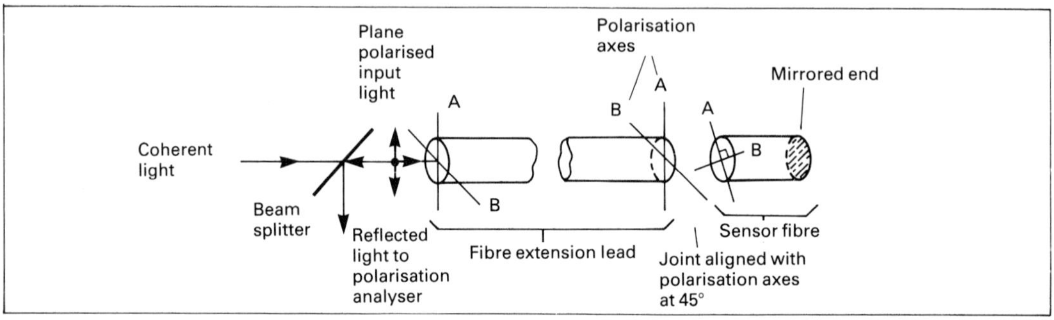

Fig. 7 A polarimetric temperature sensor

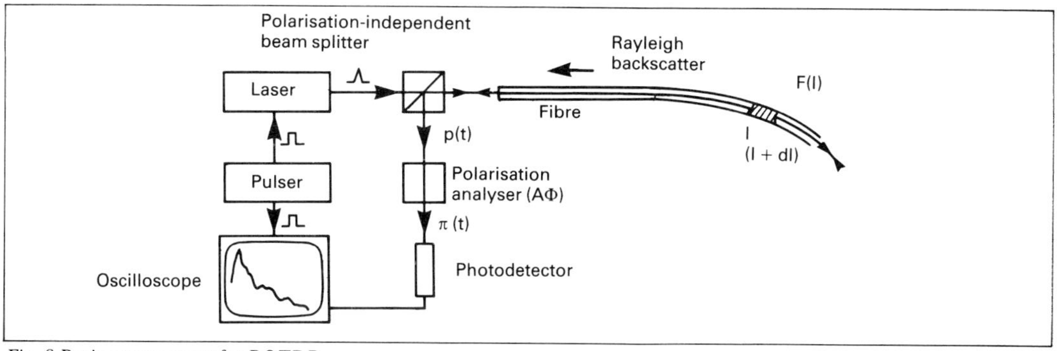

Fig. 8 Basic arrangement for POTDR

Polarisation-optical time domain reflectometry (POTDR)

The first truly distributed optical fibre sensor system of any kind was, in fact, polarimetric in form. This is known as 'Polarisation-Optical Time Domain Reflectometry'[22] and is a polarimetric extension of a technique (OTDR[23]) which for some years has proved valuable for diagnostic analysis of optical fibre communications systems. In this technique (see Fig. 8), a pulse of light is launched into a monomode fibre, and the Rayleigh back-scattered light is polarisation-analysed at the launch end. Differentiation with respect to time of the polarisation information allows certain polarisation properties of the fibre to be determined as a function of position, with a spatial resolution fixed by the pulse width and the speed of the detection electronics; the resolution can fairly readily be less than 1m. It thus becomes possible to sense the spatial distribution of any measurand which is able to modulate the polarisation properties of the fibre.

A possible application of POTDR is to voltage measurement. If a conventional monomode fibre is spiralled in the electric field (see Fig 9), then POTDR can determine the distribution of the electric field along the spiral axis, via the (non-crystalline) transverse Kerr effect. The voltage may then be determined by a separately computed line integration of the field.[22]

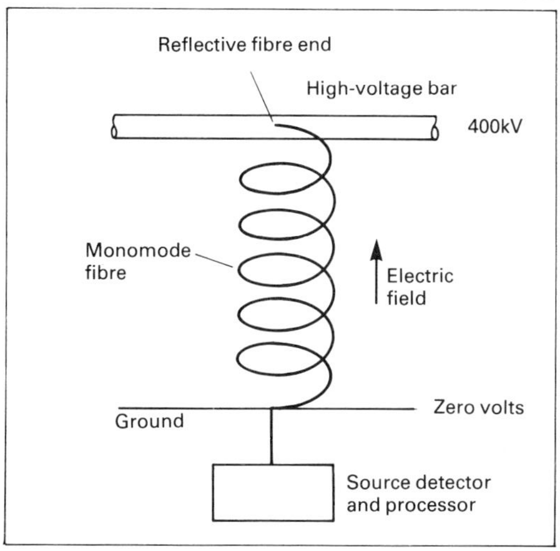

Fig. 9 Voltage measurement using POTDR

The polarisation action of any length of fibre (assumed lossless) is equivalent to a rotation of the sphere about the eigenmode diameter. This rotation will always be equivalent to a rotation about an equatorial axis in the same azimuthal plane as the eigenmode diameter, followed by a rotation about the polar axis, or vice versa. Thus, the fibre will always be equivalent to a linear retarding element and a rotator element (see Fig. 10(a)) (which are *not*, it should be stressed, equal in magnitude and orientation to the values for the continuously distributed parameters δ and θ, but are, of course, related to them[24]). Now, provided that the fibre possesses reciprocal optical properties, it follows that, for a go-and-return passage through the fibre, the effect of the rotator must be cancelled. The element, on double passage, will thus always behave as a pure retarder (see Fig. 10(b).

In POTDR, the instantaneous value, at any time, of the returning light will thus always provide information sufficient to specify the equivalent retarder for the length of fibre which corresponds to that particular go-and-return time, but the information on the equivalent rotator is unavailable.

An alternative statement of this fact is that the ellipticity of the polarisation ellipse may be determined at any point in the fibre, using POTDR, but never its orientation. The broad consequence of this is that, as only two of the three parameters which characterise the polarisation properties of each POTDR resolution element can be known (the three being orientation of retarder axes, value of retardation, value of rotation), some 'a priori' knowledge of these properties must be available for POTDR to be practicable. These points are discussed in more detail in refererence 25.

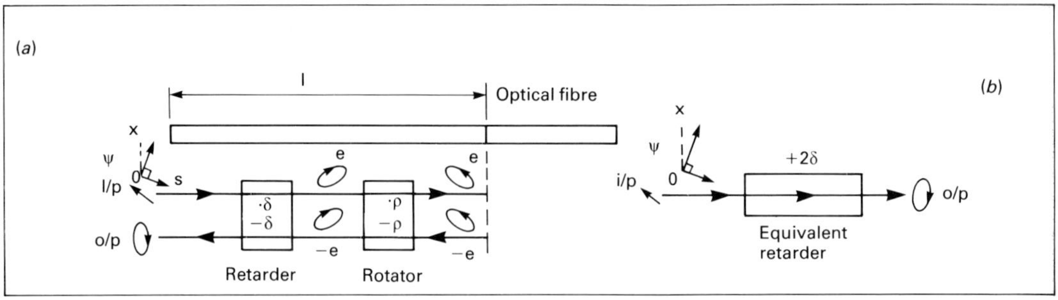

Fig. 10 Cancellation of equivalent rotator in POTDR

The assumption made so far in the discussion on POTDR has been that the fibre has reciprocal optical properties. If it has not, then the go-and-return passage is no longer equivalent to a single retarder but to a retarder-rotator pair, the rotation of the rotator being an indication of the magnitude of the non-reciprocity. Thus, in this case, POTDR is able to measure the spatial distribution of non-reciprocity along the length of the fibre. Such a facility should prove valuable for determining the source of any unwanted non-reciprocity in, for example, the optical fibre gyroscope.[25]

Use of non-linear optical effects

The relative ease with which non-linear optical effects may be made to occur in optical fibres (large intensities, long interaction lengths) offers the possibility of their use in fibre sensors. In particular, the possibility arises for the use of the polarisation dependences of the non-linear interactions in distributed sensors (although there are also many others).

The basic idea is illustrated in figure 11. A high-power pulse of polarised light is launched into the fibre and, as it propagates, the pulse alters, locally, the refractive index of the medium, as a result of one of several possible non-linear optical effects. A continuous (low power) propagation of polarised light in the opposite direction will interact with the pulse according to the relative polarisation states of the two propagations. Temporal analysis of the emerging low-power light thus allows mapping of the polarisation properties of the fibre as the pulse passes along it.

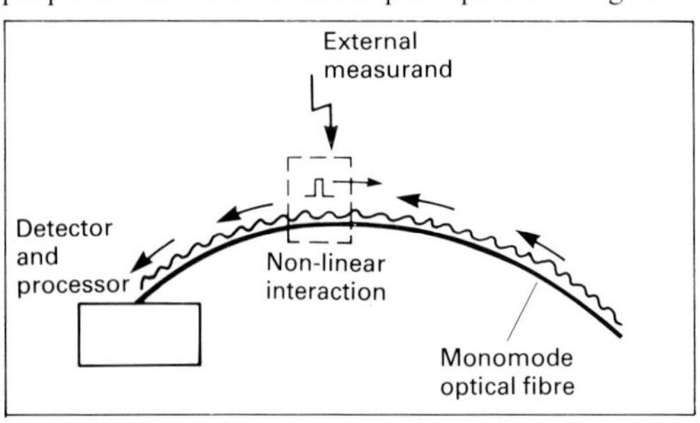

Fig. 11 Distributed sensing using a non-linear pusle/wave interaction

Although such techniques are in their infancy, two preliminary examples may be cited.

The first makes use of the optical Kerr effect, whereby the electric field of the propagating pulse introduces linear birefringence into the medium.[26] Clearly, the effect that this has on an oppositely propagating beam depends on the relative polarisation states of the two beams. (There is the possibility of using the direct temperature dependence of the Kerr effect for distributed temperature sensing, as well.[27]

The second example utilises the Raman pumping produced by the pulse to provide optical gain, at the Stokes or anti-Stokes wavelengths, for the oppositely propagating radiation. The effect has been clearly seen and the polarisation dependence has been measured[28], using the arrangement showing in figure 12.

It is difficult, however, to use, directly, these ideas for distributed sensing with presently available fibres. The main problems, as with POTDR, lie in the difficulty of interpreting a double-pass integrated polarisation state in terms of the differential fibre properties. The signal-processing problems involved are probably prohibitive. The best solution to the difficulty almost certainly lies in the provision of monomode fibres with polarisation properties carefully optimised for specific measurement applications.

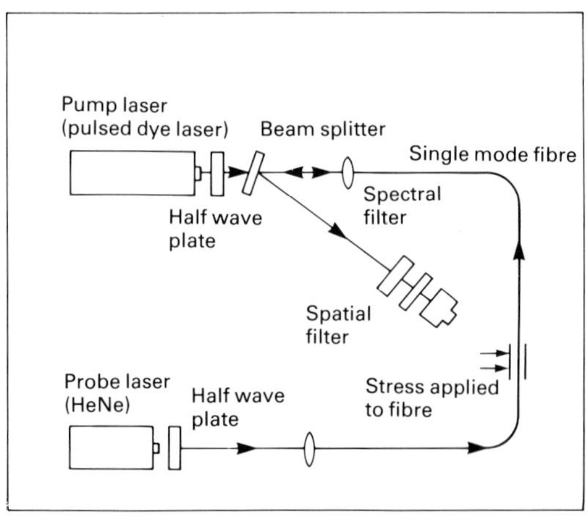

Fig. 12 Arrangements for the stimulated-Raman pulse-wave interaction

Frequency-modulated continuous-wave (FMCW) distributed sensor

The final example in this section on polarimetric distributed sensors relies neither on pulses nor on back-scatter. In fact, it only marginally qualifies as a polarimetric sensor as it does not rely on any polarisation analysis of the output light. It does, however, utilise the polarisation properties of a 'hi-bi' fibre. The method, which is illustrated in figure 13, requires a light source on which is imposed a frequency-ramped sinusoidal amplitude modulation. The light is launched exclusively into one of the two eigenmodes of a 'hi-bi' monomode fibre. If the light remains in that mode it will emerge essentially unaltered in its modulation. If a small perturbation at one point on the fibre causes coupling of some light to the other mode, the different mode velocities will give rise to a phase delay between the two modes on emergence of the light and, thus, to a modulation frequency difference, due to the frequency ramping. On mixing the two mode outputs at the detector the result will be a beat-frequency signal whose frequency value will depend directly on the delay, and which will, thus, allow identification of the position of the disturbance. In principle, also, the magnitude of the beat-frequency signal will allow measurement of the magnitude of the perturbation but, in practice, mode-coupling perturbations are notoriously inconsistent, and difficult to quantify reproducibly for coated fibres.

This idea has been demonstrated experimentally using both orthogonal polarisations and two-propagation-moded fibre.[29] Using a single-mode semiconductor laser frequency ramped up to 40GHz at 200Hz rate, it has proved possible to locate a perturbation to within 10m in a total fibre length of 300m. This technique could provide a fairly cheap, simple method of locating a few pressure points, such as might be required for a security alarm, for example.

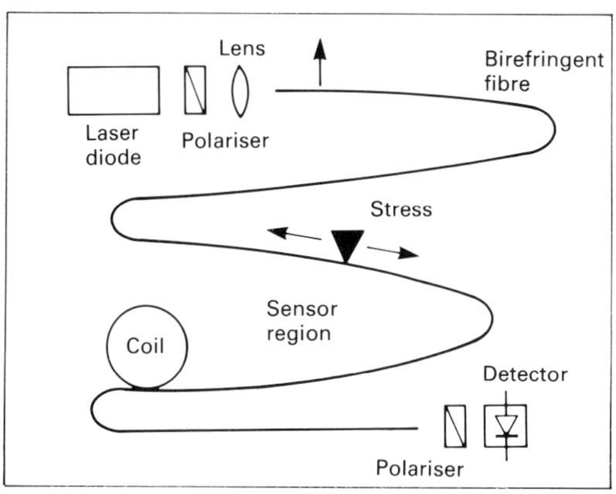

Fig. 13 Arrangement for FMCW distributed sensor.

Concluding remarks

Polarimetric monomode optical fibre techniques offer many advantages for sensing applications, especially in regard to sensitivity. The one-dimensional nature of the optical fibre as a measurement medium may be put to good use in devices which perform either line integrations or line differentiations of a measured field. Polarimetric techniques play an important role in such devices.

range of non-linear optical effects with the aid of new fibres, just emerging, which possess enhanced non-linear coefficients; and the use of acousto-optic interactions, which also present an exciting range of possibilities.[30]

Research and development progress will be limited primarily by the availability of the specialised fibres required for specific measurement functions. The special properties required may be associated with dopants, geometries or polarisation control; or with outer-fibre coatings suitable either for protection in given environments or for appropriate sensitivity and correct interfacing to specific measurands. The other areas which require close attention are those of light-source control and the processing necessary to extract the wanted information from the receive signal. In this latter context it is clear that the techniques and technology of integrated optics will have a crucial role to play. Polarimetric methods will make a major contribution to the development of optical fibre point and distributed sensor applications providing, as they do, a flexible compromise between the robust and simple on the one hand and the highly sensitive and sophisticated on the other. The range of applications which lies ahead is large and varied.

References

1. V. Vali and R.W. Shorthill, 'Fibre ring interferometer', *Appl. Opt.*, 15 (5), 1976, pp. 1099–1100.
2. J.A. Bucaro, H.D. Dardy E. Carome, 'Fibre-optic hydrophone', *J. Acoust. Soc. Am.*, 62, 1977, pp. 1302–1304.
3. A.J. Rogers, 'Optical measurement of current and voltage on power systems', *IEE J. Electr. Power Appl.*, 2 (4), 1979, pp. 120–124.
4. A.J. Rogers, 'Intrinsic and extrinsic distributed optical-fibre sensors', *Int. J. of Optical Sensors*, 1 (2), 1986, pp. 171–186.
5. J.F. Nye, Physical properties of crystals (Oxford University Press, London), 1957, chap. 14 and appendix H.
6. H. Poincaré, Théorie Mathématique de la lumière, Vol. 2, (Gauthiers-Villars, Paris), 1892, chap. 12.
7. H.G. Jerrard, 'Transmission of light through birefringent and optically active media', *J. Opt. Soc. Am.*, 44 (8), 1954, pp. 634–640.
8. M. Born and E. Wolf, Principles of optics, Pergamon Press, New York, 1959, p. 559.

9. A.M. Smith, 'Optical fibres for current measurement applications', *Opt. and Laser Technol.*, February 1980, pp. 25–30.
10. A.M. Smith, 'Optical fibre current measurement device at a generating station', *European Conference on Optical Systems and Applications (Utrecht)*, 236, pp. 352–357.
11. A.J. Barlow, J.J. Ramskov-Hansen and D.N. Payne, 'Anisotropy in spun single-mode fibres', *Electron. Lett.*, 18 (5), 1982, pp. 200–202.
12. A.M. Smith, 'Bend-induced birefringence in single-mode optical fibres', *European Conference on Optical Fibre Communications, Amsterdam*, 1979, paper 10.2.
13. A. Simon and R. Ulrich, 'Polarization optics of twisted single-mode fibres', *Appl. Opt.*, 18, 1977, pp. 2241–2251.
14. J.N. Ross, 'The rotation of the polarization in low birefringence monomode fibres due to geometric effects', *Opt. & Quantum Electron*, 16, 1984, pp. 455–461.
15. A.J. Rogers, 'Optical methods for measurement of voltage and current on power systems', *Opt. & Laser Technol.*, December 1977, pp. 273–283.
16. M.P. Varnham, D.N. Payne, R.D. Birch and E.J. Tarbox, 'Single-polarization operation of highly birefringent bow-tie fibres', *Electron. Lett.*, Vol. 19, 1983, pp. 246–247.
17. T. Hoska, K. Okamoto, T. Miya, Y. Sasaki and T. Edahiro, 'Low-loss single polarization fibres with asymmetrical strain birefringence', *Elect. Lett.*, Vol. 7, 1981, pp. 530–531.
18. M. Corke, A.D. Kersey, K. Li and D.A. Jackson, 'Remote temperature sensing using polarization-preserving fibre', *Elect. Lett.*, Vol. 20, 1984, pp. 67–69.
19. J.D.C. Jones, D.A. Jackson, M. Corke and A.D. Kersey, 'Single-mode fibre optic sensors: conventional interferometric, polarimetric and combined devices', *Proc. 'Fibre Optics 85', London*, April 1985.
20. J.P. Dakin and C.A. Wade, 'Compensated polarimetric sensor using polarization-maintaining fibre in a differential combination', *Elect. Lett.*, Vol. 20, 1984, pp. 51–53.
21. J.P. Dakin, S. Broderick, D.C. Carless and C.A. Wade, 'Operation of a compensated polarimetric sensor with a semiconductor light source', *Proc. 2nd Int. Conf. on Optical Fibre Sensors, OFS '84, Stuttgart*, September 1984, pp. 241–246.
22. A.J. Rogers, 'Polarization-optical time domain reflectometry', *Appl. Opt.*, 20 (6), 1981, pp. 1060–1074.
23. M.K. Barnoski and S.M. Jensen, 'Fibre waveguides: a novel technique for investigating attenuation characteristics', *Appl. Opt.*, Vol. 15, No. 9, 1976, pp. 2112–2115.
24. R. Clark Jones, 'A new calculus for the treatment of optical systems', *J. Opt. Soc. Am.*, 38 (8), 1948, pp. 671–685.
25. A.J. Rogers, 'Polarization properties of monomode fibres: The use of POTDR to determine spatial distributions', *Proc. Int. Conf. on Fibre-optic rotation sensors and related technologies, MIT Cambridge (Mass.)*, [Springer-Verlag 1982], November 1981, pp. 208–214.
26. J.M. Dziedzic, R.H. Stolen and A. Ashkin, 'Optical Kerr effect in long fibres', *Appl. Opt.*, 20 (8), 1981, pp. 1403–1406.
27. M.C. Farries and A.J. Rogers, 'Temperature dependence of the Kerr effect in a silica optical fibre', *Electron. Lett.*, 19 (21), 1983, pp. 890–891.
28. M.C. Farries and A.J. Rogers, 'Distributed sensing using the stimulated Raman interaction in a monomode optical fibre', *Proc. 2nd Int. Conf. on optical fibre sensors, Stuttgart*, paper 4.5, September 1984, pp. 121–132.
29. R.B. Franks and R.C. Youngquist, 'A birefringent fibre stress-location sensor', *IEE Coll. on distributed optical fibre sensors, London, Digest No. 1986/74*, paper 6, 12 May 1986.
30. A.J. Rogers, 'Acoustic wave control of optical polarization in optical fibres', *UK Patent Application 8616051*, filed 1 July 1986.
31. C.D. Hussey, R.D. Birch and Y. Fujii, 'Circularly birefringent single-mode optical fibres', *Elect. Lett.*, 22 (3), 1986, pp. 129–130.

Advances in distributed fibre optic detection and ranging

S.A. Kingsley
Fiberdyne Optoelectronics, USA

IN line with the terminology used to describe free-space optical radar systems, either light detection and ranging (LIDAR) or laser detection and ranging (LADAR), we would like to introduce a new word into the technical vocabulary, the acronym FODAR, which stands for fibre optic detection and ranging. As with free-space optical radar systems, the range information can be acquired by the use of optical time domain reflectometry (OTDR) or by optical frequency domain reflectometry (OFDR). Variations on these schemes can also exist, such as polarisation optical time domain reflectometry (POTDR), differential absorption optical time domain reflectometry (DAOTDR) and optical space domain reflectometry (OSDR). Sensors can be quasi-distributed, consisting of a number of Mach-Zehnder interferometers arranged in tandem, in either a series or parallel configuration. A technique known as coherence multiplexing may be used to isolate individual interferometers.

While this paper is essentially about ranging in optical fibre systems, much of what is discussed here is applicable to free-space ranging, with or without the use of all-fibre components. The major difference between free-space optical ranging and ranging within an optical fibre is that, in general, the fractional power that is scattered back to the transmitter is greater in the case of the latter. The sensors may be intrinsic, extrinsic and/or quasi-distributed. Single-target optical ranging systems can employ fibres and fibre optic components instead of bulk optics. Applications for such systems include simple fibre optic probes for displacement, velocity, acceleration, particle counting and aperture sizing.

Optical time domain reflectometry (OTDR)

OTDR systems are perhaps the most basic of the FODAR systems, as they make use of the propagation delay of a pulse of light to obtain the range information. Figure 1 shows a generic incoherent OTDR system. Pulses of light are sent out and the back-scatter is detected. The system can be employed for single-target ranging or for probing a distributed or quasi-distributed (sections of fibre formed into coils to increase intrinsic sensing sensitivity) fibre optic sensing system. Commercial OTDR systems have spatial resolutions of down to about 10cm of fibre.

In FODAR time-domain systems, the range of a target is given by the round-trip propagation delay:

$$\tau = \frac{2Rn}{c}, \qquad (1)$$

where R = distance to target, n = refractive index of fibre and c = velocity of light.

Opto-Electronics Inc. announced an OTDR system with 1.5mm resolution using a 150mW laser diode with an FWHM pulse width of about 60ps (PFOS picosecond fibre optic system and SE10 signal enhancer). Note that to achieve such resolutions the signal processing must be able to detect pulse shifts corresponding to a fraction of the pulse width (the classic expression for resolutions assumes that they are only considered resolved when their separation in time is equivalent to the pulse width). The quoted resolution will not be realisable when the back-scatter is a continuum, as would be the case in a distributed fibre

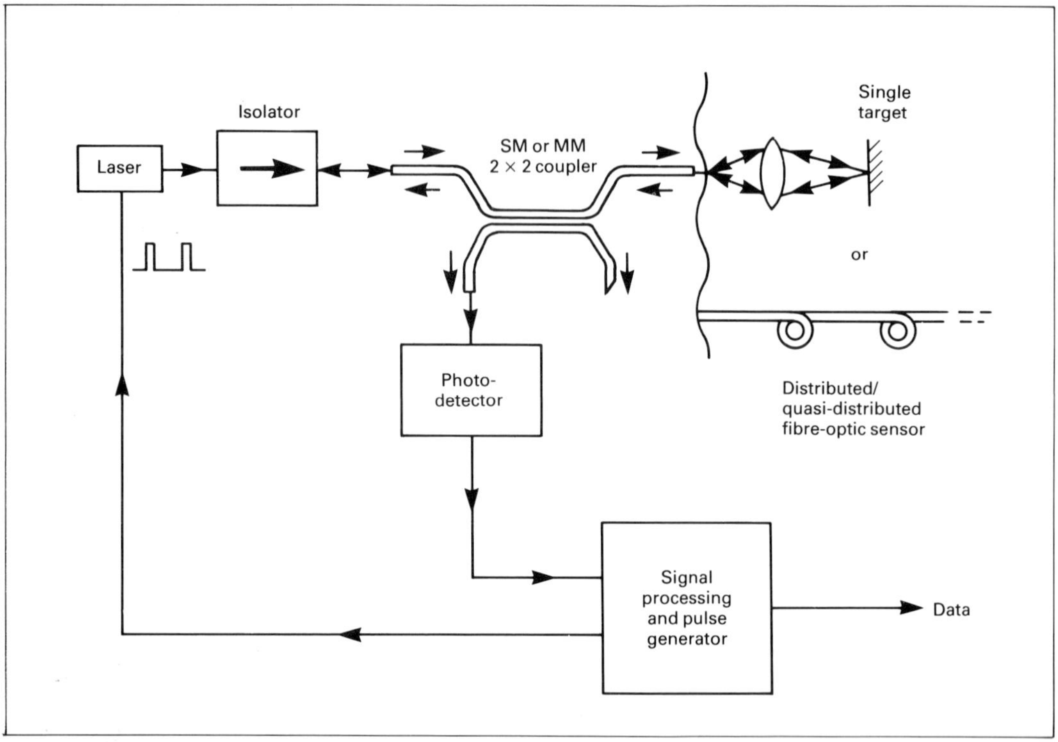

Fig. 1 FODAR system employing incoherent optical time domain reflectometry (OTDR)

optic sensing system (DFOSS). In such a system, the best resolution for a 60ps pulse would be about 6mm.

Figure 2 shows a basic coherent OTDR employing heterodyne detection. Such systems have been pioneered by Steve Wright and colleagues at STL. A major advantage of this system is that the dynamic

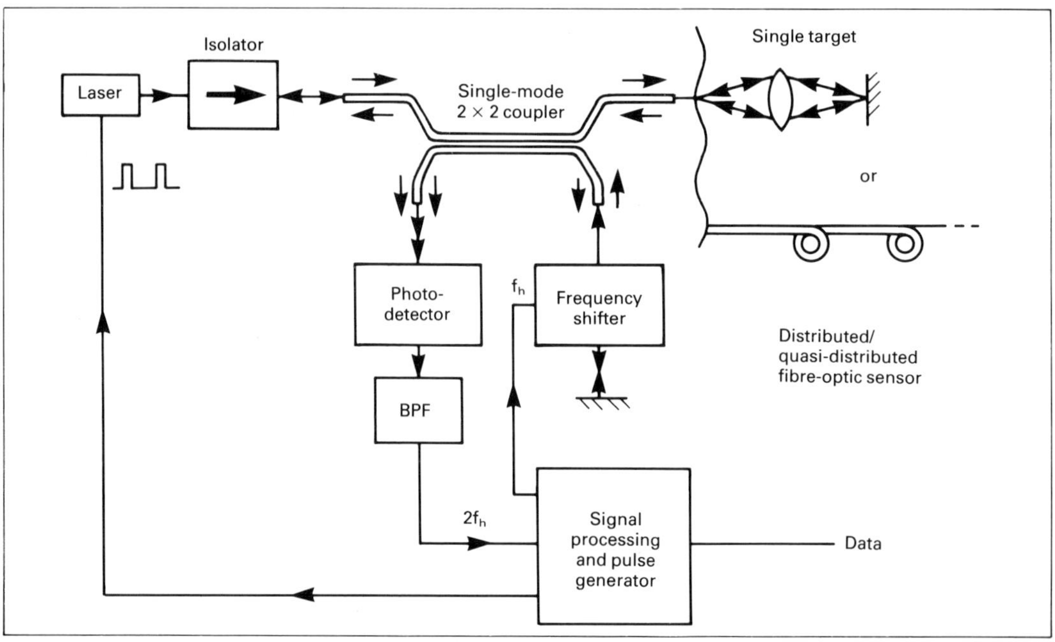

Fig. 2 FODAR system employing coherent optical time domain reflectometry (OTDR)

range and sensitivity is much higher than for incoherent OTDR due to the square-law photo-detector working in a linear fashion for homodyne or heterodyne detection systems. In other words, for each 10dB decrease in back-scattered power, the (electrical) signal from the photo-detector decreases by 10dB and not 20dB. Note that the double-pass frequency-shifter in the schematic produces a frequency-offset (hetero-dyne beat) of $2f_h$.

Optical frequency domain reflectometry (OFDR)

As with OTDR systems, OFDR systems can be divided into two basic classes: those employing incoherent techniques and those employing coherent techniques. Such systems have been and are being investigated by many research teams including D.E.N. Davies at University College London, England, and B. Culshaw at the University of Strathclyde, Scotland.

Figure 3 shows a generic incoherent OFDR system. An intensity-modulating subcarrier is frequency chirped and the back-scattered energy is mixed with a reference signal (the modulating subcarrier). The beat frequency f_b is proportional to range, viz:

$$f_b = \frac{\tau \Delta f}{T_m}, \qquad (2)$$

where Δf is the frequency deviation (chirp) and T_m is the period of the ramping waveform (assuming neligible flyback time). At a range of 1 metre, corresponding to a round-trip delay-time in an optical fibre of 10ns, and with a 10kHz ramp producing a 100GHz frequency deviation, a beat frequency of 10MHz will be produced. The isolator prevents the reflected light from disturbing the laser diode. This is particularly important in coherent systems where the return-loss from the target is low. Currently, only a few companies are manufacturing these magneto-optic devices, e.g. NEC and Optics for Research Inc., and these products are very expensive. Depending on wavelength, these Faraday rotators employ either a high-Verdet constant glass or yttrium-iron-garnet (YIG).

Probably the simplest system for single-target ranging is by cycle-counting. The range resolution of a cycle-counting system is given by:

$$\delta R = \frac{c}{4\Delta f}. \qquad (3)$$

For a deviation Δf of 100GHz, range resolutions of 0.75mm should be obtainable with simple cycle-counting. Reciprocal counting, spectral analysis and sideband-tracking techniques could be used to further increase the resolution. With one second integration times, resolution better than 10 m should be possible. Recent improvements in microwave voltage controlled oscillator (VCO) technology has led to the availability of state-of-the-art wide-deviation hyper-abrupt junction oscillators (such as the Avantek VCA-8000 VCO which can be swept at MHz rates between 8–12GHz, or the VCA-12000 which can be swept between 12–

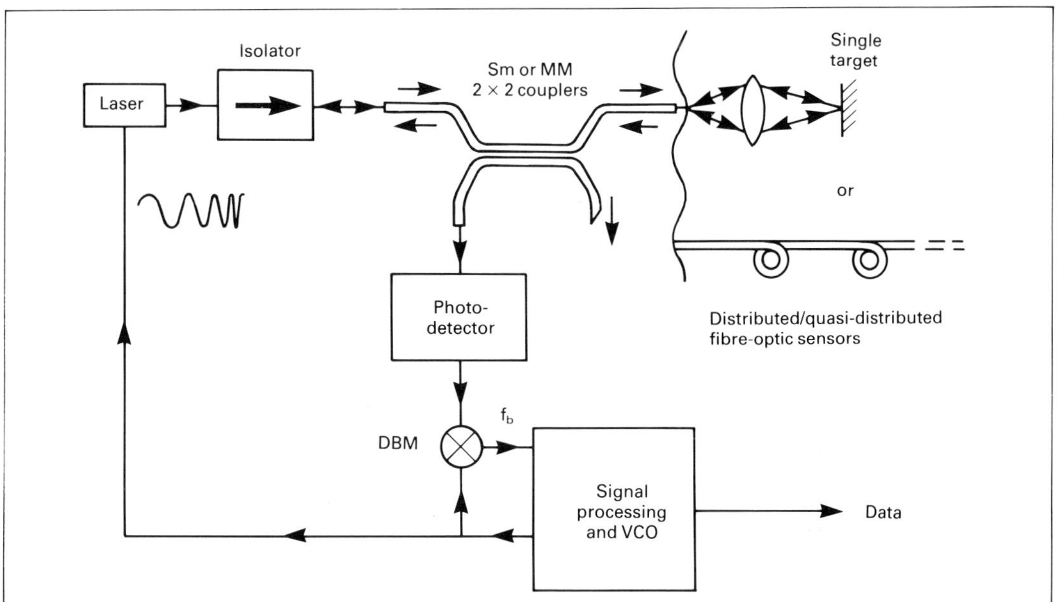

Fig. 3 FODAR system employing incoherent optical frequency domain reflectometry (OFDR)

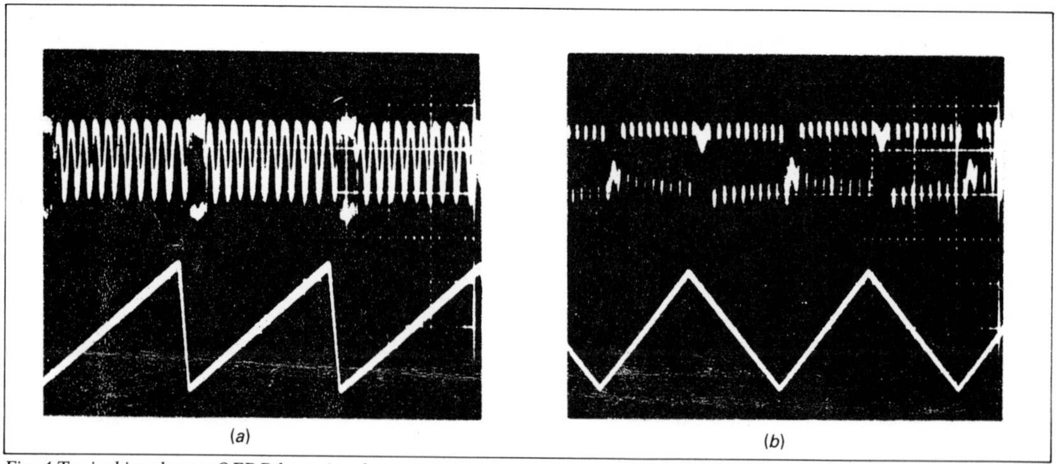

Fig. 4 Typical incoherent OFDR beat-signal responses

18GHz) that can produce very high spatial resolution in incoherent OFDR systems. Of course, state-of-the-art very high bandwidth laser diodes are also required, such as the 10GHz Ortel SL1020.

Figure 4 illustrates some typical incoherent OFDR beat signals obtained at low chirp rates from a mirror target at the end of a 100 metre fibre optic delay-line. Figure 4(a) shows the beat signal for a sawtooth ramping signal at 56kHz, while Figure 4(b) shows the recovered beat signal for a triangular-wave chirping signal at 48kHz. Frequency deviation of the VCO was approximately 10MHz.

The inherent quantisation error of cycle-counting schemes, as indicated by equation (3), means that for single target ranging, phase comparison techniques, where the phase of the detected (modulation envelope) light is compared to that of a high-frequency intensity-modulating subcarrier, generally offers much greater resolution for a given maximum (intensity) modulation frequency. State-of-the-art reciprocal counters can produce high-resolution if the overall sampling time (signal integration period) is not too short.

Figure 5 shows a basic coherent OFDR system. Here the actual frequency of the laser is chirped by applying a current ramp to the laser (thermal chirp) or, as shown in this particular schematic, by means of a double-cavity laser, where the rear cavity acts as a phase modulator. Potentially, much higher deviations are

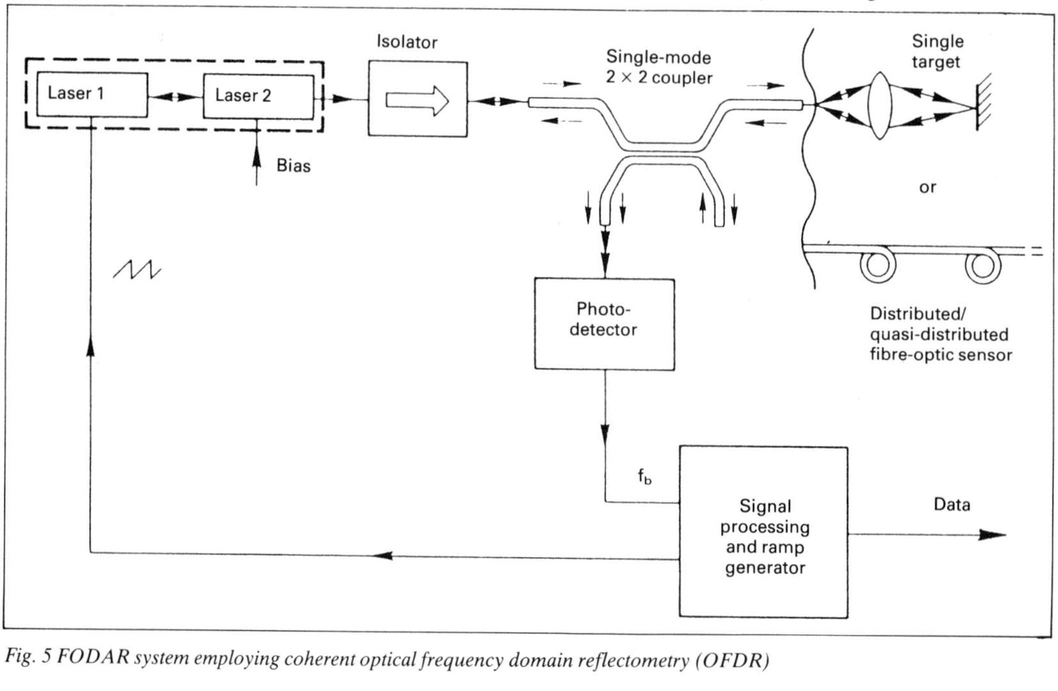

Fig. 5 FODAR system employing coherent optical frequency domain reflectometry (OFDR)

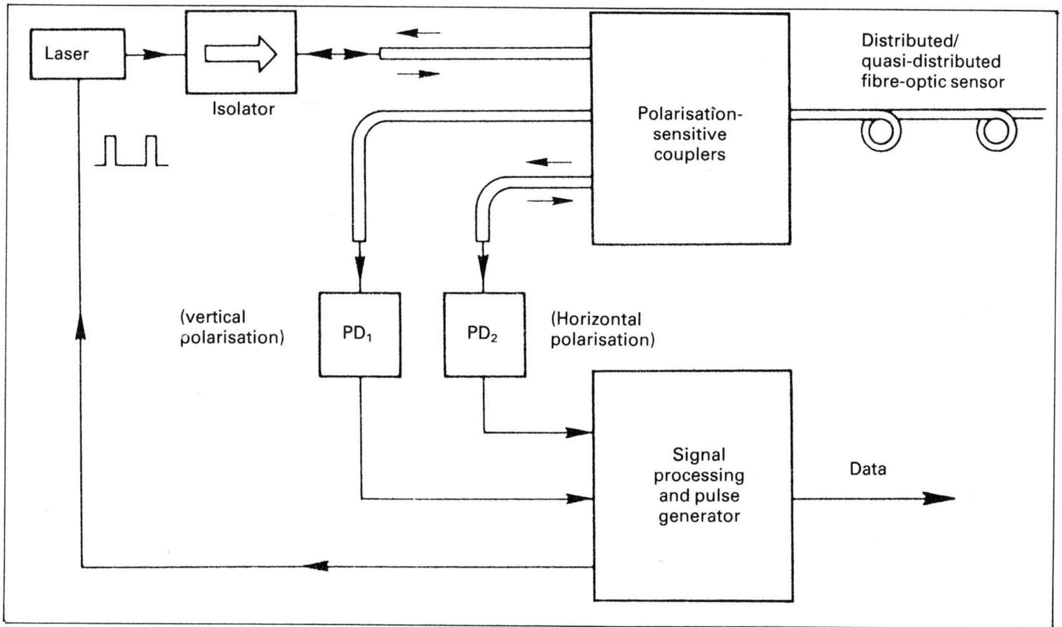

Fig. 6. FODAR system employing polarisation optical time domain reflectometry (POTDR)

possible with a coherent chirp and, hence, increased spatial resolution. The back-scattered energy is mixed with an optical reference signal in a Michelson interferometer to yield a beat signal proportional to range. Recently, a research group at the Beaverton Research Center (Defreez *et al*) have demonstrated that commercial laser diodes can be modified by ion-milling to produce electronically-tunable devices.

OFDR systems do have problems. First, as far as spectral analysis approaches are concerned, OFDR systems have high sidelobe levels and, second, the coherent systems produce beats between beats, thus making it difficult to sort out real signals from the spurious. This makes it necessary to judiciously place distributed and quasi-distributed sensors to avoid range ambiguities or phantom signals. Coherence multiplexing can be employed to suppress cross-term signals.

Polarisation optical time domain reflectometry (POTDR)

Polarisation-sensitive systems can be used in both the time and frequency domains. Figure 5 shows a POTDR system. Polarisation-sensitive systems have been extensively investigated by a group at CERL. Figure 6 shows a basic POTDR system that might be used to make a distributed temperature measurement.

Differential absorption optical time domain reflectometry (DAOTDR)

DAOTDR systems have been pioneered by E. Theocharous at CERL, and such a system is shown in figure 7. The measurement of the ratio of the back-scattered power at two wavelengths can provide the spatial distribution of the difference of the absorption coefficient of the fibre core at the two wavelengths. This type of system can be used to sense temperature in Europium-doped fibres.

The major advantage of this approach is that there is very good common-mode rejection of modulated bending-loss in the leads. The differential absorption is given by:

$$\alpha_1(R) - \alpha_2(R) = \frac{1}{2}\frac{d}{dR}\left\{\ln \frac{P_1(R)}{P_2(R)}\right\} \qquad (4)$$

where α_1 and α_2 are the absorption coefficients at closely spaced wavelengths λ_1 and λ_2, and P_1 and P_2 are the respective back-scattered powers at range R.

Non-linear optical time domain reflectometry

The research groups at CERL and Plessey have also investigated distributed systems employing non-linear phenomena in optical fibres. Figure 8 illustrates a temperature sensor using a high-power semiconductor laser and Raman ratio thermometry. Both argon and semiconductor lasers have been used in different versions of this scheme. The intensity of the Raman back-scattered light has a wavelength that is dependent on temperature. Hence, by analysing the returned light

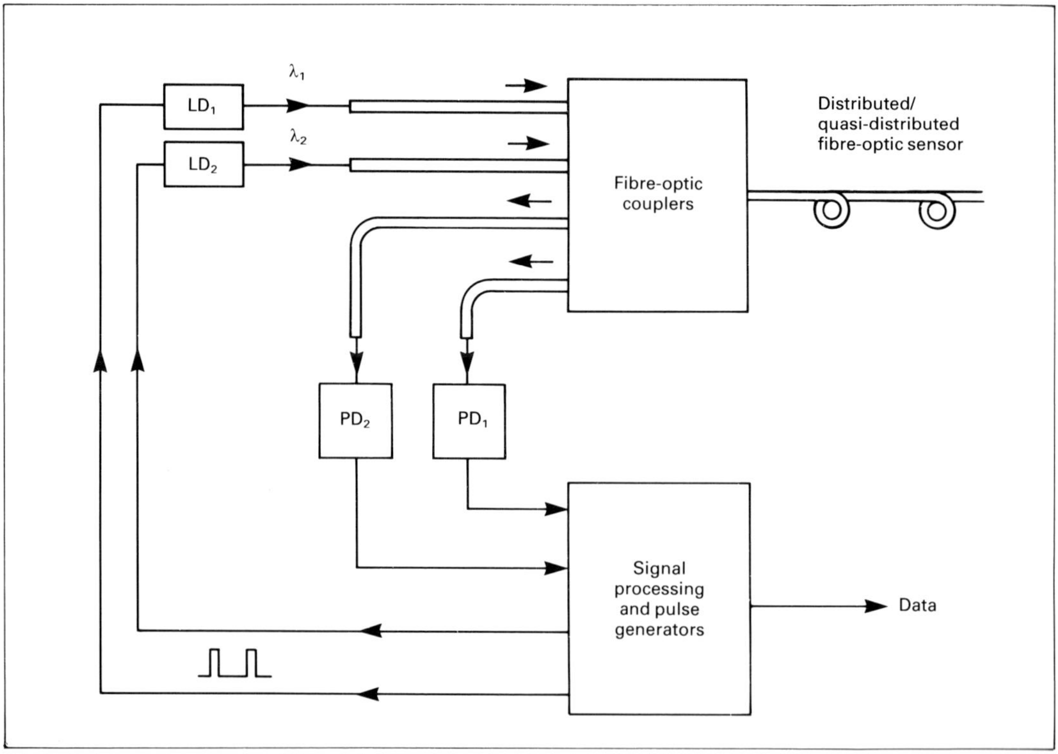

Fig. 7. Fodar system employing differential absorption optical time domain reflectometry (DAOTDR)

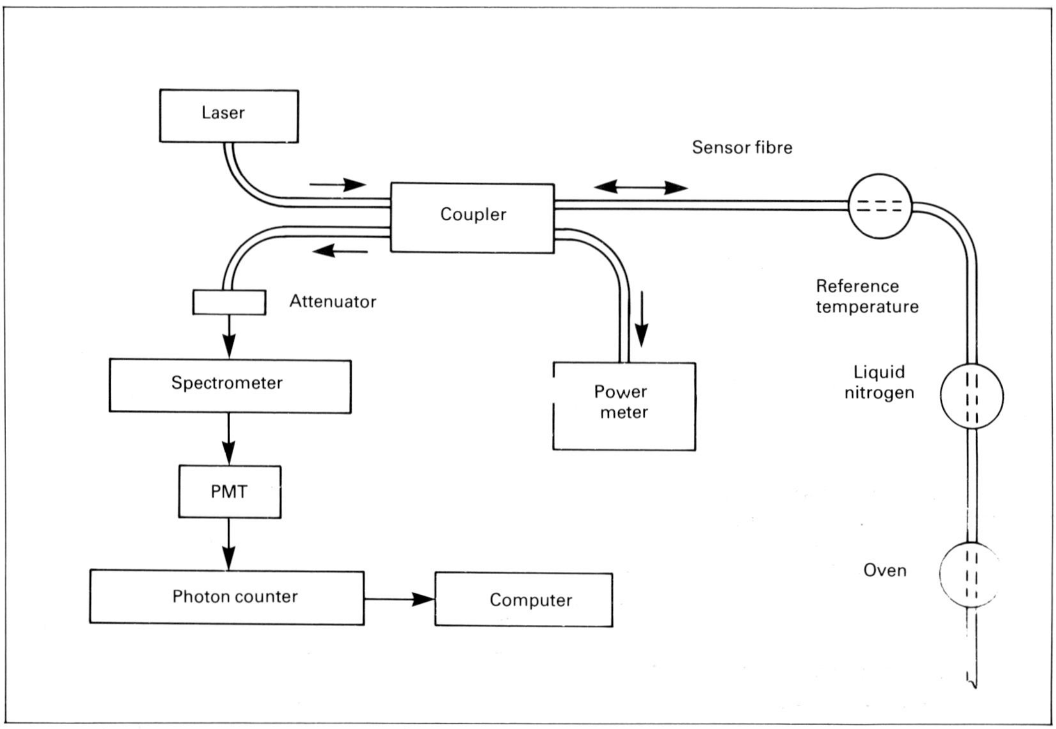

Fig. 8 Temperature sensing using non-linear OTDR

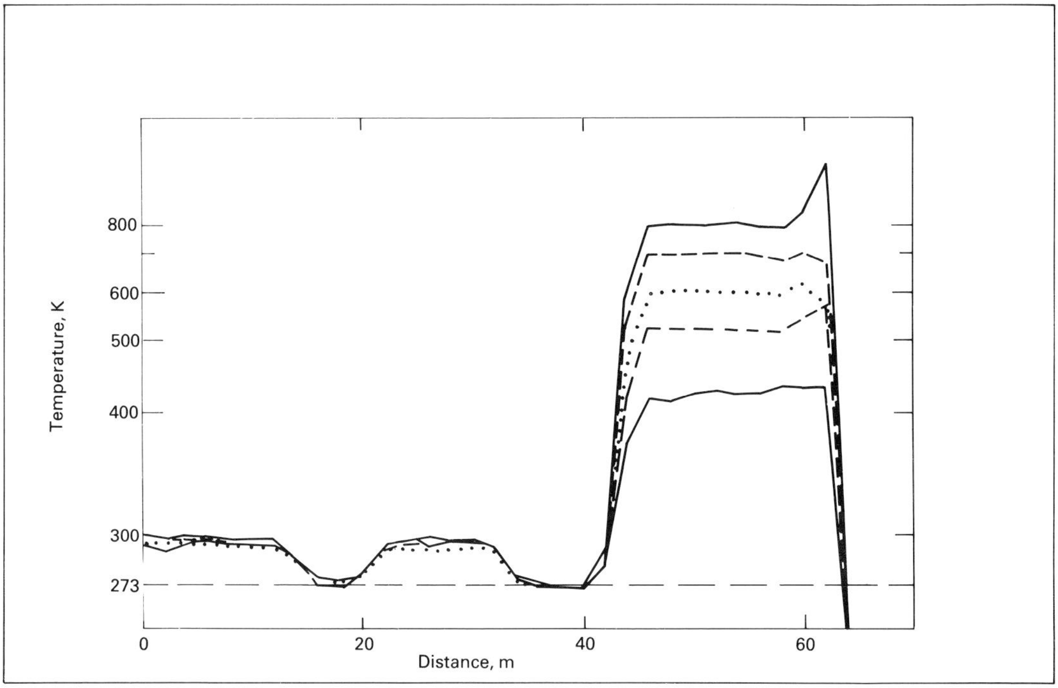

Fig. 9 Distributed fibre-optic thermometry using anti-Stokes/Stokes ratiometry (CERL)

with a spectrometer, and an OTDR system, it is possible to determine the temperature at any particular location along the fibre. This sensing technique has the major advantage that conventional communication-grade optical fibres can be used for temperature sensing. Figure 9 shows some experimental results obtained by CERL using an argon laser system.

Optical spatial domain reflectometry (OSDR)

N. Frigo *et al* at NRL have shown how very high spatial resolution may be obtained with low-bandwidth optoelectronics, by using the Faraday (magneto-optic) effect in single-mode fibres. Figure 10 illustrates their system.

The light reflected from the end of the fibre is analysed by a polarising beam-splitter. The solenoid is moved along the fibre until it passes over a region with a flaw. The reflection from flaw A will be unmodulated, while the reflection from flaw B will be polarisation-modulated via the Faraday effect. Spatial resolutions of approximately 1cm have been reported. Obviously, this approach is not very useful for a true distributed sensing system, though it can produce high spatial resolution for flaw detection if the solenoid can be slid along the fibre. The experimental system shown in figure 10 did exhibit the usual problems associated with the use of non-polarising-maintaining fibre.

Quasi-distributed Mach-Zehnder interferometers

Figure 11 shows a configuration for multiplexing quasi-distributed Mach-Zehnder interferometers. In this particular schematic, each interferometer has a differing path-length imbalance. This and other topologies have been extensively investigated by J. Shaw's group at Stanford, USA, D.E.N. Davies' group at University College London, England, and B. Culshaw at Strathclyde, Scotland. These sensors may be wavelength multiplexed or multiplexed by the use of OTDR or OFDR techniques. By making use of the limited coherence properties of semiconductor lasers, crosstalk problems between sensors may be minimised.

Discussion

Single and multiple-target ranging with fibre optics and fibre optic componentry will become a major application area for fibre optic sensors. It is perhaps not surprising that temperature has been one of the most important measurands investigated to date. See the companion paper in chapter 3 by S.A. Kingsley on distributed fibre optic hot-spot sensors. Other applications for FODAR DFOSS include stress monitoring in composite materials, electric and magnetic fields, corrosion, nuclear radiation dosimetry, chemical and biological sensing.

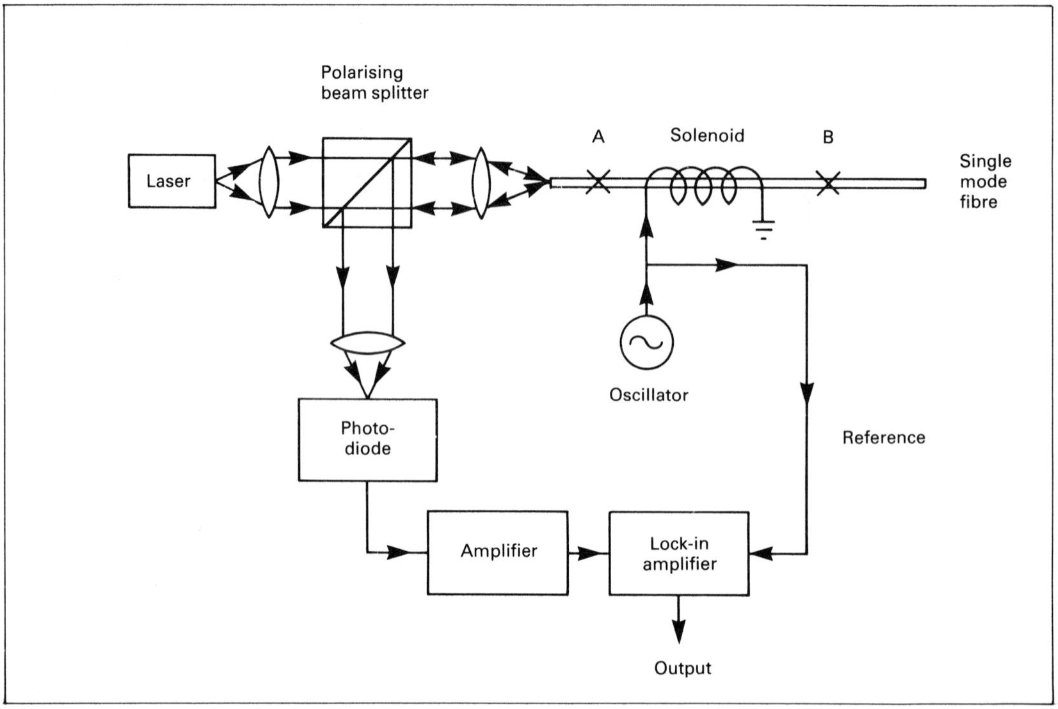

Fig. 10 Optical space domain reflectometry (OSDR)

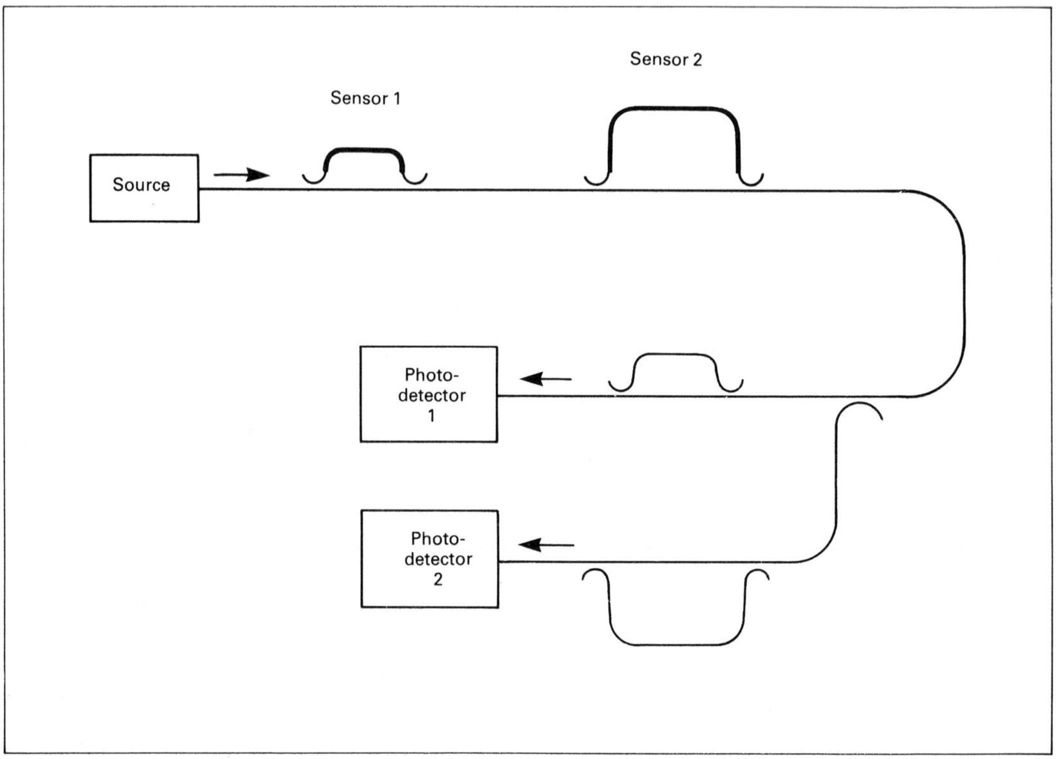

Fig. 11 FODAR system employing quasi-distributed Mach-Zehnder interferometers and coherence multiplexing (Stanford)

In this brief review it was not possible to cover every technique and system configuration published in the literature. For this reason, we have limited this overview to some of the major approaches. The extensive list of 'further reading' given at the end of this paper has been given to include other publications. To assist the reader, the list has been largely organised into sections conforming to the headings in this paper.

Concluding remarks

Distributed FODAR systems are a new class of sensors that allow intensity, phase, frequency, or spectral information to be acquired anywhere along the length of an optical fibre.

The availability of 'all-fibre' componentry will also assist in the development of this technology. Laser Doppler velocimetry devices are very similar to some of the systems described above, and make use of the same technology. Low-cost in-line fibre optic phase-modulators, frequency-shifters, polarisers, polarisation retarders, isolators and couplers will revolutionise this technology. However, in some situations, mini bulk-optics may be preferable to all-fibre componentry.

FODAR, as applied to fibre optic sensing systems, is likely to become one of the most important sensing attributes of this technology, for with it we can sense measurands over distances and areas that would not be technically and/or economically practical with any alternative technology.

Further reading

Laser radar

C.G. Bachman, *Laser Radar System and Techniques*, Artech House, 1979.

Fibre optic and distributed sensing - general

B. Bendow and S.S. Mitra (Editors), *Fibre Optics*, Plenum Press, New York and London, 1979.

B. Culshaw, *Optical Fibre Sensing and Signal Processing*, Peter Peregrinus Ltd., London. An IEE publication, 1984.

D.E.N. Davies, "Signal Processing for Distributed Optical Fibre Sensors", *Proc. 2nd International Conference on Optical Fibre Sensors, Stuttgart*, 5–7 September 1984, pp. 285–295.

A. Safaai-Jazi, C.K. Jen and G.W. Farnell, "Optical Fiber Sensor Based on Differential Spectroscopic Absorption", *Applied Optics*, Vol. 24, No. 15, 1985, pp. 2341–2345.

A.J. Rogers, "Intrinsic and Extrinsic Distributed Optical-Fibre Sensors", *Proc. SPIE 29th Annual Technical Symposium, San Diego, California, USA*, Vol. 566, 18–23 August 1985, pp. 234–242.

G. Meltz and J.R. Dunphy, "Fiber Optic Sensors for the Nondestructive Evaluation of Composite Materials", *Proc. SPIE 29th Annual Technical Symposium, San Diego, California, USA*, Vol. 566, 18–23 August 1985, pp. 159–163.

R.O. Claus, B.S. Jackson and K.D. Bennett, "Non-destructive Testing of Composite Materials by OTDR in Imbedded Optical Fibers", *Proc. SPIE 29th Annual Technical Symposium, San Diego, California, USA*, Vol. 566, 18–23 August 1985, pp. 243–248.

K.L. Belsley, J.B. Carroll, L.A. Hess, D.R. Huber and D. Schmadel, "Optically Multiplexed Interferometric Fiber Optic Sensor System", *Proc. SPIE 29th Annual Technical Symposium, San Diego, California, USA.*, Vol. 566, 18–23 August 1985, pp. 257–264.

S.A. Kingsley, "Distributed Fiber-Optic Sensors: An Overview", *Proc. SPIE 29th Annual Technical Symposium, San Diego, California, USA.*, Vol. 566, 18–23 August 1985, pp. 28–36.

V.I. Busurin, A.S. Semenov and N.P. Udalov, "Optical and Fiber-Optic Sensors (Review)", *Sov. J. Quantum Electron.*, Vol. 15, No. 5, 1985, pp. 595–621.

G.D. Pitt, P. Extance, R.C. Neat, D.N. Batchelder, R.E. Jones, J.A. Barnett and R.H. Pratt, "Optical-Fibre Sensors", *Proc. IEE*, Vol. 132, Pt. J, No. 4, 1985.

A.J. Rogers, R.J. Weiss and S. Chomet (Editors), *International Journal of Optical Sensors*, Vol 1, No. 1, January 1986.

S.A. Kingsley, "Distributed FODAR Sensors", *Presented at the 3rd Annual Hostile Environments & High Temperature Conference*, Cincinnati, Ohio, USA, 25–26 March 1986. Also presented at the IEE Colloquium on Distributed Optical-Fibre Sensors, London, England, 12 May 1986, Digest No. 1986/74.

J.P. Dakin and D.J. Pratt, "Fiber-Optic Measurement - A Comparative Study of Techniques", *Colloquium on Distributed Optical Fibre Sensors*, IEE, Savoy Place, London, England, 12 May 1986.

T.G. Giallorenzi, J.A. Bucaro, A. Dandridge and J.H. Cole, "Optical Fiber Sensors Challenge The Competition", *IEEE Spectrum*, Vol. 23, No. 9, September 1986, pp. 44–49.

S.A. Kingsley and V.D. McGinniss, "Distributed Fiber-Optic Hot-Spot Sensors", *Proc. SPIE Fiber/Lase '86*, Fiber Optic and Laser Sensors IV, Vol. 718, 22–24 September 1986.

Optical time domain reflectometry - general

D. Marcuse, *Principles of Optical Fiber Measurements*, Academic Press, chapter 5, 1981, pp. 236–254.

Y. Murakami, K. Noguchi, F. Ashiya, Y. Negishi and N. Kojima, "Maximum Measurable Distances for a Single-Mode Optical Fiber Fault Locator using the Stimulated Raman Scattering (SRS) Effect", *IEE J. Quant. Electron.*, Vol. QE-18, No. 10, 1982, pp. 1473–1477.

P. Healey, "Fading in Heterodyne OTDR", *Electronics Letters*, Vol. 20, No. 1, 1984, pp. 30–32.

B.L. Danielson, "Optical Time-Domain Reflectometer Specifications and Performance Testing", *Applied Optics*, Vol. 24, No. 15, 1985, pp. 2313–2322.

P. Healey, "Review of Long Wavelength Single-Mode Optical Fiber Reflectometry Techniques", *J. Lightwave Technology*, Vol, LT-3, No. 4, 1985, pp. 876–886.

R. Maciejko and L. Hadellis, "Two Signal Processing Enhancements for Optical Time Domain Reflectometry", *J. Lightwave Technology*, Vol. LT-4, No. 5, May 1986, pp. 538–546.

G. Ripamonti and S. Cova, "Optical Time-Domain Reflectometry with Centimeter Resolution at 10^{-15}W Sensitivity", *Electronic Letters*, Vol. 22, No. 15, 1986, pp. 818–819.

Incoherent & coherent optical time domain reflectometry

E. Theocharous, "Scattering-Optical Time Domain Reflectometry: A Technique for the Measurement of Temperature Distributions", *CEGB Memorandum, TPRD/L/AP 0021/M82.*

S.A. Kingsley, "Distributed Fiber-Optic Sensors", *Proc. of the 1984 ISA Show*, Houston, Texas, 22–25 October 1984, pp. 315–330.

D.C. Marvin and N.A. Ives, "Wide-Range Feber-Optic Strain Sensor", *Applied Optics*, Vol. 23, No. 23, 1984, pp. 4212–4217.

A.R. Mickelson, O. Klevhus and M. Eriksrud, "Backscatter Readout from Serial Microbending Sensors", *J. Lightwave Technology*, Vol. LT-2, No. 5, 1984, pp. 700–709.

C.K. Asawa, M.K. Barnoski and S. Yao, "Microbending of Optical Fibers for Remote Force Measurement", *United States Patent No. 4, 463, 254*, 31 July 1984.

S.A. Kingsley, "Fiber-Optic Sensors: Opportunities for Distributed Measurement", *InTech*, Vol. 32, No. 8, 1985, pp. 44–48.

A.H. Hartog and A.P. Leach, "A Practical Optical-Fibre Distributed Temperature Sensor", *Colloquium on Distributed Optical Fibre Sensors*, IEE, Savoy Place, London, England, 12 May 1986.

C.A. Wade, J.P. Dakin and P.B. Withers, "Multiplexed Interferometric Optical Fibre Sensors", *Colloquium on Distributed Optical Fibre Sensors*, IEE, Savoy Place, London, England, 12 May 1986.

Incoherent optical frequency domain reflectometry

R.I. MacDonald, "Frequency Domain Optical Reflectometer", *Applied Optics*, Vol. 20, No. 10, 1981, pp. 1840–1844.

Mehdi Shadaram and William L. Kuriger, "Using the Optical Frequency Domain Technique for the Analysis of Discrete and Distributed Reflections in an Optical Fiber", *Applied Optics*, Vol. 23, No. 7, 1984, pp. 1092–1095.

T.T. Tjhung, S.K. Teo, F.V.C. Mendis and B. Selvan, "Refractometry Through Optical Frequency-Domain Reflectometry", *Electronics Letters*, Vol. 21, No. 14, 1985, pp. 613–614.

H. Ghafoori-Shiraz and T. Okoshi, "Optical-Fiber Diagnosis Using Optical-Frequency-Domain Reflectometry", *Optics Letters*, Vol. 10, No. 3, 1985, pp. 160–162.

H. Ghafoori-Shiraz and T. Okoshi, "Fault Location in Optical Frequency Domain Reflectometry", *J. Lightwave Technology*, Vol. LT-4, No. 3, March 1986, pp. 316–322.

K.I. Mallalieu, R. Youngquist and D.E.N. Davies, "RF Band FMCW for Passive Multiplexing of Multimode Fibre Optic Sensors", *Colloquium on Distributed Optical Fibre Sensors*, IEE, Savoy Place, London, England, 12 May 1986.

K.I. Mallalieu, R. Youngquist and D.E.N. Davies, "FMCW of Optical Source Envelope Modulation for Passive Multiplexing of Frequency-Based Fibre-Optic Sensors", *Electronics Letters*, Vol. 22, No. 15, 1986, pp. 809–810.

F. Faridian, M.G. Somekh and I. Sakai, "New CW Imaging Mode for Scanning Acoustic Microscope", *Electronics Letters*, Vol. 22, No. 15, 1986, pp. 800–802.

Coherent optical frequency domain reflectometry

W. Eickhoff and R. Ulrich, "Optical Frequency Domain Reflectometry in Single-Mode Fiber", *Appl. Phys. Lett.*, Vol. 9, No. 39, 1981, pp. 693–695.

I.P. Giles, D. Uttam, B. Culshaw and D.E.N. Davies, "Coherent Optical-Fibre Sensors with Modulated Laser Sources", *Electronics Letters*, Vol. 19, No. 1, 1983, pp. 14–15.

D. Uttam, I.P. Giles, B. Culshaw and D.E.N. Davies, "Remote interferometric sensors using frequency modulated laser sources" *Proc. 1st International Conference on Optical Fibre Sensors*, London, IEE Conference Publication No. 221, 26–28 April 1983, pp. 182–184.

D. Uttam, I.P. Giles and B. Culshaw, "Closed Loop FM Interferometric Remote Optical Fibre Sensor", *Proc. SPIE Technical Symposium East '83*, Fiber Optic and Laser Sensors, Vol. 412, pp. 256–260.

S.A. Kingsley and D.E.N. Davies, "OFDR Diagnostics for Fibre and Integrated-Optic Systems", *Electronics Letters*, Vol. 21, No. 10, 1985, pp. 434–435.

G. Beheim and K. Fritsch, "Remote Displacement Measurements Using a Laser Diode", *Electronics Letters*, Vol. 21, No. 3, 1985, pp. 93–94.

S.A. Kingsley and D.E.N. Davies, "OFDR Diagnostics for Fiber/Integrated Optic Systems and High Resolution Distributed Fiber-Optic Sensing", *Proc. SPIE 29th Annual International Technical Symposium*, San Diego, California, USA, Vol. 566, 18–23 August 1985, pp. 265–275.

D. Uttam and B. Culshaw, "Precision Time Domain Reflectometry in Optical Fiber Systems Using a Frequency Modulated Continuous Wave Ranging Technique", *J. Lightwave Technology*, Vol. LT-3, No. 5, 1985, pp. 971–977.

I. Sakai, "Multiplexing Systems for Optical Fibre Sensors Using a Frequency Modulated Source", *Colloquium on Distributed Optical Fibre Sensors*, IEE, Savoy Place, London, England, 12 May 1986.

R.B. Franks and R.C. Youngquist, "A Birefringent Fibre Stress Location Sensor", *Colloquium on Distributed Optical Fibre Sensors*, IEE, Savoy Place, London, England, 12 May 1986.

R.K. Defreez, J. Puretz, R.A. Elliot, J. Orloff and L.W. Swanson, "CW Operation of Widely and Continuously Tunable Micromachined-Coupled-Cavity Diode Lasers", *Electronics Letters*, Vol. 22, No. 17, 1986, pp. 919–921.

Polarisation optical time domain reflectometry

A.J. Rogers, "Polarisation Optical Time Domain Reflectometry", *Electronics Letters*, Vol. 16, No. 13, 1980, pp. 489–490.

A.J. Rogers, "Polarization-Optical Time Domain Reflectometry: A Technique for the Measurement of Field Distributions", *Applied Optics*, Vol. 20, No. 6, 1981, pp. 1060–1074.

A.J. Rogers, "Polarization Properties of Monomode Optical Fibres: The Use of POTDR to Determine Spatial Distributions", *Proc. 1st International Conference on Fiber-Optic Rotation Sensors*, MIT, Cambridge, Mass, USA, 9–11 November 1981, Springer-Verlag, Vol. 32, 1982.

J.N. Ross, "Birefringence Measurement in Optical Fibers by Polarization-Optical Time-Domain Reflectometry", *Applied Optics*, Vol 21, No. 19, 1982, pp. 3489–3495.

M. Nakazawa, M. Tokuda and N. Uchida, "Analyses of Optical Time-Domain Reflectometry for Single Mode Fibers and of Polarization Optical Time-Domain Reflectometry for Polarization-Maintaining Fibers", *Optics Letters*, Vol. 8, No. 2, 1983, pp. 130–132.

D.G. Lubnau, "Polarization Backscatter Analysis of Field Distributions using Fiber Optics", *Applied Optics*, Vol. 22, No. 3, 1983, pp. 377–378.

M. Corke, A.D. Kersey, K. Liu and D.A. Jackson, "Remote Temperature Sensing Using Polarisation-Presserving Fibre", *Electronics Letters*, Vol. 20, No. 2, 1984, pp. 67–69.

A.J. Rogers and M.C. Farries, "Distributed Sensing Using Stimulated Raman Interaction in a Monomode Optical Fibre", *Proceedings of 2nd International Conference on Optical Fibre Sensors*, Stuttgart, paper 4.5, 5–7 September

1984, pp. 121–132.

A.J. Rogers, "Polarisation Optics for Monomode Fibre Sensors", Special Issue on Gyroscopes and Monomode Optical Components, *IEE Proc.*, Vol. 132, Pt. J, No. 5, October 1985, pp. 303–308.

Differential absorption optical time domain reflectometry

E. Theocharous, "Differential Absorption Distributed Thermometer", *Proc. 1st International Conference on Optical Fibre Sensors*, London England, IEE Conference Publication No. 221, 26–28 April 1983, pp. 10–12.

E. Theocharous, "Distributed Sensors Based On Differential Absorption", *Colloquium on Distributed Optical Fibre Sensors*, IEE, Savoy Place, London, England, 12 May 1986.

Non-linear optical time domain reflectometry

M. Nakazawa and M. Tokuda, "Measurement of the Fiber Loss Spectrum Using Fiber Raman Optical-Time-Domain Reflectometry", *Applied Optics*, Vol. 22, No. 12, 1983, pp. 1910–1914.

L.B. Jeunhomme, *Single-Mode Fiber Optics: Principles and Applications*, Marcel Dekker Inc., New York & Basel, Chapter 9, 1983, pp. 257–268.

K. Mochizuki, "Optical Fiber Transmission Systems Using Stimulated Raman Scattering: Theory", *J. Lightwave Technology*, Vol. LT-3, No. 3, 1985, pp. 688–694.

J.P. Dakin, D.J. Pratt, G.W. Bibby and J.N. Ross, "Distributed Optical Fibre Raman Temperature Sensor Using a Semiconductor Light Source and Detector", *Electronics Letters*, Vol. 21, No. 13, 1985, pp. 569–570.

J.P. Dakin, D.J. Pratt, G.W. Bibby and J.N. Ross, "Temperature Distribution Measurement Using Raman Ratio Thermometry", *Proc. SPIE 29th Annual Technical Symposium*, San Diego, California, USA, Vol. 566, 1985, pp. 249–253.

F. Cisternino, B. Costa, M.M. Rao and B. Sordo, "Tunable Backscattering for Single-Mode Optical Fibers in 1.1–1.6μm Spectral Region", *J. Lightwave Technology*, Vol. LT-4, No. 7, 1986, pp. 884–888.

Optical spatial domain reflectometry

N.J. Frigo, A. Dandridge and C.A. Villarruel, "An Optical Space Domain Reflectometer Based on the Faraday Effect", *J. Lightwave Technology*, Vol. LT-4, No. 3, March 1986, pp. 256–258.

Quasi-distributed Mach-Zehnder interferometers

S.A. Al-Chalabi, B. Culshaw and D.E.N. Davies, "Partially Coherent Sources in Interferometric Sensors", *Proc. 1st International Conference on Optical Fibre Sensors*, London, IEE Conference Publication No. 221, 26–28 April 1983, pp. 132–135.

J.L. Brooks, R.H. Wentworth, R.C. Youngquist, M. Tur, B.Y. Kim and H.J. Shaw, "Coherence Multiplexing of Fiber-Optic Interferometric Sensors", *J. Lightwave Technology*, Vol. LT-3, No. 5, 1985, pp. 1062–1072.

R.H. Wentworth and H.J. Shaw, "Expected Noise Levels for Interferometric Sensors Multiplexed Using Partially Coherent Light", *Proc. SPIE 29th Annual Technical Symposium, Fiber Optic and Laser Sensors III*, San Diego, California, USA, Vol. 566, 18–23 August 1985, pp. 212–217.

J.L. Brooks, M. Tur, B.Y. Kim, K.A. Fesler and H.J. Shaw, "Fiber-Optic Interferometric Sensor Arrays with Freedom from Source Phase-Induced Noise", *Optics Letters*, Vol. 11, No. 7, July 1986, pp. 473–475.

Distributed sensor systems design

THIS second chapter covers the details of many of the basic concepts used in networked fibre sensors.

The first paper, by A.J. Rogers, covers sensors which exploit the changes in the polarisation of light that can occur in response in external influences, and describes a number of ingenious methods in some detail.

The second paper, by N.E. Jolley *et al*, is strictly a step-wise-distributed or multiplexed scheme which monitors optical phase modulation arising from mechanical strain in fibres. The main novel feature is the signal/noise advantage of coherent detection with a strong pump signal derived from the original light source.

The third item, by Dakin *et al*, represents a novel means of locating thermal or mechanical disturbances on a continuous length of fibre, using the 'Shupe effect', normally considered to be a disadvantage when present in optical fibre gyroscopes.

The fourth paper, by E. Theocharous, is again a multiplexing system for point sensors, and is interesting because it represents the first publication of attenuation modulation for step-wise continuous temperature monitoring.

The fifth contribution, by Bernard *et al*, and the sixth, by Everard *et al*, show how the concept of pseudo-random source modulation, well-known in conventional radar systems, can be exploited to improve the signal/noise ratio of distributed sensors when the system is limited by the peak-power limitations of the source.

The seventh paper, by Desforges *et al*, shows a number of novel multiplexed sensors that can be tagged on to a continuous (unbroken) fibre line. These are in number of configurations that, first, allow light to leave the fibre and then a fraction of the light to be recoupled back in, after transmission through the external sensor.

The eighth and ninth papers, by Franks *et al* and Kurosawa *et al*, respectively, show similar methods for the location of mechanical disturbance on an optical fibre, based on coupling between forward-travelling modes. Due to frequency chirping of the source, coupled into each mode at the launch end, there is a frequency difference between the light in each mode depending on the location. Hence, the beat frequency, at the output detector, is a function of the coupling position. The paper by Kurosawa *et al* is subsequent to that of Franks *et al*, but has been considered worth including due to the differences in mathematics and presentation of results.

The tenth and final paper in this chapter, by A.C. Boucouvales *et al*, considers the design of special D-shaped fibres for sensors. The designs of other special fibres for the sensing of a single parameter are treated in later sections.

Distributed optical fibre sensors for the measurement of pressure, strain and temperature

A.J. Rogers
King's College London, UK

OPTICAL fibre methods of measurement sensing offer many important advantages for industrial use:[1,2] the fibre is a flexible, insulating, dielectric medium which can readily be installed in industrial plant without significant disturbance of the measurement environment; the range of measurands which is accessible to measurement by optical fibre techniques is very large, since the propagation of light within an optical fibre is sensitive to a wide variety of physical influences external to it.[3,4]

Optical fibre distributed measurement sensing[5] is a technique which utilises the one-dimensional nature of the optical fibre as a distinct measurement feature. It is possible, in principle, to determine the value of a wanted measurand continuously as a function of position along the length of a suitably-configured optical fibre, with arbitrarily large spatial resolution. The normal temporal variation of the distribution is determined simultaneously.

Such sensing systems are normally referred to as fully-distributed systems, to distinguish them from the quasi-distributed systems, which possess the capability of sensing the measurand only at a number of discrete, predetermined points.

The fully-distributed facility opens up an enormous number of possibilities for industrial application. For example, it would allow the spatial and temporal strain distributions in large critical structures such as multi-storey buildings, bridges, dams, aircraft, pressure vessels, electrical generators, etc. to be monitored continuously. It would allow the temperature distributions in boilers, power transformers, power cables, aerofoils, office blocks, etc. to be determined, and thus heat flows to be computed. Electric and magnetic field distributions could be mapped in space so that electro-magnetic design problems would be eased and sources of electro-magnetic interference would be quickly identifiable.

There are two important definable reasons why one would wish to obtain the information afforded by distributed optical fibre measurement sensors. The first is that of providing continuous monitoring so as to obtain advance warning of any potentially damaging condition in a structure, and thus to allow alleviative action to be taken in good time. The second is that this spatial and temporal information allows a much deeper understanding of the behaviour of large (or even quite small) structures, with many implications for improvements in their basic design.

Conventional industrial measurement sensor technology does not provide this facility. When measurand distributions of any kind are vital in a given situation the solution usually is to festoon the structure with a multitude of thermocouples or strain gauges, or whatever. This then presents problems of multiplexing, logging and calibration and, in any case, relies on the choice of positions for each of the many sensors being the correct one — a choice which cannot properly be made without a prior knowledge of the very distribution one is seeking to measure. This 'solution' is thus expensive, tedious and usually broadly inadequate.

The optical fibre can be readily installed in industrial plant (retrospectively if necessary), produces minimal disturbance of the measurement environment, is cheap, passive and electrically insulating, acts as its own telemetering channel, can easily be rearranged in accordance with acquired knowledge and allows a choice of any or all measurement points along its length within the limits of the spatial resolution interval. If such a technique can be made to work satisfactorily for a number of measurands, a new dimension appears in the field of industrial measurement.

The purpose of this present analysis is two-fold: it is, first, to define the principles and practical limitations of distributed optical fibre sensors, in order to assess the potential importance of the technique for industrial measurement; and, second, it is to point the way for a particular research strategy which might successfully spearhead the advance into provision of a range of practical, usable industrial systems.

The range of industrial measurands is large, the number of physical measurement methods is great. In order that the thrust of the paper should not be too diluted, it has been decided to focus the study on essentially two measurands: strain (τ) and temperature (T). Closely associated with strain are the parameters force (F) and pressure (p), since it is they which lead to the strain. Consequently we are effectively considering $\bar{p}$, $\bar{F}$, σ and $\bar{T}$. These are, of course, industrial parameters of almost universal importance.

Finally, it should perhaps be emphasised that the analysis will consider in detail only the 'fully' distributed linear systems described above, since these are by far the most versatile. It will not consider what are variously called quasi-distributed, multiplexed or multi-point array optical fibre sensor systems, where individual (either intrinsic or extrinsic) fibre sensors are addressed by means of an interrogating (not necessarily linear) fibre network. For details of these sorts of systems the reader is referred to other literature.[6–9]

Principles of distributed optical fibre measurement

In order to investigate the principles on which distributed optical fibre measurement sensors (DOFS) are based it is convenient to construct a model which embraces all possible forms of DOFS systems. This will allow an investigation of the broad behaviour of these sensors and will establish the relationships between the various practical, characterising system parameters. These relationships will facilitate the decisions which must be made about the trade-offs necessary in the design of the DOFS to meet a particular application requirement, and also allow quantitative comparison to be made between differing systems.

We shall, however, restrict the treatment to back-scatter systems. These are systems in which a propagating light pulse is back-scattered by inhomogeneities in the fibre, so as to provide a signal, at the front end of the fibre, whose time resolution provides the required information on the spatial position along the fibre. This restriction imposes no significant loss of generality as the extension to forward-scatter systems (e.g. 25) is straightforward. It does, however, offer the advantage of a considerable simplification of the algebra.

Systems parameters

The prime characterising parameters for any DOFS arrangement may be defined as follows (see Figs. 1 and 2):

Spatial resolution (δl). This is the minimum length of fibre over which sensible change in the spatial variation of the measurand can be detected. The spatial distribution of the measurand is thus presented effectively as a series of values, each value corresponding to its magnitude averaged over one spatial resolution length. The spatial resolution is measured in metres.

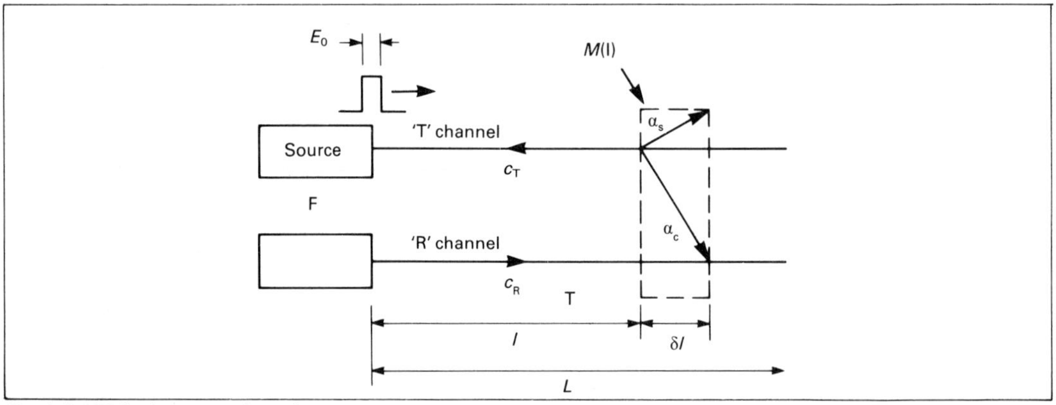

Fig. 1 DOFS back-scatter schematic

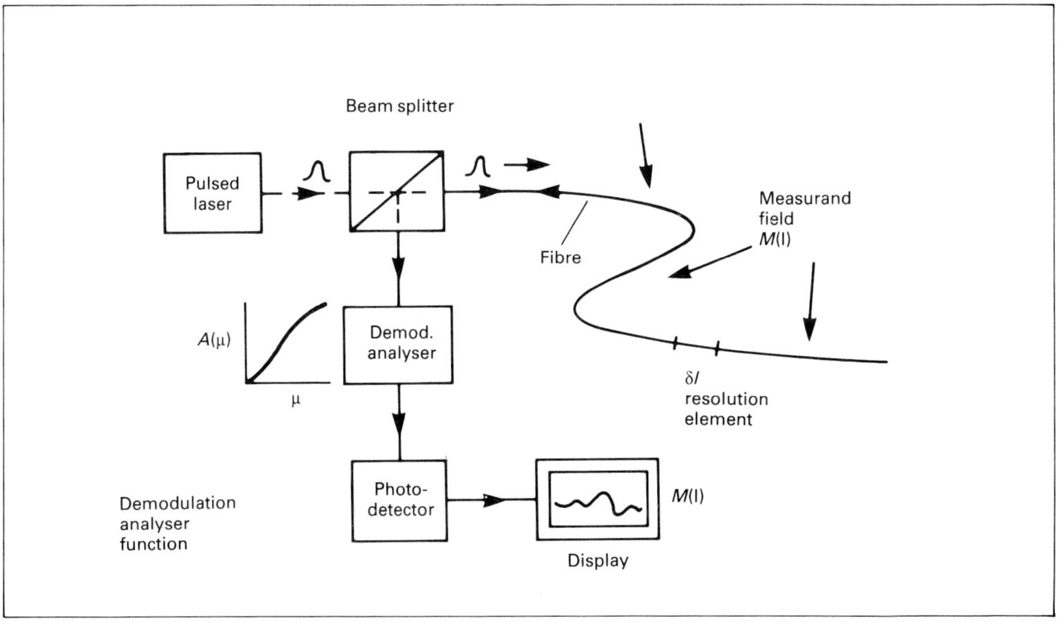

Fig. 2 Back-scatter arrangement for DOFS

Fibre path length (L). This is the total length of fibre over which sensible measurements can be made. 'Sensible' in this sense usually means 'within the specified accuracy'. It is measured in metres.

Sensitivity (S). The sensitivity of the DOFS is defined as the change of optical power at the detector input which is produced by unit change of measurand field per unit length of fibre. Its units will clearly depend on those of the measurand but typical examples are watts per microstrain per metre, watts per degree Kelvin per metre, etc.

Measurement bandwidth (B). The measurement bandwidth is the bandwidth over which the measurand can be measured over the complete length L of the fibre. Its units are hertz.

System bandwidth (W). This bandwidth is the one which the detector must possess in order to operate the complete system as a distributed sensor. Again, its units are hertz.

Noise level. This is the RMS fluctuation of the magnitude of the output signal, within the system bandwidth. Its units will be volts or amperes, depending on the system.

Dynamic range (D). The dynamic range is the ratio of maximum to minimum values of the measurand which allows each to be measured to the required accuracy. Its units are usually decibels.

Fibre properties. The system characterisation must include a specification of the fibre which it utilises. This specification should contain:

- Attenuation spectrum.
- Dispersion spectrum.
- Modal type (i.e. monomode, multimode, graded index, etc.).
- Geometry (core size, cladding size, numerical aperture, etc.).
- Coatings (the protective materials on the outer surface of the cladding).

We may proceed now to examine the basic relationships which exist between the above parameters.

The system equation

Consider the arrangement shown in figure 1. An optical fibre of total length L lies under the influence of a measurand field $M(l)$. Consider the element of fibre lying between distances l and $l+dl$ from the front end F of the fibre. A pulse of radiation with energy E_0 and duration ΔT is launched into the fibre at F and travels to the element through channel T (transmit) with a velocity C_T. While propagating within the element at l:

- A fraction of the energy ($\alpha_s dl$) is either absorbed within it, and/or is scattered out of the fibre.
- A fraction of the energy ($\alpha_c dl$) is coupled into a channel R (receive) which then returns to F at velocity C_R and is detected by the photo-detector.

- The energy comes under the influence of $M(l)$, which is the average value of the measurand field between l and $l+\mathrm{d}l$.

Let us assume that the properties of the fibre are uniform, i.e. are independent of position l, in the absence of a measurand M; that the effect of M on C_T and C_R is negligible; and that α_s, $\alpha_c \ll 1$. All of these assumptions are very reasonable in practice.

Let us first consider the situation in the absence of the measurand field $M(l)$. We may write for the fraction of the energy lost from T during passage through $\mathrm{d}l$:

$$\frac{\mathrm{d}E_T(l)}{E_T(l)} = -(\alpha_s+\alpha_c)\,\mathrm{d}l$$

and thus:

$$E_T(l) = E_0 \exp\left[-(\alpha_s+\alpha_c)l\right] \qquad (1)$$

Hence, the energy which enters the R channel from the element is:

$$\mathrm{d}E_R(l) = \alpha_c E_T(l)\,\mathrm{d}l = \alpha_c E_0 \exp\left[-(\alpha_s+\alpha_c)l\right]\mathrm{d}l$$

Now this energy is further attenuated during passage back through the R channel over distance l so that it arrives at F with magnitude:

$$\mathrm{d}E_R(2l) = \alpha_c E_0 \exp\left[-(\alpha_s+\alpha_c+\alpha_s'+\alpha_c')l\right]\mathrm{d}l \qquad (2)$$

where α_s' and α_c' refer to the R channel. For convenience let us set:

$$(\alpha_s+\alpha_c+\alpha_s'+\alpha_c') = 2\alpha \qquad (2a)$$

The energy given by equation (2) will arrive back at F after a time t, where:

$$t = l\left(\frac{1}{C_T}+\frac{1}{C_R}\right) = \frac{2l}{c} \qquad (3)$$

We can use this linear relationship to effect the transformation:

$$\mathrm{d}E(2l) \rightarrow \mathrm{d}E(t)$$

according to:

$$\mathrm{d}E_R(t) = \alpha_c E_0 \exp(-\alpha ct).\left(\frac{c}{2}\right)\mathrm{d}t$$

But $\mathrm{d}E_R(t)/\mathrm{d}t$ is the power, $p(t)$, say, averaged between t and $(t+\mathrm{d}t)$, and hence, finally:

$$p(t) = \frac{c\alpha_c E_0}{2}\exp(-\alpha ct) \qquad (4)$$

Giving the measurand received power as a function of time, starting from the launch of the pulse.

Equation (4) requires some further discussion. First, it is clear that the equation is strictly applicable only if the pulse is of infinitesimal duration ($\Delta T \rightarrow 0$). The practical implication of this restriction is that the detector system for the receive channel must make a measurement of $p(t)$ averaged over a time which is significantly greater than ΔT. If this were not so the output signal would depend upon the pulse shape. This is undesirable for two reasons: first, it would necessitate troublesome deconvolution and, second, it would render the output sensitive to pulse-to-pulse variations. It follows, then, that the detector response time, τ, must be greater than the pulse width, and thus we have:

$$\tau(=1/W) > \Delta T$$

Now the detector response time will determine the spatial resolution δl, since the latter is defined as that distance over which the measurand is averaged by the system. Since the effective pulse velocity is c we have:

$$\delta l = c\tau/2$$

In order to get some feel for the order of these quantities, it may be noted that for a light pulse with propagation velocity c equal to 2×10^8 ms^{-1}, and a typical detector response time of 1ns (10^{-9} s), we have δl=0.1m. For this situation we should thus require that the optical pulse width be significantly less than 1ns, say 300ps.

Consider now the effect of the measurand field $M(l)$. This modulates, in some way, the radiation while it is within the element $\mathrm{d}l$. The modulation may be via modification of amplitude, frequency, phase or polarisation state of the optical waves. The total energy in the pulse may or may not be altered by the modulation but, in any case, we may assume that the energy is, in fact, only altered after the radiation, on re-emerging from F, passes through the demodulating analyser A (see Fig. 2). The action of A is thus seen, effectively, as being a conversion of the measurand modulation to an energy variation.

Let us suppose that the modulation which $M(l)$ produces between l and $(l+\mathrm{d}l)$ can be represented by a dimensionless parameter which takes the form:

$$\mathrm{d}\mu(l) = \gamma\delta l M(l) \qquad (5)$$

where γ is an identifiable constant which depends directly on the physical effect being used to make the measurement and determines the modulation produced by the measurand per unit length of fibre.

We now assume that the relevant physical effect is reciprocal (i.e. is independent of direction of optical propagation) and is the same for both T and R channels. We may then integrate (5) to obtain the total modulation resulting from go-and-return passage to l as:

$$\mu(2l) = 2\gamma \int_0^l M(l)\,\mathrm{d}l$$

or, using equation (3):

$$\mu(t) = 2\gamma \int_0^{ct/2} M(l)\,\mathrm{d}l \qquad (6)$$

Now suppose that we use a demodulating analyser for $\mu(t)$ which is such as to transmit a fraction of the incident power $p(t)$, this fraction being dependent on μ and thus allowing demodulation to occur. We represent this fraction as $A(\mu)$ and refer to it as the analyser function (see Fig. 2). The power transmitted through the demodulating analyser to the photodetector will thus be:

$$\pi(t) = A[\mu(t)]\,.\,p(t) \qquad (7)$$

Now from our definition of the sensitivity S and equation (5) we have that:

$$S = \frac{\partial^2 \pi}{\partial M \partial l} = p(t)\frac{\partial}{\partial M}\left(\frac{\partial A}{\partial \mu}\frac{\partial \mu}{\partial l}\right) = p(t)\frac{\mathrm{d}A}{\partial \mu}\gamma \qquad (8)$$

Also from equation (7) we may write:

$$\mu(t) = A^{-1}\left(\frac{\pi(t)}{p(t)}\right) = 2\gamma \int_0^{ct/2} M(l)\,\mathrm{d}l \qquad (9)$$

using equation (6). (Here A^{-1} refers to the inversion of the function A, not its reciprocal).

Differentiating equation (9) we have:

$$M(l) = \frac{1}{c\gamma}\cdot\frac{\mathrm{d}}{\mathrm{d}t}\left[A^{-1}\left(\frac{\pi(t)}{p(t)}\right)\right] \qquad (10)$$

showing that $M(l)$ can be determined provided that $\pi(t)$ and $p(t)$ are known together with a knowledge of the function A, and the constants c, γ. We shall call equation (10) the system equation.

The application of this equation will be illustrated by applying it to two practical examples.

Example 1: Temperature-dependent absorption. Let us suppose that we wish to measure a spatial temperature (θ) distribution, and that the physical effect to be used is the dependence on temperature of the absorption coefficient (α_a) of the fibre material. In this case the modulation brought about by $\theta(l)$ is indistinguishable from that brought about by other absorption and scattering processes, so we must measure $p(t)$ in the absence of $\theta(l)$ (or for a known $\theta(l)$) and trust that it remains constant. (An alternative approach is to infer $p(t)$ by simultaneous measurement at a wavelength for which there is no dependence on θ, but for which the other dependences are closely similar.)

Suppose that the temperature-dependent component of the absorption coefficient is linear with temperature so that:

$$\mathrm{d}\mu(l) = \frac{\mathrm{d}E_\theta(l)}{E(l)} = -\alpha_\theta \theta(l)\,\mathrm{d}l$$

where $\mathrm{d}E_\theta(l)$ is the loss of energy due to the temperature change $\theta(l)$. We can thus, from equation (5), note that:

$$\gamma = -\alpha_\theta$$

Hence, using the same arguments which led to equations (1) and (2) we have that the energy which arrives back at F from the element $\mathrm{d}l$ is given by:

$$\mathrm{d}E'_R(2l) = \alpha_c E_0 \exp(-\alpha ct)\exp[-\alpha_\theta \theta(l) ct]\,\mathrm{d}l$$

and thus:

$$\pi(t) = p(t)\exp[-\alpha_\theta \theta(l) ct]$$

The function A defined in equation (7) is thus, clearly, the exponential function, and A^{-1} the natural logarithm. Hence, we have:

$$\frac{1}{c\gamma}\cdot\frac{\mathrm{d}}{\mathrm{d}t}\left[A^{-1}\left(\frac{\pi(t)}{p(t)}\right)\right] = \frac{1}{c\gamma}\,.\,c\gamma\theta(l) = \theta(l)$$

in accordance with equation (10), as required.

Example 2: Strain-dependent birefringence. The second illustration is concerned with the measurement of spatial distribution of strain (or pressure) via the change in refractive index which occurs in a fibre when it is subjected to physical strain. (The phenomenon is known as the elasto-optic effect.) The refractive index will change for the optical polarisation direction which lies parallel to the strain vector, relative to the orthogonal direction. The resultant difference in velocities (very small compared with their absolute values) for the two polarisation directions is known as linear birefringence, and leads to an optical phase difference between the two directions, per unit length of fibre, which is proportional to the magnitude of the strain. Thus, we may write for the phase difference introduced by the element lying between l and $(l+\mathrm{d}l)$:

$$\mathrm{d}\phi(l)[\equiv \mathrm{d}\mu(l)] = b\sigma(l)\,\mathrm{d}l$$

where $\sigma(l)$ is the strain averaged over the element and b

is a constant which may be identified with the γ of equation (10).

Also we have for the total phase difference over l:

$$\phi(l) = b \int_0^l \sigma(l)\,dl$$

and

$$2\phi(l) \rightarrow \phi(t);\ t = 2\frac{l}{c}$$

Hence $\pi(t) = p(t)\ A[\phi(t)]$ where A now refers to the transfer function of a polarising analyser (e.g. polaroid or Wollaston prism) and has the general form:

$$A(\phi) = \sin 2\phi$$

Hence it is clear that, in this case, $p(t)$ can always be measured before polarisation analysis takes place, $\pi(t)$ can be measured after its action, and we can then readily perform the operation:

$$\frac{1}{cb}\cdot\frac{\mathrm{d}}{\mathrm{d}t}\left[A^{-1}\left(\frac{\pi(t)}{p(t)}\right)\right] = \frac{1}{cb}\cdot\frac{\mathrm{d}}{\mathrm{d}t}\left[2b\int_0^{ct/2}\sigma(l)\,\mathrm{d}l\right] = \sigma(l)$$

again, in accordance with equation (10), as required.

System relationships

The system equation (10) defines the basic relationship between the required distribution $M(l)$ and the measured quantities $\pi(t)$ and $p(t)$. We shall now examine the implications which this relation has for the sensitivity, accuracy, bandwidth and dynamic range of the measurement.

The quantity $A(\mu)$ in equation (7) is obtained by taking the ratio of two measured power levels. Thus the error $\mathrm{d}A$ in A may be written:

$$\mathrm{d}A = A\frac{\mathrm{d}\pi}{\pi} - A\frac{\mathrm{d}p}{p}$$

Any variations in power level which correlate for π and p clearly will cancel (e.g. pre-demodulation power variations due to source fluctuations). The uncorrelated components will be due to contributions from thermal noise and dark current noise in the photo-detector circuit, and from the photon noise (quantum noise) in the received optical signal. If the two photo-detection systems are closely similar we may write:

$$\mathrm{d}A = \frac{A}{\pi}(\Delta\pi_q + \Delta n) + \frac{A}{p}(\Delta p_q + \Delta n)$$

where Δn represents the sum of thermal noise and dark current noise, the suffix q refers to the quantum noise component, and where we have added all the errors to represent a worse-case error, $\mathrm{d}A$, in A.

Now the quantum noise components are given by:[10]

$$\frac{\Delta\pi_q}{\pi} = (r_\pi\tau)^{-1/2};\ \frac{\Delta p_q}{p} = (r_p\tau)^{-1/2}$$

where r_π, r_p are the mean photon arrival rates and $\tau(=1/w)$ is the photodetector response time. For a spatial resolution δl we require that $\tau < 2\delta l/c$ and thus, in the optimum case:

$$\frac{\Delta p_q}{p} = \left(\frac{h\nu c}{2p\delta l}\right)^{1/2};\ \frac{\Delta\pi_q}{\pi} = \left(\frac{h\nu c}{2\pi\delta l}\right)^{1/2}$$

where h is the quantum constant and ν the optical frequency. We may relate these via equation (7) to obtain:

$$\frac{\Delta\pi_q}{\pi} = A^{-1/2}\frac{\Delta p_q}{p}$$

Thus we now have for $\mathrm{d}A$:

$$\mathrm{d}A = \left(\frac{h\nu c}{2p\delta l}\right)^{1/2}(A + A^{1/2}) + \frac{\Delta n}{p}(A+1) \qquad (11)$$

Let us now write equation (10) in the form:

$$M(l) = \frac{1}{c\gamma}\cdot\frac{[\mu(t+\tau) - \mu(t)]}{\tau}$$

so that the standard error in $M(l)$ may be written:

$$\mathrm{d}M = \frac{1}{c\gamma\tau}\mathrm{d}(\mathrm{d}\mu)$$

Now $\mathrm{d}A = A'(\mu)\mathrm{d}\mu$ and thus, if $\mathrm{d}A$ is to be the standard error:

$$\mathrm{d}A = A'(\mathrm{d}\mu)\,\mathrm{d}(\mathrm{d}\mu)$$

where A' represents the derivative of A. Hence:

$$\mathrm{d}M = \frac{1}{c\gamma\tau}\cdot\frac{\mathrm{d}A}{A'(\mathrm{d}\mu)}$$

and, thus, substituting for $\mathrm{d}A$ from (11) and using $\tau = 2\delta l/c$ we have:

$$\mathrm{d}M = \frac{1}{2\gamma\delta l}\left(\frac{h\nu c}{2p\delta l}\right)^{1/2}\cdot\frac{(A + A^{1/2})}{A'} + \frac{\Delta n}{p}\cdot\frac{(A+1)}{A'} \qquad (12)$$

Equation (12) represents the error in M in terms of the mean power received from the element δl for a given A, p and γ.

Probably the most useful initial piece of information we may derive from equation (12) is the percentage accuracy with which we may measure M when its magnitude is small. For this we may reasonably assume a linear function A of the form:

$$A(\mathrm{d}\mu) = \frac{\mathrm{d}A}{\mathrm{d}\mu}\mathrm{d}\mu = A'\,\mathrm{d}\mu = g\,\mathrm{d}\mu,$$

say (taking $A(0) = 0$), where g is a constant under these 'small signal' conditions. If M is small it follows that both $\mathrm{d}\mu$ and A will be small (A is a fraction and has a maximum value of unity, corresponding to the largest value of M which is measurable; in this case we have $A \ll 1$) and hence:

$$\frac{(A+A^{1/2})}{A'} \approx \left(\frac{\mathrm{d}\mu}{g}\right)^{1/2}$$

and

$$\frac{(A+1)}{A'} \approx \frac{1}{g}$$

Now for the element $\mathrm{d}l$ we have $\mathrm{d}\mu(l) = \gamma M(l)\delta l$ from (5) and thus equation (12) becomes:

$$\mathrm{d}M = \frac{1}{2\gamma\delta l}\left[\left(\frac{h\nu c}{2p\delta l}\right)^{1/2}\left(\frac{\gamma m\delta l}{g}\right)^{1/2} + \frac{\Delta n}{pg}\right]$$

giving (using the expression for sensitivity S from equation (8)):

$$\frac{\mathrm{d}M}{M}\,(=\eta_{(\mathrm{s})},\ \text{say}) = \left(\frac{h\nu c}{8MS\delta l^2}\right)^{1/2} + \frac{\Delta n}{2MS\delta l} \qquad (13)$$

If, on the other hand, M is large (and hence $A \approx 1$), but conditions, by the photo-detector noise (rather than the

$$\frac{(A+A^{1/2})}{A'} \approx \frac{(A+1)}{A'} \approx \frac{2}{g_{(\mathrm{L})}}$$

and thus we find:

$$\frac{\mathrm{d}M}{M} = \eta_{(\mathrm{L})} = \frac{1}{MS_{(\mathrm{L})}\delta l}\left[\left(\frac{h\nu cp}{2\delta l}\right)^{1/2} + \Delta n\right] \qquad (14)$$

We shall call equations (13) and (14) the operating equations for small and large M, respectively.

Now the above values for $\mathrm{d}M/M$ (η_L and η_S) refer, of course, to just one propagating pulse. The accuracy of either measurement can be improved by averaging over a number, N, of such pulses. The improvement takes the form:

$$\eta_N = \eta/N^{1/2} \qquad (15)$$

However, an increasing value of N inevitably reduces the measurement bandwidth B, since one pulse must pass completely through the fibre before the next may be launched. Thus we have:

$$B = c/2NL \qquad (16)$$

where L is the total fibre length, as previously defined.

Hence we also find from equations (15) and (16):

$$\frac{B}{\eta_N} = \frac{c}{2\eta N^{1/2}L} \qquad (17)$$

and equation (13) becomes the operating equation for small M:

$$\eta_N = \left(\frac{BL}{2c}\right)^{1/2}\left[\left(\frac{h\nu c}{2MS\delta l^2}\right)^{1/2} + \frac{\Delta n}{MS\delta l}\right] \qquad (18)$$

The dynamic range of the measurement system refers to the ratio of the maximum to minimum allowable values of the measurand, for any given resolution element.

The maximum value under any circumstances must be that which leads to $A = 1$ (though other practical limitations may dictate a lower value). The minimum value will be the lowest value for which equation (13) provides the required measurement accuracy. If we assume that the detection system is limited, under these conditions, by the photo-detector noise (rather than the signal quantum noise, or 'shot noise') then we find that for N pulses:

$$M_{\min} = \frac{\Delta n}{2N^{1/2}S\delta l\eta}$$

where η is the single phase accuracy. The value of $M_{\max}$ will be given by the condition that $A = 1$. Hence:

$$M_{\max} = \frac{1}{g\gamma\delta l}$$

and thus the dynamic range will be given by:

$$D = \frac{M_{\max}}{M_{\min}} = 2N^{1/2}\frac{p\eta}{\Delta n}$$

and with $N = c/(2BL)$ we have:

$$D = \left(\frac{2c}{BL}\right)^{1/2}\cdot\frac{p\eta}{\Delta n} \qquad (19)$$

We now have a set of system relationships: equations (13), (14), (17), (19).

Usable physical effects for pressure, strain and temperature measurement

Definition of 'standard' system

In the preceding section the relationships between the primary characterising system parameters for DOFS were derived. In this section we shall use some of these relationships to compare the potential usefulness of the various physical effects by which pressure, strain and temperature may be measured with DOFS.

In order to do this we shall define a 'standard' requirement for the measurement of these quantities and also we shall assume certain 'typical' characteristics for the components and devices which are to be used.

Suppose that the basic requirement is to measure the parameters to an accuracy of 1% with a spatial resolution of 0.1m over a total length of 100m. Suppose also that for this task we decide on the following sub-system characteristics.

Source. Pulsed semiconductor laser

frequency, ν	$=3.3\times 10^{14}$ Hz ($\lambda=900$nm)
pulse energy, E_0	$=1$nJ
pulse width, ΔT	$=0.5$ns
velocity of light in the fibre, c	$=2\times 10^{8}$ ms^{-1}

Detector. Avalanche photo-diode

bandwidth, W	$=1$GHz
responsivity, ρ	$=50$AW^{-1}
dark current, i_d	$=50$nA
load resistor, R	$=100\Omega$
capacitance, C	$=2$pF
noise equivalent power	$=2\times 10^{-14}$ WHz$^{-1/2}$

(From the above parameters, the sum of thermal and dark current noise powers, Δn, is found to be ~ 10nW).

Method. We shall make the comparison using the back-scatter arrangement of figure 2.

In the arrangement considered, the pulse propagates in the forward direction through the fibre to constitute the T channel and a fraction of the light is Rayleigh back-scattered continuously to the source end, to constitute the R channel. (This arrangement is commonly used for measuring the distribution of attenuation in communications fibres and is known as optical time domain reflectometry (OTDR).[11] One of the consequences of this choice of arrangement is that it defines the optimum scattering coefficient of the fibre: it is required to maximise the received power from the far end of the fibre. Thus, from equation (4) we have:

$$p(2L)=\frac{c\alpha_c E_0}{2}\exp(-2\alpha L)$$

If we assume that α is due solely to scattering (i.e. no absorption) and that the fraction of the scattered light in T which is coupled into the receiver channel R is Σ then we may write:

$$p(2L)=\frac{c\Sigma\alpha_s E_0}{2}\exp(-2\alpha_s L) \qquad (20)$$

This has a maximum value when:

$$\alpha_s=\frac{1}{2L}=5\times 10^{-3}\ \mathrm{m}^{-1}\ (\approx 20\ \mathrm{dB\ km}^{-1})$$

when $L=100$m. This is the value for α_s which we shall use for our system; we shall also use a value for Σ of 5×10^{-3}, which is a typical value for silica fibre.[12] We have yet to decide on the efficiency of the light launch into the fibre. This will depend upon the type of fibre which is used. For simplicity we shall assume that we can launch all of the 1nJ of energy from the semiconductor laser into a multimode fibre and only 100pJ into a monomode fibre. These are fair estimates for presently-available sources and fibres. We can now use all the above information to write equations (13) and (15) in the reduced form:

$$\eta_N=\frac{dM}{M}=0{\cdot}01=\frac{1}{N^{1/2}}\left(\frac{0{\cdot}074}{(Mfg\gamma)^{1/2}}+\frac{1}{2Mfg\gamma}\right) \qquad (21)$$

where $f=10$ for multimode fibre, $f=1$ for monomode fibre and $p(2L)\approx 10^{-7}f$ watts (from equation (20)).

Thus we now need to determine γ and g for the various physical effects we shall consider, and hence to calculate the minimum value of M which can be measured to 1% accuracy. Clearly, the lower this value the more generally useful will be the effect.

Finally, since the light from a single pulse will require 1μs to return from the far end of 100m of fibre, the single-pulse measurement bandwidth will be ~1MHz.

Strain and temperature dependence

We take the complex refractive index of the fibre material to have the form:

$$n=n'-\mathrm{i}n''$$

and we require, for a fibre of length l:

$$\frac{\partial(nl)}{\partial(\theta,\sigma)}=\frac{\partial(n'l)}{\partial(\theta,\sigma)}-\mathrm{i}\frac{\partial(n''l)}{\partial(\theta,\sigma)} \qquad (22)$$

where θ is the mean temperature over the length l, and σ is the strain. Pressure and strain will be related by a linear expression of the form:

$$P=K\sigma \qquad (23)$$

where K is an appropriate elastic modulus.

Temperature. Attenuation. The second term in equation (22) represents the variation in attenuation coefficient and becomes, for temperature alone:

$$\frac{\partial(n''l)}{\partial\theta}=\frac{n''\partial l}{\partial\theta}+\frac{l\partial n''}{\partial\theta} \qquad (24)$$

The first of the two terms in equation (24) relates to the thermal expansion of the fibre material, and, in general, will be too small to cause a significant variation in the attenuation. The second term represents a variation of the attenuation coefficient itself with temperature and it is known[36] that this can be large when the fibre is doped with rare-earth elements. The largest reported temperature coefficient occurs with holmium-doped fibre,[35] the measured value being ~ 1% per degree at 900nm wavelength. Let us, therefore, take this figure for our calculations.

We have seen from the section 'Temperature-dependent absorption' that γ may be identified with $-\alpha_\theta$ where:

$$d\mu = \alpha_\theta \theta \, dl$$

and thus, with $\alpha = 5 \times 10^{-3}\ m^{-1}$ we have:

$$\gamma(= -\alpha_\theta) = -5 \times 10^{-5}\ m^{-1}\ K^{-1} \text{ (i.e. 1\% of } d\text{)}.$$

We have also identified the function A in equation (7) as:

$$A(\mu) = \exp(-\mu)$$

and thus $A' = -exp(-\mu) \approx -1$ if μ is small. Thus $g_{(s)} = -1$.

Now for pure attenuation measurements we may use multimode fibre. No phase or polarisation information is to be extracted. Thus we have $f = 10$. Hence we have: $\gamma = -5 \times 10^{-5}\ m^{-1}\ K^{-1}$; $g = 1$; $f = 10$; and equation (21)

$$\theta_m \approx \frac{3 \times 10^5}{N^{1/2}}$$

In other words, 10^{11} pulses would need to be averaged in order to measure a temperature of 1K to 1% accuracy. Since the measurement for one pulse requires 1μs, the total time required is 10^5 s or about 28 hours. This is clearly too long to be practicable. The conclusion is that this method must be limited in its applicability to the DOFS type of measurement to larger spatial resolution lengths, to shorter total lengths and to accuracies less demanding than 1%.

Phase. The first term in equation (22) represents a phase variation of the propagating light and, for temperature, may be written:

$$\frac{\partial(n'l)}{\partial\theta} = \frac{n'\partial l}{\partial\theta} + \frac{l\partial n'}{\partial\theta} \tag{25}$$

Again, measurements have shown that the second of the two terms in equation (25) is dominant.[13] For a temperature change θ we may thus write the consequent phase change over the element dl as:

$$\Delta\phi(\theta) = \frac{2\pi}{\lambda}\frac{\partial n'}{\partial\theta}\theta \, dl$$

(assuming that $\partial n'/\partial\theta$ is independent of θ).

Hence, from equation (5), we may identify γ as:

$$\gamma = \frac{2\pi}{\lambda}\frac{\partial n'}{\partial\theta}$$

and reported values show that $\gamma \sim 75$ radians $K^{-1}\ m^{-1}$ in silica at 900nm wavelength. The demodulation of the phase variation will be performed with the aid of polarisation analysers which provide the form of A:

$$A(\Delta\phi) = \sin 2\Delta\phi$$

Hence $A' = 2$ when $\Delta\phi$ is small, giving $g = 2$.

Finally, for effective phase demodulation we must use monomode fibre, since a multimode pattern will randomise the phase distribution across the fibre aperture, and hence $f = 1$. Thus: $\gamma = 75$ radians $K^{-1}\ m^{-1}$ $g = 2$; $f = 1$; and equation (21) yields:

$$\theta_m \approx 1/N^{1/2}$$

which is a very acceptable result; even for $N = 1$, 1% accuracy is achieved with ~ 1K.

The conclusion for temperature measurement, then, is that phase modulation, via the temperature dependence of refractive index, can provide viable DOFS technology, while the temperature dependence of the absorption coefficient, in general, cannot.

Strain. Attenuation. For this case we have:

$$\frac{\partial(n''l)}{\partial\sigma} = \frac{n''\partial l}{\partial\sigma} + \frac{l\partial n''}{\partial\sigma} \tag{26}$$

The first term in this equation denotes the increase in absorption resulting from the length change which accompanies the strain and is usually quite negligible. The second term refers to the variations of attenuation coefficient with strain and, while this is normally very small for a bulk material, it can be rendered large in a fibre by arranging for strain to cause scatter loss. This may be done, for example, by placing the fibre in a corrugated structure which is then subject to a stress (see Fig. 3). The stress causes the gap in which the fibre is laid to change in dimension, resulting in microbending in the (multimode) fibre, causing loss of the higher order propagation modes as the stress increases. The characteristics of an arrangement such as this depend quite strongly on the materials and the geometry, but we may estimate orders of magnitude. A power variation ~ 5% per μm of displacement may be expected.[14] Thus for a 250μm gap with our optimized

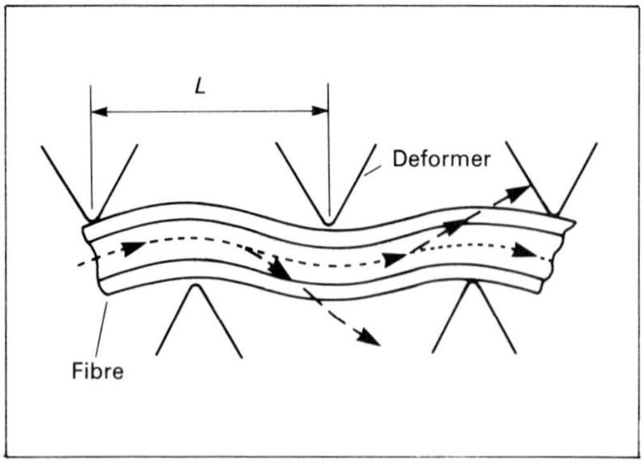

Fig. 3 Microbend strain sensing

fibre we may write:

$$d\mu = \alpha_\sigma \sigma \, dl$$

with α_σ $(=\gamma)=0.05$ neper m^{-1} strain^{-1}.

Since we are again considering an attenuation modulation, A will take an exponential form, and thus $g=1$ for small $d\mu$.

We use multimode fibre (of necessity) and hence $f=10$. Thus: $\gamma=0.05$ neper m^{-1} strain^{-1} $g=1; f=10$; and we find, using equation (21):

$$\sigma_m = 300/N^{1/2}$$

Thus for a strain of 1% we require 10^9 pulses and a measurement time of 1000 seconds for 1% accuracy.

The conclusion is that this method can only work well when fairly large strains/displacements are to be measured, and when the requirements in regard to spatial resolution and/or accuracy of measurement are relaxed considerably from our stated requirements.

Phase. The relevant equation for this case is:

$$\frac{\partial(n'l)}{\partial\sigma} = \frac{n'\partial l}{\partial\sigma} + \frac{l\partial n'}{\partial\sigma}$$

The first of these terms refers to the phase change consequent upon any increase in fibre length which occurs as a result of the strain; the second term refers to the change of refractive index of the fibre material when it is strained, i.e. to the photo-elastic effect. For the latter the situation is complicated by the fact that we should distinguish between radial and longitudinal strain, and remember that these are related via Poisson's ratio.

To simplify matters for the purpose of our order of magnitude calculations, however, we shall assume that only longitudinal stress is present. In this case the longitudinal strain σ_z is related to the radial strain σ_r by:[15]

$$\sigma_r = -\frac{(1-\nu)}{2\nu}\sigma_z$$

where ν is Poisson's ratio. We may write the phase change for the length dl in the form:[15]

$$\Delta\phi(\sigma_z) = \frac{2\pi\delta l}{\lambda}\left[\sigma_z + \frac{n^2}{2}(P_{11}+P_{12})\frac{(1-\nu)}{2\nu}\sigma_z + P_{12}\sigma_z\right]$$

where P_{11} and P_{12} are the photo-elastic constants for silica.

Inserting the known values for the constants we find:

$$\Delta\phi(\sigma_z) = 1{\cdot}66 \times 10^7 \, \sigma_z \delta l$$

and hence $\alpha_\sigma(=\gamma)=1.66\times10^7$ radians m^{-1} strain^{-1}.

We are now using phase demodulation so that $g=2$, and we must use monomode fibre, so that $f=1$. Using, again, equation (21) we find:

$$\sigma_m = 5 \times 10^{-6}/N^{1/2}$$

Hence, even with a single pulse (1 μs response time) 5 microstrain can be measured to 1% accuracy, and this can thus be considered to be a practically viable method.

Pressure/stress

The effect of pressure on a medium is to produce strain. Thus the effectiveness of a DOFS in the measurement of pressure is directly related to its ability to produce strain in the measurement fibre.

We have seen that strains $\sim 5\times10^{-6}$ may be measured to 1% accuracy at MHz over 100m with 0.1m spatial resolution, using a phase modulation of the light by the measurand. It follows that we can measure, with the same performance characteristics, pressures which produce this order of strain. Young's modulus for SiO_2 at room temperature is 7.3×10^{10} N m^{-2}. Thus a strain of 10^{-6} is produced by a pressure of $\sim 10^5$ N m^{-2}, or 1 bar. It will be remembered that this was for single pulse operation and can readily be reduced by up to two orders of magnitude while retaining an acceptable measurement bandwidth.

Concluding remarks

The broad conclusions which the analysis of the previous sections has produced are thus that for good spatial resolution ($\sim$ 0.1m), long measurement lengths ($\sim$ 100m) and large measurement bandwidths ($\sim$1MHz), phase modulation methods must be used). Attenuation modulation methods have their advantages (multimode fibre, simple demodulators) but offer markedly inferior performance characteristics, in that, first, they are much less sensitive and therefore require long detector integration times for acceptable accuracy, and, second, the performance characteristics depend upon the actual temperature distribution. (If, for example, the temperature is very high at the launch end of the fibre, the attenuation will be large there, thus reducing the optical signal for the remainder of the fibre length). Attenuation methods are thus seen as valuable only in special cases.

The major disadvantage of the phase modulation methods lies in the increased complexity of the detector necessary for phase demodulation.

A general difficulty for both methods resides in the fact that most systems which have been devised rely on back-scatter from an optical pulse for the spatial information, and the back-scatter levels are intrinsically low (a factor $\sim 10^{-5}\,m^{-1}$ down on forward propagation) so that signal-to-noise problems in the detector are inevitably severe, even when optimised fibre is used.

Another general difficulty is that the fibres which are available for use are sensitive to a range of external effects via their compositions, geometries and coatings, and for an effective measurement of a specific field it is necessary to have available fibres which possess enhanced sensitivity to that field. There is also, of course, a necessity to ensure that the fibre material (including the coating) is compatible with the measurement environment. Neither of these materials problems has yet been very seriously addressed.

Finally, the analysis has provided the essential relationships between the system parameters; these will allow the system designer to effect quantitative trade-offs when attempting to meet specific applications requirements.

Examples of practical distributed measurement systems which have been studied are to be found in references 14 and 16–27.

Component Technology

In this section we shall review the state-of-the-art in respect of the various components which are necessary for DOFS, and discuss the limitations which their performances impose. Some extrapolation of present trends will also be included.

Sources

The light source in any practical industrial system has to be either a semiconductor light emitting diode (LED) or laser diode (LD). Only the packaged semiconductor module can be expected to possess the ruggedness and durability which is required in the quite hostile environments associated with the majority of industrial applications. In addition it enjoys the advantage of requiring low operating voltage.

The primary demands made of the light source in the context of our proposed line of research for DOFS are those of output power and coherence.

We have noted that, for back-scatter systems, the performance depends critically on the energy which can be delivered in a pulse of width $\sim$ 1ns (for 0.1m spatial resolution). The figure of 1nJ which we have used for our calculations in the section 'Usable physical effects for pressure, strain and temperature measurement' can now readily be achieved using commercially available devices. Coherence of the source is required for any DOFS method which utilises phase modulation. Clearly the coherence length must be at least as large as the spatial resolution if the phase information is to be extracted satisfactorily. Thus we require a coherence length $> \sim$ 0.2m and a coherence time $\sim$ 1ns. This translates into a source spectral linewidth of $<$ 1GHz, and this can only be achieved with single-longitudinal mode semiconductor lasers (SMLSs). (Linewidths as low as 30MHz have been reported with these devices.[28])

To satisfy the two requirements simultaneously (1nJ in 1ns, 1GHz linewidth) implies a device which is close to the state-of-the-art. The conclusion, therefore, is that the source requirement is a demanding one but that it can now just be met. Further developments, which are taking place primarily for high-bandwidth coherent optical fibre communications systems, will allow the demands to be met more easily.

Detectors

For detection of the optical radiation we again turn towards a semiconductor device, and for the same reasons as for the source.

We can use either a PIN photo-diode or an avalanche photo-diode. For back-scatter we would normally choose the latter for its increased sensitivity. Its disadvantages are that it requires a relatively high voltage (100–200V) for its operation and that it is intrinsically noisier than the PIN device. The noise figure must be traded-off carefully against the sensitivity, for any given application.

Both types of device are quite readily capable of the $\sim$ 1GHz bandwidths which are required for the DOFS function. A large range of devices of both types is available commercially, in both silicon and germanium structures, and the choice of detector thus presents relatively very few difficulties.

Fibres

Until relatively recently very little attention has been given to the special needs of optical fibre sensors by those who research and develop fibre fabrication techniques. Even now, most of the developments favourable to the sensor function are seen to be more significant in application to telecommunications, since the prospects there are perceived to be much more commercially profitable.

There are three aspects to fibre composition which are important in the provision of optical fibre sensors and these are common to both DOFS and point systems.

The first is the fibre geometry. This is especially important in relation to phase devices (i.e. interferometric and polarimetric devices) where controlled polarisation characteristics are desirable for the fibre. Considerable progress has been made in past years on this front with low birefringence,[29] high linear birefringence,[30,31] high circular birefringence,[32] and 'polarising' fibres of various designs having become available. Such fibres are, of course, especially valuable for DOFS since, as we have seen, it is the phase modulation type of system which is most likely to provide the generally-applicable solution to the DOFS problem.

Another geometrical approach which should be mentioned is that where dual cores are positioned within the same cladding.[33] Such a structure allows the measurand to control the coupling between the two guides and thus effectively physically separates the T and R channels. This has distinct advantages in detection.

The second important aspect of fibre fabrication is that of dopant composition. Clearly the inclusion of dopants can render the fibre's optical properties specifically sensitive to a particular measurand, and thereby ease very considerably the DOFS design. Some progress has been made also in this area, particularly in regard to temperature measurement, as has been discussed. A number of rare-earth elements (e.g. Nd, Er, Yb, Tb, Ho, Pr) have been tried[34] with interesting resulting values for the temperature coefficient of absorption, especially in the case of Holmium.[35,36] Other properties of these fibres are presently under intensive investigation. These include the non-linear properties, such as those which might lead to enhancement of the electro-optic effects, Brillouin and Raman scattering coefficients, etc.[36] These last developments are unlikely to have any direct impact on the measurement fibre itself, as, even with enhancement, the non-linear effects are likely still to require much higher peak optical power levels than can readily be provided by semiconductor lasers, taking into account the required DOFS performance. They might very well, however, give rise to a range of active fibre components (e.g. phase/frequency modulators, polarisers, amplifiers, switches, etc.) which could prove invaluable in the signal processing functions necessary for optimal extraction of the measurand information in DOFS.[37]

There is very much more that could be done in this second, dopant area. Dopants could lead to some controlled anisotropy in the fibre, for example, so as to enhance elasto-optic effects for strain/pressure measurement and so as to allow enhancement of measurand-dependent mode coupling in the fibre.

For the third and final aspect of fibre composition, we turn to the fibre coatings. Scarcely any attention at all has yet been devoted to the coating problems which are thrown up by the sensor requirements.

This problem has the following three aspects. First, it is necessary to ensure that the coating can survive any mechanical attack to which it might be subjected in the measurement environment; second, the coating must allow optimum interaction of the measurand with the fibre property which is being modulated in the measurement action; and third, that interaction must remain constant over long periods of time if frequent, inconvenient (and possibly unallowable) recalibration of the system is not to be required. The coatings problem is not, therefore, at all an easy one. In fact it is not so much one problem as a range of problems, representing the range of measurement tasks and applications.

In this area, and in the dopant area, there is a requirement for a concerted attack by materials scientists and engineers.

Having pointed out various difficulties in the area of fibre fabrication it must be said, however, that a good deal can still be done with existing fibres, as regards DOFS. In particular, suitable 'hi-bi' fibre probably already exists, though no doubt it will require some modification for an optimised system, especially in regard to its coating. But the subject of optical fibre sensors has progressed to its present state largely by using fibres designed for telecommunications systems. Perhaps they will become widely accepted for industrial application only when the special problems appropriate to such application are properly addressed and solved.

Other components

Other components which are needed for DOFS are devices such as source drivers and modulators, directional couplers, polarisers and fast switches.[38] The fibre versions will be most convenient by far for DOFS use, since there are potential advantages in regard to ruggedness, optical alignment, compactness and cost. They will benefit, undoubtedly, from advances in the study of non-linear optical effects in fibres, and one can be fairly optimistic about the prospects of an 'all-fibre' DOFS technology in the medium term.

Concluding remarks

The conclusions from this section are that almost all of the component problems are associated with the provision of the specialised fibres needed for the individual measurement functions. A possible alleviation of these might be obtained by developing techniques which allow a few specially fabricated fibres to be used for a range of measurands, in association with electronic processing functions which provide the required discrimination amongst the measurands.

Other components such as sources, detectors, modulators, switches, etc., sometimes require state-of-the-art technology, but the required performances have, at least, all been demonstrated, and performance margins will inevitably increase as time progresses.

Overall concluding remarks

Conclusions have been drawn at the end of the various sections, where appropriate, but the main ones are concisely re-stated here, for convenience:

- Fully-distributed DOFS systems are much more versatile than quasi-distributed systems.
- A generalised system analysis has derived a set of relationships between the important system parameters for DOFS.
- System parameters have been defined for a system regarded as of general industrial usefulness. This is 100m in total length, has 0.1m spatial resolution and measures to 1% accuracy over all resolution lengths. For such a system to work in back-scatter in respect of temperature and strain measurement, it is concluded that phase measurand-modulation techniques must be used.
- Existing optoelectronic components are capable of meeting the demands made by DOFS, but much more attention needs to be directed towards the fabrication of specialised fibres.
- The industrial economics of DOFS for temperature and strain measurement should prove favourable if the above performance can be achieved.

Acknowledgements

The work reported in this paper was performed originally as part of a study for the UK Optical Sensors Collaborative Association (OSCA), and I am very grateful to them for permission to publish it. I am also very grateful to Dr John Wykes of the British Coal Corporation for his constructive criticisms of the paper.

References

1. T.G. Giallorenzi *et al.* 'Optical-fibre sensor technology', *IEEE J. Quantum Electron.*, QE-18, No. 4, April 1982, pp. 626–65.
2. A.L. Harmer 'Principles of optical fibre sensors and instrumentation', *Proc. Symposium on Optical Sensors and Optical Techniques in Instrumentation*, Institute of Measurement and Control, London, November 1981.
3. A.J. Rogers 'Optical measurement of current and voltage on power systems', *IEE J. of Elec. Power Appns*, 2, No. 4, 1979, pp. 120–4.
4. B. Culshaw '*Optical-fibre Sensing and Signal Processing*' (Peter Peregrinus, London 1984).
5. A.J. Rogers 'Distributed optical-fibre sensors', *J. Phys. D.* 19, 1986, pp. 2237–55.
6. A.R. Nelson *et al.*, 'Passive multiplexing system for fibre-optical sensors', *Appl. Opt.*, 19, No. 17, 1980, pp. 2917–20.
7. B. Culshaw 'Distributed and multiplexed fibre-optic sensor systems', *Proc. NATO Adv. Study Inst.* Erice, Sicily, published by Martinus Nijhoff, May 1986.
8. T. Ormond 'Wavelength multiplex-demultiplex devices emerge from fibre-optic research', *Electronic Development News*, 4th February 1981, pp. 73–4.
9. R. Kist 'Fibre-optic sensors for networks', *Proc. 4th Int. Conf. on Optical-fibre Sensors, OFS '86*, Tokyo, 1986, pp. 209–25.
10. See, for example, A. Yariv '*An Introduction to Optical Electronics*', 2nd ed., (Holt, Rinehart & Winston), 1977, pp. 305–9.
11. M.K. Barnoski *et al.*, 'Optical time domain reflectometry', *Appl. Opt.*, 16, 1977, pp. 2375–9.
12. E.-G. Neumann 'Optical time-domain reflectometer: comment', *Appl. Opt.*, 17, 1978, p. 1675.
13. J.P. Dakin 'Temperature measurement using intrinsic optical-fibre sensors', *Int. J. of Optical Sensors*, 1, No. 2, 1986, pp. 101–14.
14. A.J.A. Bruinsma *et al.*, 'Fibre-optic strain measurement for structural integrity monitoring', *Proc. 2nd Int. Conf. on Optical-fibre Sensors*, Stuttgart, 1984, pp. 399–402.
15. B. Culshaw 'Optical Fibre Sensing and Signal Processing', (Peter Peregrinus, London), 1984, p. 91.
16. G. Hartog 'A distributed temperature sensor based on liquid-core fibres', *J. of Lightwave Technology*, LT1, No. 3, 1983, pp. 498–509.
17. M.C. Farries *et al.*, 'Distributed temperature sensor using ND^{3+}-doped optical fibre', *Electr. Lett.*, 22, No. 8, 1986, pp. 418–9.
18. J.P. Dakin *et al.*, 'Distributed anti-Stokes Raman thermometry', *Proc. 3rd Int. Conf. on Optical-fibre Sensors*, San Diego, post-deadline paper, 1985.
19. M.K. Barnoski and S.M. Jensen, 'Fibre wave-guides: a novel technique for investigating attenuation characteristics', *Appl. Opt.*, 15, No. 9, 1976, pp. 2112–5.
20. A.J. Rogers 'Polarization-optical time domain reflectometry: a technique for the measurement of field distributions', *Appl. Opt.*, 20, No. 6, 1981, pp. 1060–74.
21. J.N. Ross 'Birefringence measurement in optical fibres by POTDR', *Appl. Opt.*, 21, No. 19, 1982, pp. 3489–95.
22. A.J. Rogers 'Polarization properties of monomode optical fibres: use of POTDR to determine spatial distributions', *Proc. Int. Conf. on Fibre-optical Rotation Sensors and Related Technologies*, MIT Cambridge, MA, (Springer Verlag, 1982), November 1981, pp. 208–14.
23. R.C. Youngquist *et al.*, 'A Two-moded Distributed/Multiplexed Sensor System' (Private communication).
24. A.J. Rogers 'Optical fibres in communications and measurement', *CEGB Research* No. 16, August 1984, pp. 3–17.
25. M.C. Farries and A.J. Rogers, 'Distributed sensing using stimulated Raman interaction in a monomode optical fibre', *Proc. 2nd Int. Conf. on Optical-fibre Sensors*, Stuttgart, Paper 4.5, September 1984, pp. 121–32.
26. E. Theocharous 'Differential absorption distributed

thermometer', *Proc. 1st Int. Conf. on Optical-fibre Sensors*, London, IEEE Conf. Publ. No. 221, April 1983, pp. 10–12.

27. I.P. Giles, D. Uttam, B. Culshaw and D.E.N. Davies, 'Coherent optical-fibre sensors with modulated laser sources', *Electr. Lett.* 19, No. 1, 1983, pp. 14–5.
28. K. Peterman and G. Arnold, 'Noise and distortion characteristics of semiconductor lasers in optical-fibre communications systems', *IEEE J. Quantum Electron.*, QE-18, No. 4, 1983, pp. 543–55.
29. A.J. Barlow *et al.*, 'Production of single-mode fibres with negligible intrinsic birefringence and polarization mode dispersion', *Electr. Lett.*, 17, 1981, pp. 725–6.
30. R.D. Birch *et al.*, 'Fabrication of polarization maintaining fibres using gas-phase etching', *Electr. Lett.*, 18, 1982, pp. 1036–8.
31. R.B. Dyott 'Preservation of polarization in optical-fibre waveguides with elliptical cores', *Electr. Lett.*, 15, pp. 380–2.
32. C.D. Hussey *et al.*, 'Circularly birefringent single mode optical fibres', *Electr. Lett., 22, 1986, pp. 129*–30.
33. G. Meltz *et al.*, 'Multi-wavelength twin-core fibre-optic sensor', *4th Int. Conf. on Optical-fibre Sensors, OFS '86*, Tokyo, Paper 3.6, 1986, pp. 67–70.
34. S.B. Poole *et al.*, 'Fabrication of low-loss optical fibres containing rare-earth ions', *Electr. Lett.*, 21, 1985, pp. 737–9.
35. M.C. Farries (private communication).
36. W.A. Gambling 'New fabrication techniques for optical-fibre sensors and lasers', *Int. J. of Optical Sensors*, 1, No. 4, 1986, pp. 307–16.
37. A.J. Rogers 'Polarization optics for monomode fibre sensors', *Proc. IEE, pt J*, 132, No. 5, 1985, pp. 303–8.
38. B. Culshaw 'Trends in fibre-optic sensors', *Int. J. of Optical Sensors*, 1, No. 4, 1986, pp. 327–47.

A coherent time multiplexed distributed sensing system

N.E. Jolley and R.E. Epworth
STC Technology Ltd, UK

S. Wright
Hewlett-Packard Laboratories, UK

OPTICAL fibre sensors using interferometric techniques offer very high sensitivity because of their ability to measure changes in fibre length to a small fraction of an optical wavelength. A number of separate interferometers on a single 'highway' fibre can be separately interrogated using time multiplexing. The resulting optical system is very similar to that of a coherent reflectometer, and we report here on a multiplexed sensing system based on the optical system of a commercial coherent reflectometer (STC-OFR-6).[1]

In such a system, the perturbed and reference optical paths have very different lengths and, hence, system performance depends critically on the phase noise of the optical source. The sensing system that we describe has been operated with a number of different sources, and we compare here their relative merits. In particular we present measurements showing the extremely low phase noise of a Rayleigh back-scatter line narrowed semiconductor source.[2]

Source evaluation

Source phase noise can be measured using a delayed self-heterodyne Mach-Zehnder interferometer as shown in figure 1. This consists of Mach-Zehnder formed using fused couplers with a Bragg cell and suitable length of fibre in one arm. Thus the light source is mixed with a delayed frequency shifted version of itself in a photo-detector. In this system the delay was 100m, simulating a 50m sensing element length. If the laser has no phase noise, the detector output would be a pure spectral line at the Bragg cell driver frequency. Laser phase noise appears as a broadening of the carrier at the offset frequency. Thus the phase noise spectral density at a particular frequency may simply be measured on a spectrum analyser.[3] This measurement system has been used to determine the phase noise of a 1152nm Helium-Neon laser and various semiconductor sources.

For the measurements the source and measurement systems were environmentally isolated as far as possible, using sound-proofing and an isolation table. This was more difficult for the HeNe laser because significant heat has to be removed from the tube – an obvious problem with this source.

Results and discussion

A back-scatter line narrowed (BSLN) semiconductor laser has already been shown to have better short term coherence than a gas laser.[2] Figure 2 shows the measured phase noise of the HeNe laser and a BSLN inverted rib waveguide (IRW) laser. For both lasers the noise consists of a wideband noise floor and a low frequency l/f component. The HeNe laser has the lowest wideband noise floor, but below 2000Hz the line narrowed IRW laser has the lowest noise.

Figure 3 shows the comparison between the phase noise of the line narrowed IRW laser and a recently published result for active feedback control of the phase noise, using a Fabry-Perot interferometer as a frequency discriminator.[4] The results here show that the line narrowed IRW laser provides a robust,

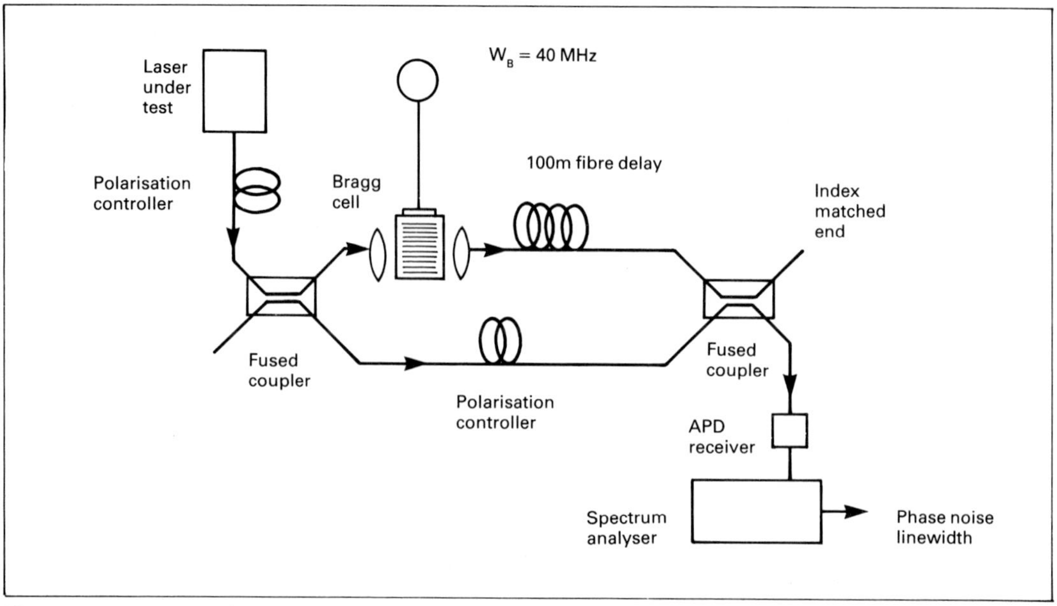

Fig. 1 Delayed self-heterodyne measurement

extremely low phase noise source without the need for excessive acoustic isolation, or complex active-feedback phase noise control circuitry.

Distributed sensing system

A time multiplexed multi-element sensing system is shown in figure 4. A pulse of coherent light is partially reflected from a large number of equally spaced reflective splices along the length of the fibre. The reflected optical pulses are coherently detected by combining with some optical reference, converting them to electrical pulses at the Bragg cell driver frequency. Processing, to derive the phase difference between adjacent reflectors, can be applied on either the electrical or optical signals. Thus, the system can sense a large number of independent elements, with no sensitivity to perturbations to the optical fibre downlead.

We have used an optical system in which the optical reference is a cw local oscillator. This approach has advantages over other configurations.[5] The use of local oscillator detection provides a large increase in sensitivity as the system is only limited by source phase noise, not receiver noise. In addition, the coherent detection process produces a signal current proportional to the electric field of the reflected optical signal, and so the dynamic range problem of the electrical signal is considerably reduced. Thus, if the reflectance of A and B are reduced by an order of magnitude the current signal is reduced by a factor of three.

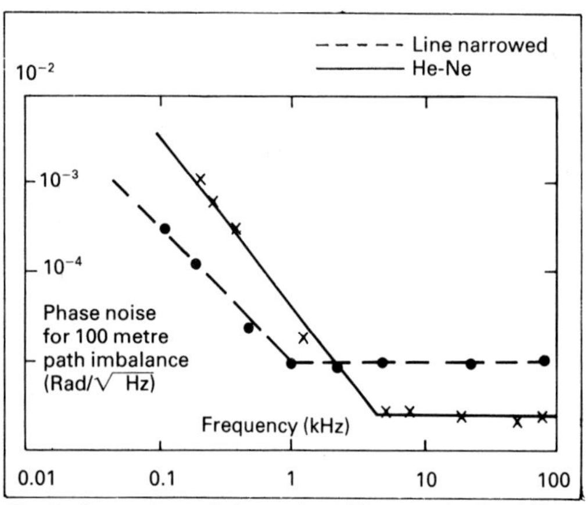

Fig. 2 Comparison of phase noise of line narrowed and HeNe laser

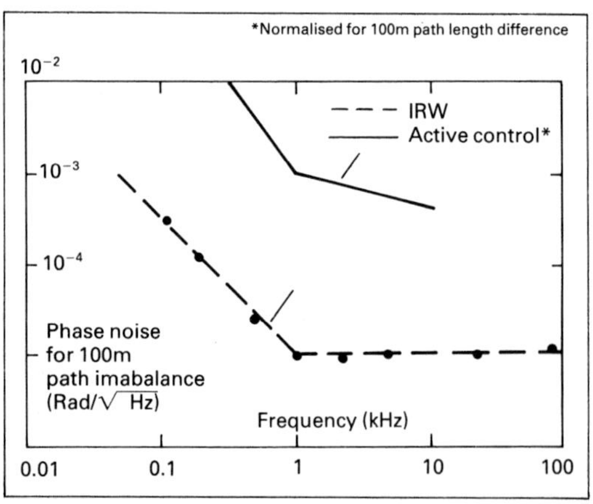

Fig. 3 Comparison of phase noise of IRW and controlled laser

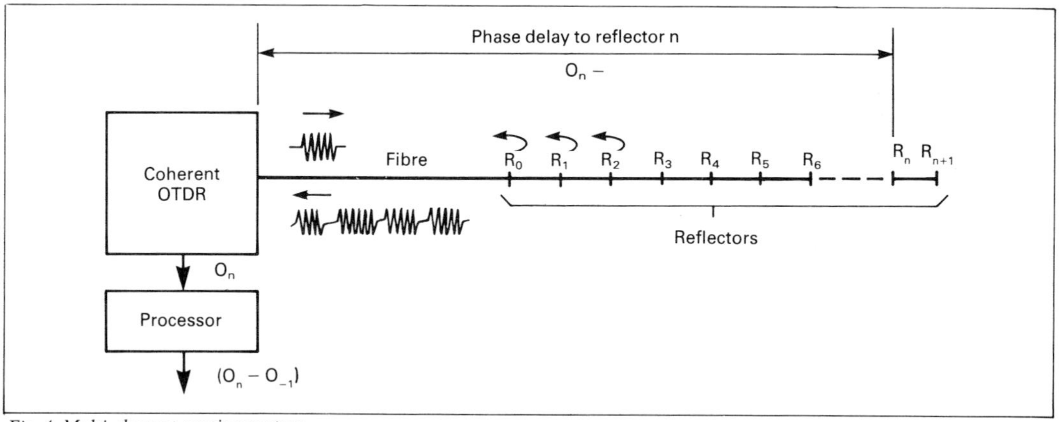

Fig. 4 Multi-element sensing system

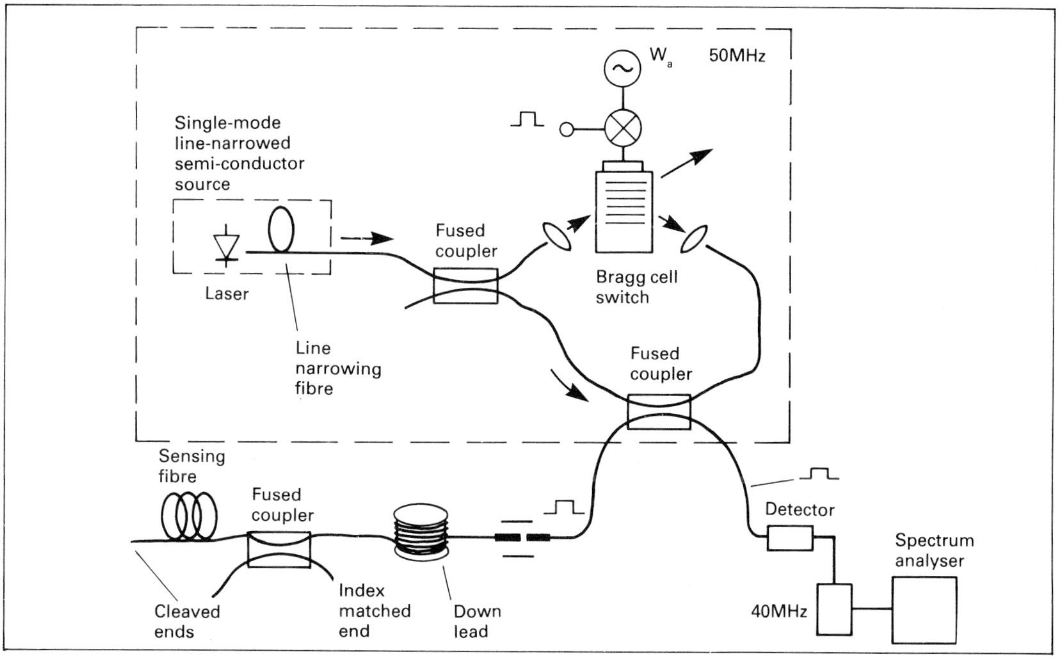

Fig. 5 Time multiplexed sensor sensitivity measurement

Measurements and discussion

In order to measure the effects of phase noise in such a sensor, the measurement system shown in figure 5 was set up. For these experiments a coupler and single sensing fibre with cleaved ends were used to simulate a single sensing element with reflective splices.

Figure 6 shows the measured sensitivity of the system. The peak launch power was 40 W (mean 1 W). The reflection coefficients at the cleaved ends were 4%. Some improvement in sensitivity can be achieved by providing acoustic isolation of the back-scatter fibre from its environment. The system still detects phase changes of less than one milliradian at

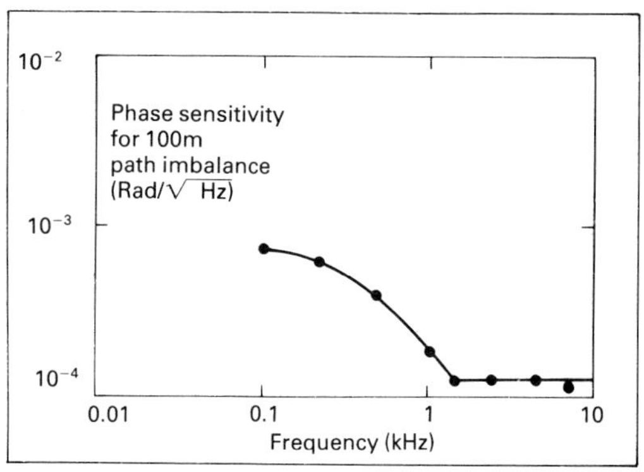

Fig. 6 System phase sensitivity with line narrowed IRW laser

100Hz, and above 2000Hz it approaches one tenth of this figure — a change of one part in 10^{12} in the sensing fibre length. This sensitivity is achieved with an optical system that has already been developed into a robust, compact form for other applications.

Acknowledgements

The authors would like to thank STC for permission to publish this paper, and A. Robertson and R. Jones for helpful discussion and assistance.

References

1. D.F. Smith, J.P. King, P. Timpson, M. Gray, S. Wright, 'Coherent OTDR for fault location in long wavelength optical systems', *IEE International Conference on 'Measurements for Telecommunication Systems', London,* 27–28 November 1985.
2. R.E. Epworth, D.F. Smith and S. Wright, 'A practical 1.3μm semiconductor source with significantly better short term coherence than a gas laser', *Proceedings of the 10th ECOC, Stuttgart,* 1984, pp. 132–133.
3. P. Gallion, F.J. Mendienta and R. Leconte, 'Single frequency laser phase-noise limitation in single-mode optical-fibre systems with correlated fields', *Journal of the Optical Society of America,* Vol. 72, No. 9, September 1982.
4. J.P. Dakin and P.B. Withers, 'A study of phase noise reduction of semiconductor lasers for sensor applications', *AGARD Conference reprint No. 383, Guided Optical Structures in a Military Environment, Istanbul, Turkey,* September 1985, pp. 12.1–12.7.
5. J.P. Dakin and C.A. Wade, 'Optical fibre hydrophone array—recent progress', *Proceedings of the 2nd OFS, Stuttgart,* 1984, pp. 375–379.

A novel distributed optical fibre sensing system enabling location of disturbances in a Sagnac loop interferometer

J.P. Dakin
York Ltd., UK

D.A.J. Pearce
Madge Networks, UK

A.P. Strong and C.A. Wade
Plessey Research, UK

THE interferometric detection of induced phase modulation of light propagating in an optical fibre forms the basis of a family of highly sensitive transducers, which includes hydrophones,[1] magnetometers,[2] dynamic thermometers,[3] and gyroscopes.[4] In most of these sensors the observed phase modulation corresponds to the effect of the measurand integrated over the interaction length of the fibre, and the distribution of the measurand along the fibre cannot usually be determined. Few techniques have been published for the location of disturbances on a continuous fibre, although in a recent report the FMCW technique was used to locate a stress-induced modal coupling point on high-birefringence fibre.[5]

This paper presents a new type of disturbance location sensor which is based on a modified Sagnac loop interferometer and exploits an effect which has been identified previously as a source of error in optical fibre gyroscopes.

Operating principle

There is currently considerable interest in the development of fibre gyroscopes, which are based on the Sagnac interferometer (see Fig. 1). Light from the optical source is split into two beams which propagate in opposite directions around a loop of fibre. Rotation

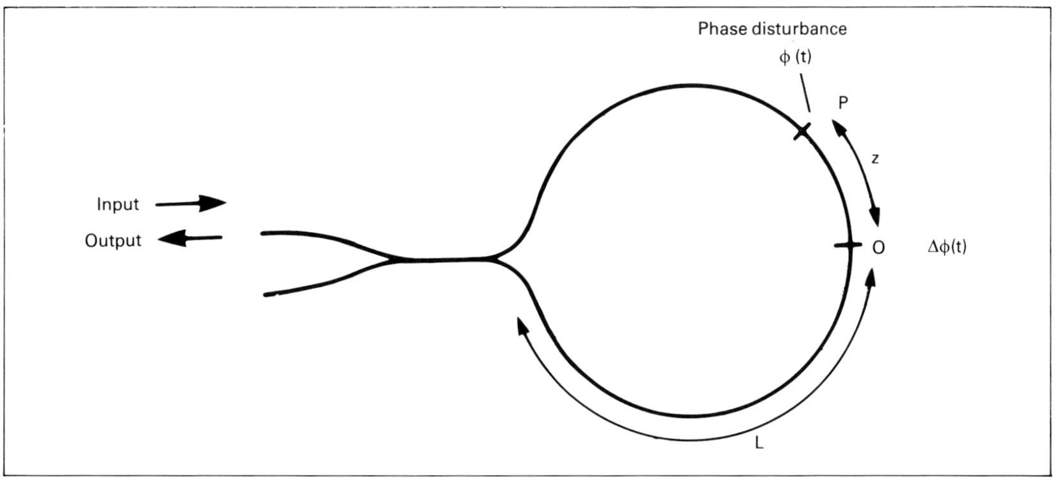

Fig. 1 Sagnac loop interferometer

of the loop gives rise to a phase difference between the counter-propagating beams, and a corresponding fringe shift is observed at the output of the interferometer. The Sagnac interferometer is also sensitive to other stimuli such as variations in temperature or in mechanical stress in the fibre and, if the intention is to measure rotation rate, these effects are regarded as sources of error which can limit the accuracy of fibre gyroscopes. Seen from another point of view, however, these effects can be exploited to form a new sensor.

The effects of changing temperature and strain on a Sagnac interferometer have been described and evaluated.[4,6] Referring to figure 1, consider a simple Sagnac interferometer having a loop of length $2L$ on which a phase modulation $\phi(t)$ is applied at point P, which is located a distance z along the fibre from 0, the centre of the loop. The resulting phase difference between the counter-propagating beams is given by:

$$\Delta\phi(t) = \phi(t - (L - z)/v) - \phi(t - (L + z)/v) \quad (1)$$

where v is the velocity of light in the fibre. If $2z/v$ is small, Taylor's approximation can be used, yielding:

$$\Delta\phi(t) = [\phi(t) - (L - z)/v\,(\mathrm{d}\phi/\mathrm{d}t)] - [\phi(t) - (L+z)/v\,(\mathrm{d}\phi/\mathrm{d}t)] \quad (2)$$

neglecting higher order terms. This simplifies to:

$$\Delta\phi(t) = (2z/v)\,(\mathrm{d}\phi/\mathrm{d}t) \quad (3)$$

It can be seen that the observed fringe shift is directly proportional to the distance z from the centre of the loop. If $\phi(t)$ is measured independently, for example by interfering one of the counter-propagating beams with a coherent reference, it can be differentiated and substituted into equation (3) to yield z, the location of the phase disturbance.

Experimental arrangement

An experimental sensor was constructed to establish the feasibility of locating disturbances by this method (see Fig. 2). It consisted of a Sagnac interferometer loop combined with a Mach-Zehnder interferometer, both utilising the same 1150nm HeNe laser source. The Sagnac loop contained approximately 200m of 1300nm single-mode fibre closely wound in a single layer on an 11cm diameter cylindrical former. A pseudo-heterodyne detection scheme[7] was used, and the sinusoidal phase modulation necessary for correct bias was applied via a piezo-electric fibre stretcher located near one end of the fibre coil. The output from the Sagnac interferometer was received on a photo-diode, D2, and the resulting photo-current was amplified and processed, as in reference 7 to produce an output voltage proportional to the phase difference between the two beams $\Delta\phi$.

In the Mach-Zehnder interferometer, light from one of the counter-propagating Sagnac beams, split off from the loop using a 3dB fibre coupler, was combined directly with light direct from the laser. This enabled

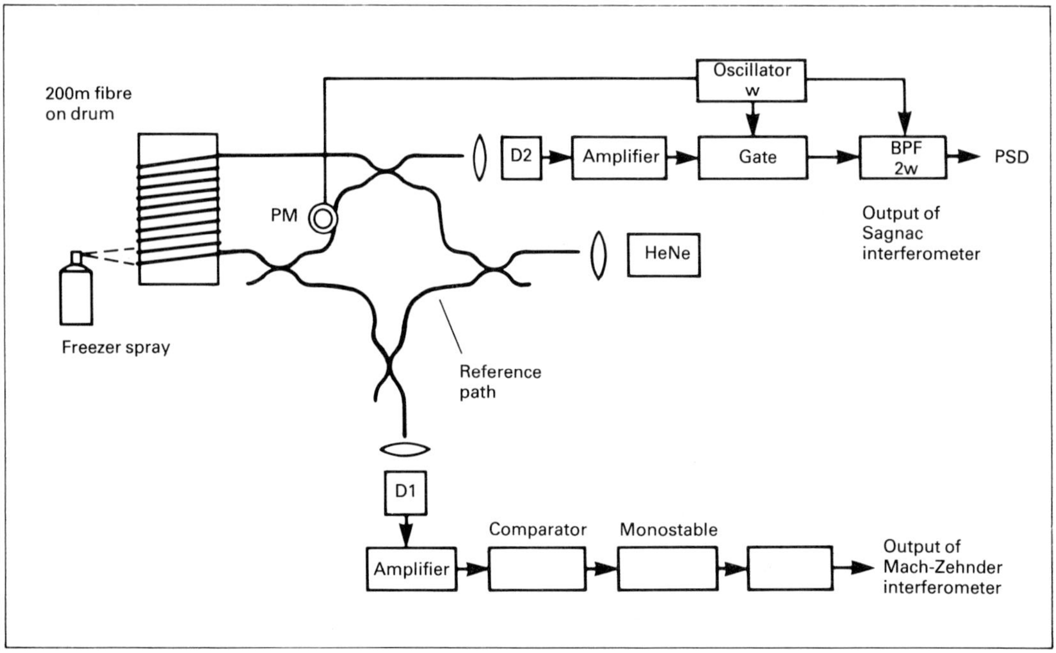

Fig. 2 Experimental arrangement

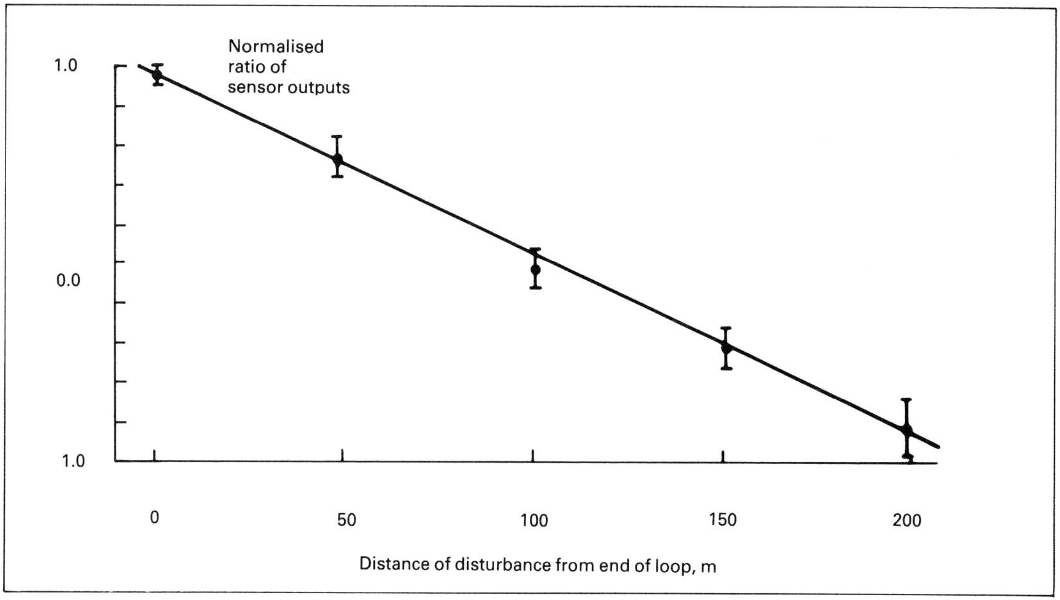

Fig. 3 Response of sensor to disturbances at different locations

any phase disturbance $\delta(t)$ to be detected and quantified. For simplicity, a fringe counting arrangement was used to measure ϕ: the optical receiver was AC coupled and was followed by a comparator which produced sharp edges at zero-crossing points. A monostable followed by a low-pass filter converted this signal into a low voltage proportional to $|d\phi/dt|$.

In this experiment, the equipment produced two outputs: one, from the Sagnac interferometer, proportional to $(2z/v).d\phi/dt$; and the other from the Mach-Zehnder interferometer, proportional to $d\phi/dt$.

Experimental results

Phase disturbances were induced in the fibre by spraying small areas of the Sagnac coil with freezer spray. This produced a rapid cooling of a section of the fibre, followed by a rather slower return to ambient temperature, and resulted in a corresponding sawtooth phase disturbance. Disturbances were induced at five different places on the fibre drum, and in each case the two outputs were recorded as a function of time. In figure 3 the ratio of the two outputs is plotted against the location of the disturbance along the 200m fibre loop, estimated from distance measurements on the drum. Each point is the mean of five measurements, and the error bars indicate the spread of the individual readings. A straight line has been fitted to the data, since, from equation (3), the dependence is expected to be linear. The relatively large scatter of readings in this experiment, indicated by the error bars, is believed to result from the method of inducing the phase disturbance by cooling, with the fibre wound on a drum, and a considerable improvement in resolution is expected to be possible.

Concluding remarks

We have demonstrated a disturbance location sensor which is based on detecting the non-reciprocal phase shift induced by a localised stimulus on a Sagnac loop interferometer. In the feasibility experiment described, the sensor detected thermal disturbances with a resolution of approximately ±20m in a total fibre length of 200m, and a substantial improvement in resolution is expected to be possible. The sensor thus shows considerable potential for locating thermal or mechanical disturbances and may have a wide range of applications in areas such as intruder detection, vehicle location or detection of leaks or cryogenic liquids.

References

1. J.A. Bucaro, H.D. Dardy and E.F. Corome, J. Acoust. Soc. Am. Vol. 62, 1977 pp. 1302–1304.
2. A. Dandridge, A.B. Tveten, G.H. Siegel, Jr., E.J. West and T.G. Giallorenzi, *Elec. Lett.* Vol. 16, 1980, pp. 408–409.
3. L.S. Schuetz, J.H. Cole, J Jarzynski, N. Lagakos and J.A. Bucaro, *App. Opt.* Vol. 22, 1983, pp. 478–483.
4. R.A. Bergh, H.C Lefevre and H.J. Shaw *Light Tech*, Vol. 2, 1984, pp. 91–107.
5. R.B. Franks, W. Torruellas and R.C. Youngquist, SPIE *Fibre Optic Sensors*, Vol. 586, 1985, pp. 84–89.
6. D.M. Shupe, *App. Opt.* Vol. 19, 1980, pp. 654–655.
7. A.D. Kersey, A.C. Lewin and D.A. Jackson, *Elec. Lett.* Vol. 20, 1984, pp. 368–370.

Distributed sensors based on differential absorption

E. Theocharous
BP Research, UK

THE slope of an optical time domain reflectometry (OTDR) trace at a particular point is a measurement of the attenuation that light experiences in propagating twice through that point. A number of methods have recently been proposed which employ OTDR for the measurement of an external field distribution along the length of a fibre[1,2,3,4]. However, parameters like microbending can influence the OTDR trace so some referencing is required to remove any ambiguities in the interpretation of the OTDR trace. The purpose of this contribution is to describe the differential absorption optical time domain reflectometry (DAOTDR), a technique which relies on the dependence of the spectral absorption characteristics of the fibre core on an external influence, such as temperature, in order to monitor its distribution along the fibre length.

Assume that a doped optical fibre is available, such that the impurity content causes an absorption band around λ_A. Assume further that an external field, such as temperature, changes the spectral profile of the band but the field has little effect on the absorption coefficient at a nearby wavelength λ_R which is away from the band. An OTDR trace corresponding to λ_A shows strong dependence on the field, whereas for wavelength λ_R it is not significantly modulated.

The OTDR theory states that when a pulse of light, of power P_0 and duration τ, is launched into an optical fibre, the back-scattered power returning from a distance r inside the fibre is given by;[5]

$$P_{(r)} = \tfrac{1}{2} P_0 \tau V_g B_{(r)} S_{(r)} \exp\left\{-2\int_0^r [\alpha_{(\epsilon)} + B_{(\epsilon)}]\, d\epsilon\right\} \qquad (1)$$

where V_g is the group velocity of the pulse, B is the scattering coefficient, S is the numerical aperture factor and α is the absorption coefficient. If wavelengths λ_A and λ_R are denoted by subscripts A and R respectively, then dividing the corresponding expression for the two wavelengths:

$$\frac{P_{A(r)}}{P_{R(r)}} = \frac{P_{0A}}{P_{0R}} \frac{\tau_A}{\tau_R} \frac{V_{gA}}{V_{gR}} \frac{B_{A(r)}}{B_{R(r)}} \frac{S_{A(r)}}{S_{R(r)}} \exp\left\{-2\int_0^r \left\{[\alpha_{A(\epsilon)} - \alpha_{R(\epsilon)}] + [B_{A(\epsilon)} - B_{R(\epsilon)}]\right\} d\epsilon\right\} \qquad (2)$$

If λ_R is chosen to be close to λ_A then the values of the parameters P_0, τ, V_g, B and S are not expected to vary significantly between the two wavelengths. Furthermore, provided their values are known, the ratios on the RHS of equation (2) can be replaced by numerical factors. Equation (2) then simplifies:

$$\alpha_{A(r)} - \alpha_{R(r)} = \tfrac{1}{2} \frac{d}{dr}\left\{ \ln \left[\frac{P_{A(r)}}{P_{R(r)}}\right]\right\} \qquad (3)$$

Hence the measurement of the ratio of the back-scattered power at two wavelengths can provide the spatial distribution of the difference of the absorption coefficients of the fibre core at the two wavelengths.

Figure 1 shows the various stages in the extraction of the external field profile from the range resolved DAOTDR return signals. A uniform disturbance is assumed between distances r_2 and r_3 inside the fibre which causes the absorption coefficient at the active wavelength to increase whereas λ_R remains unaffected.

The most important requirement for the practical

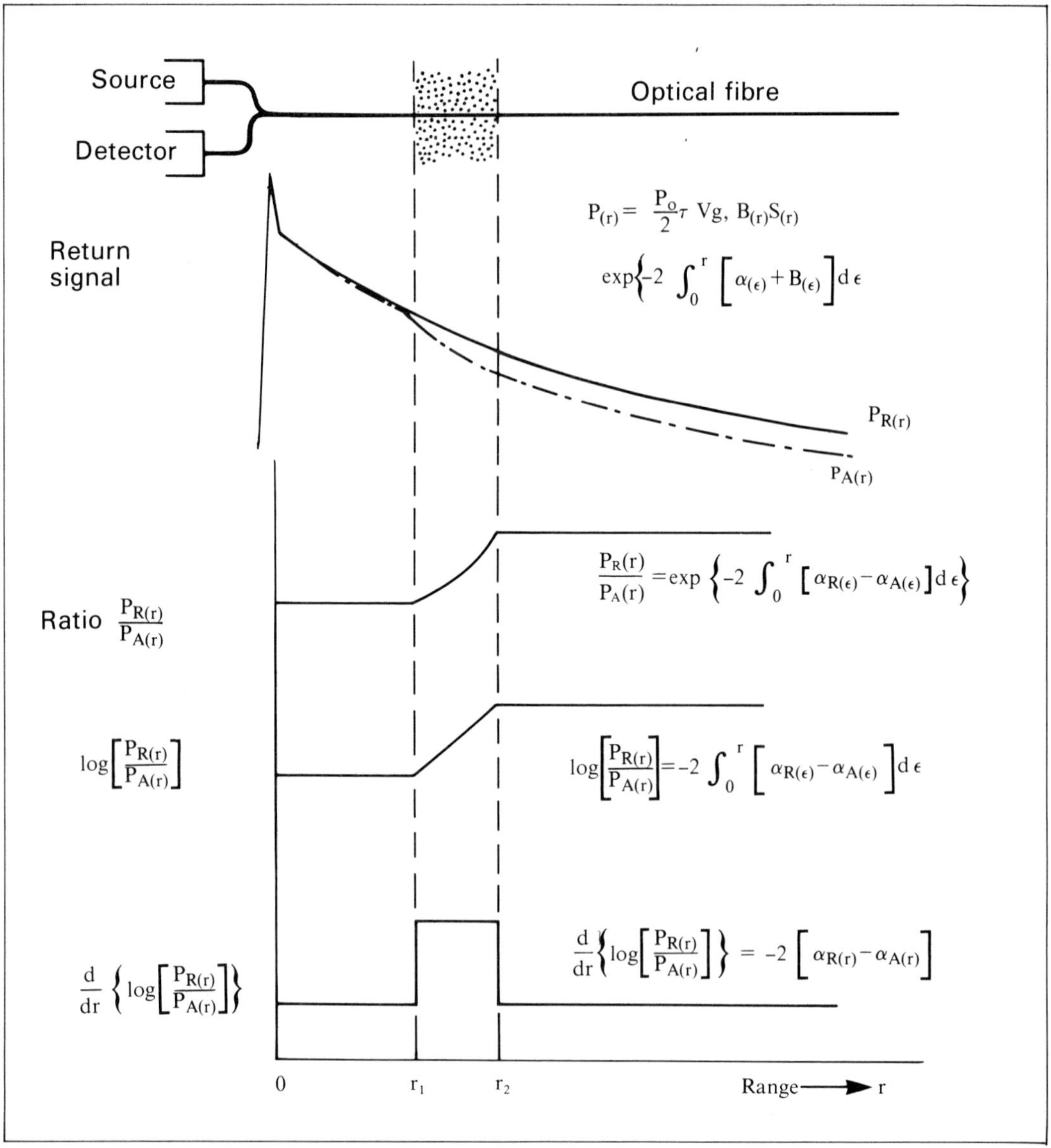

Fig. 1 The principle of DAOTDR

demonstration of DAOTDR is the availability of a suitable optical fibre. External conditions such as temperature, pressure, electric and magnetic fields are known to affect the spectra of materials. In the case of gases these effects are well documented. Atoms in solids are approximately ten times closer to each other than in a gas at normal temperature and pressure. The spectral bands of impurity atoms in a crystal lattice are relatively sharp. However, in an amorphous (glass) lattice inhomogeneous broadening, due to the different conditions that exist at different points in the lattice, dominates. The modulation of the spectral profile of doped glasses by an external influence may not be readily observed. The modulation of the spectra of solids by pressure[6] and magnetic field[7] have only been observed in doped crystals. In doped glasses only temperature has been observed to modify their spectra.[8] This effect is particularly strong in semiconductors and, to a lesser extent, in glasses doped with rare earths. The ruby glasses are essentially semiconductor doped glasses and their absorption spectra exhibit strong temperature modulation. However, the transmission of commercially available ruby glasses is not high enough to be drawn into an optical fibre. On the other hand, optical fibres doped with rare earth ions (Nd^{3+}, Er^{3+}, Tb^{3+} and Eu^{3+}) have been fabricated with attenuation figures of <2dB/km away from the absorption bands.[9] It was also shown[10] by the same group that the OTDR trace of a Nd^{3+} doped fibre at 904nm can provide the temperature profile along the fibre.

Concluding remarks

A comparison of DAOTDR with other distributed sensor techniques indicates that apart from requiring special fibres, DAOTDR, based on attenuation measurements, seriously attenuates the back-scattered signal compared to a scattering type technique. On the other hand, DAOTDR monitors the Rayleigh component of the back-scattered signal which is over 30dB greater than the Raman component. Furthermore, scattering techniques can only measure temperature whereas DAOTDR can, at least in theory, be adapted to monitor other parameters.

An interesting development of DAOTDR is the multiplexing of a number of differential absorption point sensors along the length of an ordinary communications fibre.[11] Although this technique provides measurements only where a sensor is located, its spatial resolution and performance is superior to DAOTDR.

Acknowledgements

The author would like to thank the Central Electricity Generating Board and the British Petroleum Research Centre for permission to publish this paper.

References

1. A.J. Rogers, 'Polarization optical time domain reflectometry', *Applied Optics*, Vol. 5, 1981, pp. 1060–1074.
2. A.H. Hartog and D.N. Payne, 'Remote sensing of temperature distribution using an optical fibre', *Proc. ECOC 8, Cannes*, 1982, pp. 215–220.
3. E. Theocharous, Measuring apparatus and method, *Patent application PCT/GB84/00136.*
4. J.P. Dakin, D.J. Pratt, G.W. Bibby and J.N. Ross, Distributed anti-stokes ratio thermometry, *Proc. OFS 3, San Diego*, 1985, Postdeadline paper PDS-3.
5. E.G. Neumann, 'Optical time domain reflectometry: comment', *Applied Optics*, Vol. 17, 1978, p. 1675.
6. R.G. Munro, G.J. Piermarini and S. Block, 'Model lineshape analysis for the ruby R lines used for pressure measurement', *Journal of Applied Physics*, Vol. 57, pp. 165–169.
7. K.H. Hellwege, 'Spectroscopy of crystalline rare earth compounds', *Quantum Electronics Conference, Paris*, Vol. 1, 1963, pp. 623–644.
8. A.J. Holland and W.E.S. Turner, 'An investigation into the variation with temperature of the spectral transmission of glasses', *Journal of the society of glass technology*, Vol. 25, 1941, pp. 164–220.
9. S.B. Poole, D.N. Payne and M.E. Fermann, 'Fabrication of low loss optical fibres containing rare-earth ions', *Electronics Letters*, Vol. 21, 1985, pp. 737–738.
10. M.C. Farries, M.E. Fermann, A.P. Leach, S.D. Poole and D.N. Payne, 'Distributed temperature sensing using Nd^{3+} doped optical fibres', *Electronics Letters*, Vol. 22, 1986, pp. 418–419.
11. E. Theocharous, 'Differential absorption distributed termometer', *Proceedings of the First International Conference on Optical Fibre Sensors, London*, 1983, pp. 10–12.

Correlation optical time domain reflectometry for high resolution distributed fibre-optic sensing

J-J. Bernard and E. Depresles
CGE Research Centre, France

THE optical fibre is by nature a one-dimensional measurement medium and is well suited to all distributed sensing methods using the spatial distribution of the measurands.

In fully distributed systems, any point of the fibre can be simultaneously interrogated by the well-known optical reflectometry methods.

Conventional optical time domain reflectometry (OTDR) is rapidly limited in spatial resolution because of time response of electronics limited to about 10ns corresponding to a few-metre resolution.

Furthermore, compared to the 0.8–0.9μm wavelength range, in the 1.3–1.6μm region high power pulsed semiconductor lasers do not exist yet. Thus, it becomes necessary to use low CW power laser diodes, such as those developed for telecommunication systems.

Correlation technique tackles the problem of limited source power by improving the mean launched power, not the peak power as in other techniques, and leads to high resolution by using high bit-rate pseudorandom sequences.

Principle of the correlation OTDR

Correlation technique has been described previously for OTDR application.[1] Figure 1 shows the principle of the COTDR.

The laser diode is intensity modulated by a pseudorandom bit sequence $S_1(t)$ having the duration $T_o = (2^n - 1)T_b$ where T_b is the bit length. In order to avoid simultaneous measurement at two different points, T_o is chosen to be at least equal to $2L/v$, where L is the whole length of the tested line and v is the propagation speed of the light in the fibre.

The back-scattered signal is detected by a photodiode, amplified and then multiplied by a time-shifted sequence $S_2(t-\theta)$. The shift time $\theta = 2z/v$ corresponds to a round-trip propagation time up to the abscissa z of the measured point. The product signal is integrated during a time $T = kT_o$, multiple of T_o. The cross-correlation of the two sequences is then:

$$R(\theta) = \int_0^T S_1(t)\, S_2(t-\theta)\, dt$$

Figure 2 shows the shape of the theoretical cross-correlation function. It appears as a triangle of base equal to $2T_b$ and of height equal to $NT_o/2$ where $N = 2^n - 1$ (n corresponds to the number of flip-flops of the shift register which generates the sequence). Outside the interval $-T_b$, $+T_b$ the cross-correlation function is equal to -1.

Comparison between conventional and correlation OTDRs

In order to give some comparative figures between these two methods we will consider the worst case of the analysis of a matched break (without any reflected energy) on a single-mode fibre. The use of a cw laser diode allows to launch about 0.25mW in the test fibre, that is −6dBm, taking into account various losses (coupling, connectors, etc.)

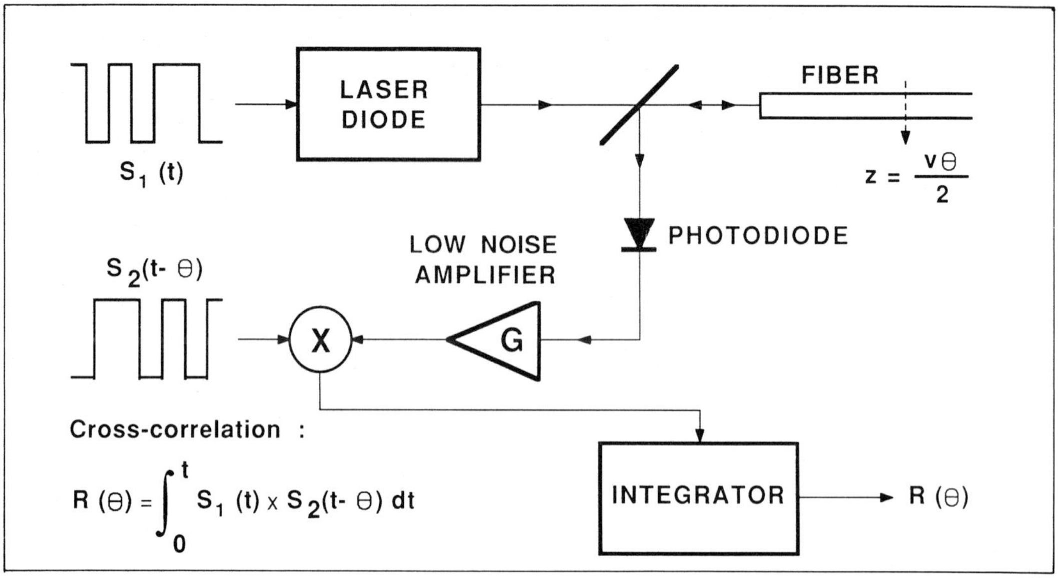

Fig. 1 Principle of the Correlation Reflectometry

The back-scattered signal level at the input of the fibre is at 78 – 80dB below the launched power level, for 1ns width pulses corresponding to a spatial resolution of 10cm. In the case of short length fibres (some hundred metres), the fibre attenuation can be considered as negligible for the used wavelengths (1.3 to 1.6μm). The optical power received by the photo-diode is then about -85dBm.

Localisation of a matched break requires a minimum signal/noise ratio of 5dB at this point. Taking a noise equivalent power (NEP) of 3.10^{-9}W in a bandwidth of 1GHz, that is −55dBm, the detection of a break by the conventional pulse technique is possible only with an averaging process providing a signal/noise ratio gain of 35dB, which leads to an average number of 10.[7] Measurements of a 1km-long fibre requires a time interval between consecutive pulses of 10μs, leading to an averaging time of the order of 2 minutes. A sampling frequency of at least 10GHz is then necessary for the acquisition of the 10,000 analysed points of the 1km-long fibre that seems to be difficult to realise today with available electronic circuits.

Correlation technique, very useful in radar systems, improves the mean power while maintaining the same peak power and without degrading spatial resolution (formally identical to the pulse compression technique). It has previously been used for medium dynamic range portable reflectometers designed for field operation.[2,3,4]

Spatial resolution is determined by the width (FWHM) of the cross-correlation peak, corresponding to the bit duration of the sequence. A resolution of

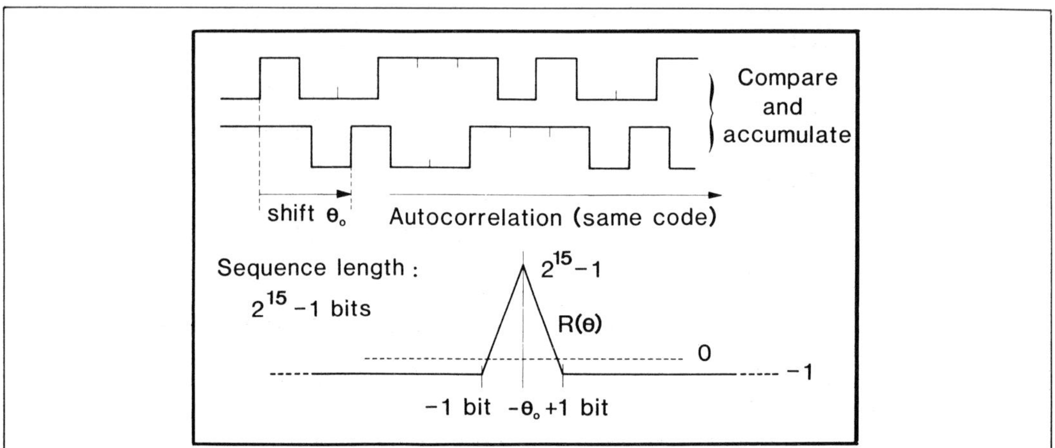

Fig. 2 Maximal sequence autocorrelation function

10cm requires a bit duration of 1ns, thus a clock frequency of 1GHz. The signal/noise ratio gain provided by the correlation process is, at most, 10 log (BT/8), where B is the sequence bandwidth and T its duration.[5]

Taking a sequence of 32kbits ($2^{15}-1$) and a bit duration of 1ns, the sequence duration T is 32μs and its bandwidth 1GHz. The maximum gain is then 36dB compared to the level back-scattered by a single pulse of 1ns duration. This allows to preserve, with a received power of −85dBm, the minimum signal/noise ratio of 5dB required for a matched break detection.

But this method, the analysis of the 10,000 points of a 1km-long fibre cannot be achieved in less than 320ms. Taking the same measurement time as for the pulse method (2 minutes), the correlation method provides the possibility of an averaging process over about 375 samples, corresponding to an extra gain of the order of 13dB.

Through those elements, we can see that the correlation technique theoretically gives much better performances than the conventional pulse method in the case of high resolution short distance reflectometry.

Description of the correlation reflectometer

The schematic diagram of the correlation reflectometer is shown in figure 3. It consists of an InGaAsP/InP buried-heterostructure (BH) laser, coupled to a single-mode fibre allowing a mean launched power of about 750μW with a possible 'ECL' modulation of +700μW.

The injection current of the laser is modulated by a pseudo-random bit sequence (PRBS) with a clock frequency of 200Mb/s corresponding to a spatial resolution of 50cm.

The optical power is then launched into the test fibre through a single-mode fibre coupler, made by the fused biconical taper technique. The back-scattered signal is detected by an integrated InGaAs PIN-FET receiver. After a high gain amplification (51dB), the signal is multiplied to the shifted PRBS. The product signal is then time-integrated, providing the cross-correlation signal between the emitted signal and the signal back-scattered from the analysed point in the fibre. This time-integrated signal is proportional to the optical power loss between the signals back-scattered from the fibre input and from the explored distance.

The delay time varying by steps of 5ns from zero to 2L/C, the cross-correlation signal provides the temporal signature of the fibre, that is its attenuation image including reflection peaks generated by the sensors distributed along the fibre.

The analogue signal coming out from the correlator is then digitalised and sent to a microcomputer through an IEEE-488 bus in order to be stored and processed.

Evaluation of the one-way dynamic range

As previously explained, location of a matched break requires a signal/noise ratio of at least 5dB. The one-way dynamic range is defined by:

$$DR = 0.5\ (S/N - 5dB).$$

Figure 4 shows the back-scattering curve obtained on a 40m-long fibre.

Signal/noise ratio is about 10dB, corresponding to a one-way dynamic range of 2.5dB, that allows to detect and locate a matched break at the end of a 5km long fibre having a 0.5dB/km attenuation.

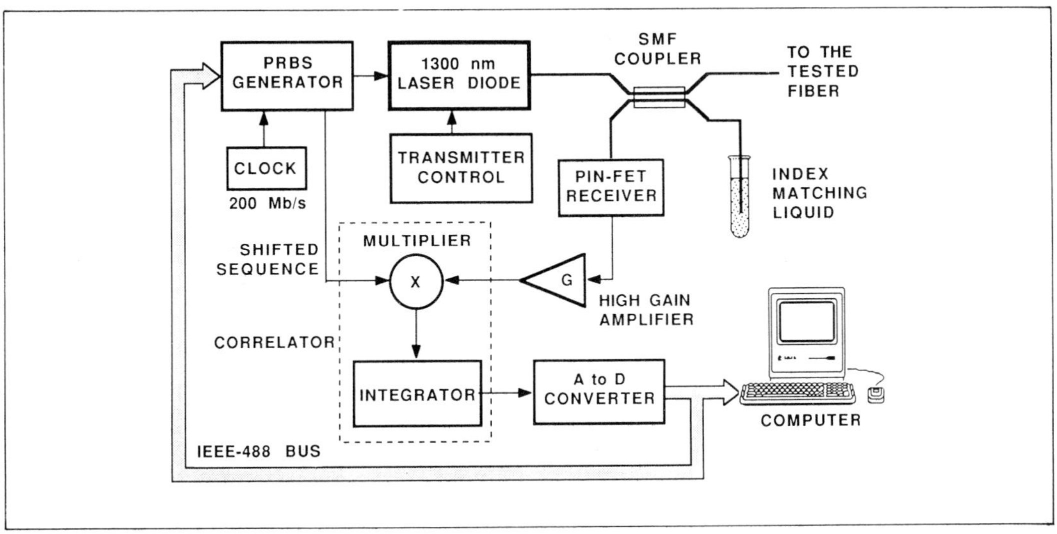

Fig. 3 Schematic diagram of the correlation reflectometer

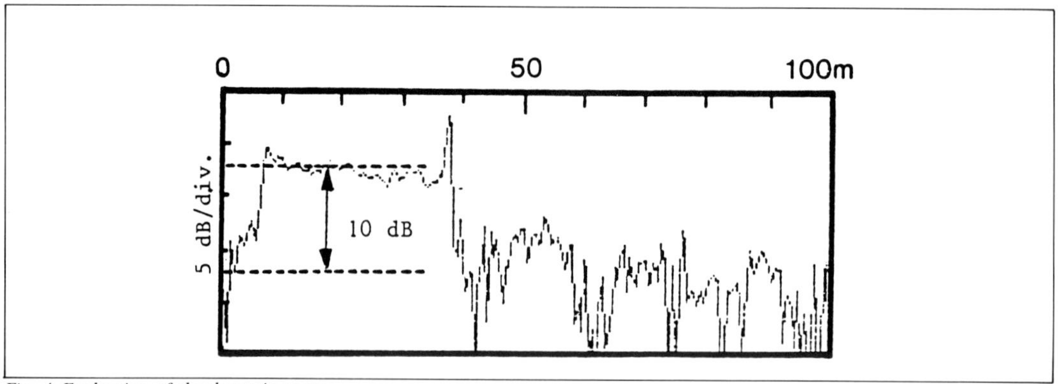

Fig. 4 Evaluation of the dynamic range

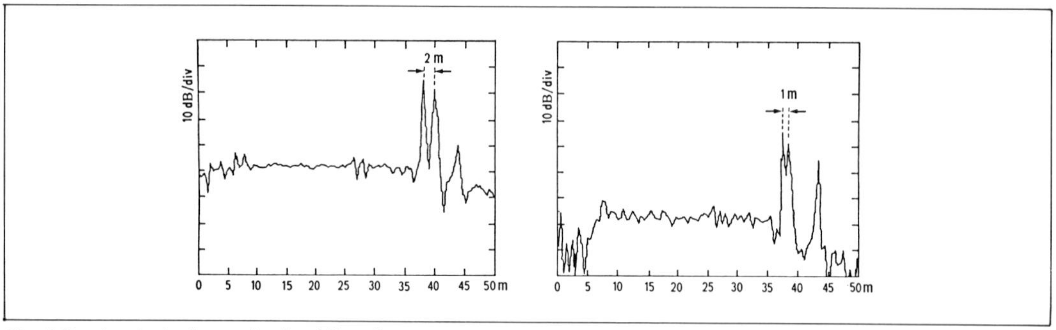

Fig. 5 Results obtained on a simulated line of two sensors

Results obtained on a simulated line of sensors

In order to evaluate the spatial resolution of the COTDR, a short single-mode fibre line of 45m length including two reflecting defects spaced by 1 or 2 metres has been constituted, simulating a line of two distributed sensors. These defects have been done by connecting short pieces of fibre through a liquid, the refractive index of which did not well match the fibre core index, to generate some reflected energy to each connection point.

Results obtained on this line, using a 200Mb/s, $2^{10}-1$ bits PRBS, are given in figure 5 where we can see two defects located at about 40m and spaced by 2 metres (a) and by 1 metre (b), giving so the resolving power of the instrument.

Spatial resolution is 50cm, corresponding to the interval between two consecutive points.

Then, a 400m long single-mode fibre line has been realised by connecting with the same method 6 short pieces of fibre (with lengths from 40 to 100m) as shown in figure 6. The reflected energy that occurred at each connection point stimulated a line of 7 distributed sensors (including the reflection occurring at the end of the line).

Figure 7 shows the energy back-scattered from this line, where each reflection is clearly set off, demonstrating the measurement possibilities of this technique.

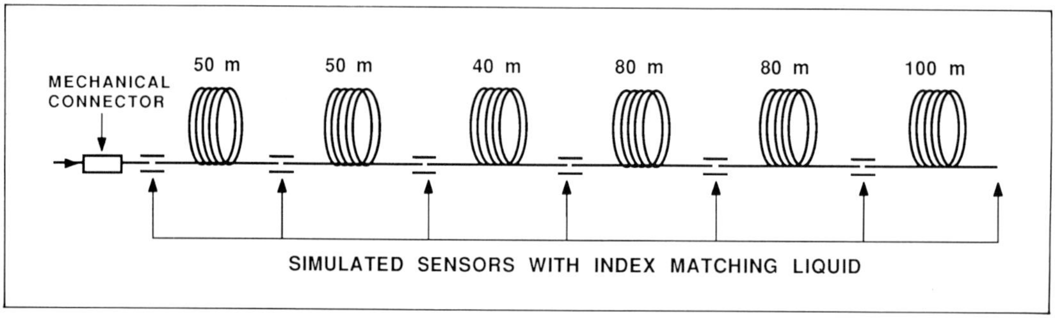

Fig. 6 Simulation of a 7 sensors line

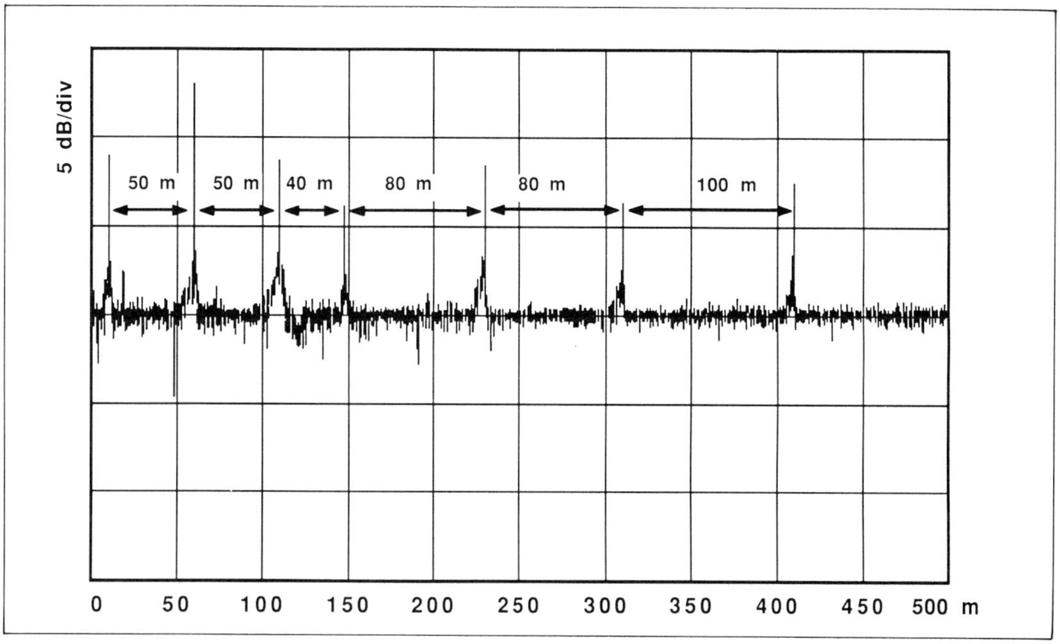

Fig. 7 Result obtained on the 7 sensors line

Spatial resolution improvement

Resolution improvement requires the realisation of higher bit-rate PRBS. The main difficulty in realising a 10cm resolution correlation reflectometer is a technological problem. In fact PRBS with bit-rate of more than 300Mb/s can no more be realised by using silicon ECL circuits.

Design and realisation of high speed functions as sequence generator, time-shifted sequence driver, A-D converter and laser driver, require GaAs technology. But today a very few functions are commercially available. A 2^7–1 bits pseudo-random sequence has been realised at 1.12Gb/s in the prospect of a future 10cm resolution reflectometer.

Concluding remarks

A 50cm resolution 1.3μm OTDR using the correlation technique has been realised for the test of distributed single-mode fibre sensors. This technique leads to an acceptable compromise between spatial resolution, dynamic range and measurement time, and offers better performance than the conventional pulse technique for high resolution tests of short single-mode fibre lengths.

Furthermore, a 1.12 Gb/s pseudo-random sequence has been realised with GaAs circuits for a future 10cm resolution reflectometer.

Acknowledgements

The authors are grateful to DRET for its support to this work.

Distributed optical fibre temperature sensor using spread-spectrum techniques

J.K.A Everard and R. Thomas
King's College London, UK

THE demand for a large number of sensors to monitor complicated structures is increasing and distributed sensors offer a practical solution. It has been shown that optical fibres can be used both as the sensor and as the communications medium. However, because the optical signals are usually very low in power, the time taken to obtain accurate results is often very long. In this paper we present a Raman based distributed temperature sensor using spread spectrum pseudorandom sequences. This technique enables both high spatial resolution and rapid response times to be obtained simultaneously.

In conventional optical time domain reflectometers (OTDR) an optical pulse is launched into the fibre and the back-scattered light is detected in a sampling receiver. To avoid sampling more than one point at the same time the maximum repetition rate of the transmitting pulser must be limited to $c/2L$ where c is the velocity of light in the fibre and L is the length of the fibre. To obtain high resolution the launched pulse width gets narrower and considerable signal averaging is required to obtain a good signal to noise ratio.

Experiment

The new prototype optical fibre temperature sensor,[1] shown in figure 1, solves the problem by launching a

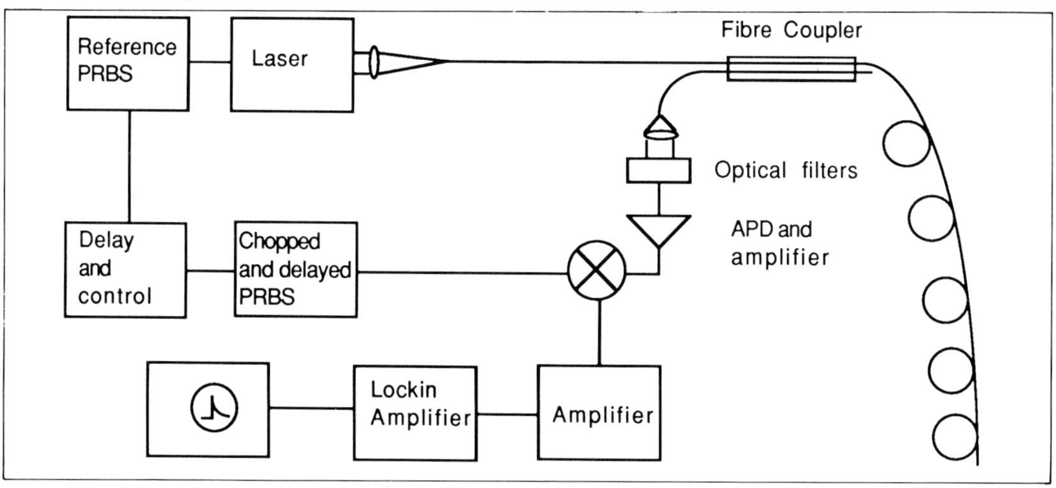

Fig. 1 Distributed optical fibre sensor

semiconductor laser beam modulated with a pseudo-random bit sequence (PRBS)[2] into an optical fibre where the average launch power into the fibre is 5mW. The back-scattered light is detected via a fibre coupler which is built with a 1km pigtail to prevent unwanted reflections. The optical signal is then filtered, using a Fabry-Perot stack to measure the Stokes and the anti-Stokes Raman lines, and detected using an avalanche photo-diode (APD) transimpedance amplifier combination. The output signal is then multiplied by a delayed version of the pseudorandom signal which, because of the autocorrelation function, allows the spatial information to be obtained without ambiguity. The autocorrelation function for a maximal length sequence consists of a peak for zero delay and a flat background correlation of $-1/m$ where m is the number of bits in the sequence before it repeats. In this system the average launch power is therefore equal to half the peak power. It should be noted that in repetition rate limited OTDR the average back-scattered power is proportional to $P\Delta L_2$, whereas in PRBS OTDR the average back-scattered power is proportional to $P\Delta L$, where P is the peak launch power and ΔL is the resolution length. There is, therefore, considerable advantage to be gained using PRBS OTDR for high-resolution, high repetition, rate systems.

In the system described here a $2^{15}-1$ maximal length pseudorandom sequence is used which repeats after 32767 bits. This gives a theoretical two-way optical dynamic range of 45dB. It should be noted however that to obtain this dynamic range a very pure sequence has to be generated and, therefore, the pulses need to be accurate both in pulse amplitude and zero crossing position. In this system the pulses were generated using 15 shift registers and an exclusive OR gate made using fast TTL. The output of each of the shift registers contains the same sequence with different delays. It is, therefore, essential to ensure that these signals do not couple to the output. Charge storage in the transistors causes mark to space modulation of the pulses which degrades the correlation function. These problems were solved by using ECL clean up circuitry after the TTL and by transmitting all the signals using balanced lines to obtain high common mode rejection.

If one examines the autocorrelation function when applied to a distributed back-scatter system the following equation can be written:[3]

$$F(\tau) = \int_{-m/2}^{+m/2} f(t-\tau_x)$$

$$\times [f_1(t-\tau_1) + \ldots + f_x(t-\tau_x) + \ldots + f_n(t-\tau_n)]\, dt$$

Where $F_{(\tau)}$ is the output, $f_x(t-\tau_x)$ is the wanted signal and $f_n(t-\tau_n)$ (for $n \neq x$) are the unwanted terms, m is the sequence length and n is the number of bits in the fibre. As the optical back-scatter emanates from all points in the fibre then, when the wanted term $\int f(t-\tau_x) \times f_x(t-\tau_x)$ is normalised to one, all the other terms are multiplied by $-1/m$ and added together. This unwanted term produces a large component which could be similar in value to the wanted signal. This problem is solved by chopping the delayed sequence between the wanted delay inside the fibre to a delay outside the fibre and by making the sequence length considerably longer than the length of the fibre. This also has the added advantage of producing an AC signal out of the multiplier which is easy to detect and does not suffer from DC drift.[3] This signal is detected in a lock-in amplifier. The reference delay is scanned through the fibre and the output of the lock-in is digitised and displayed.

Each point can be observed simultaneously by using multiple correlators[1] or digitising the waveform and multiplying the waveforms digitally. However, this becomes more difficult at high repetition rates.

To obtain temperature measurements, optical filters are placed in front of the photo-detector to measure the Raman back-scatter[4] which consists of two broad spectral lines on either side of the Rayleigh back-scatter. These lines are the Stokes (lower frequency sideband) and anti-Stokes (higher frequency sideband).

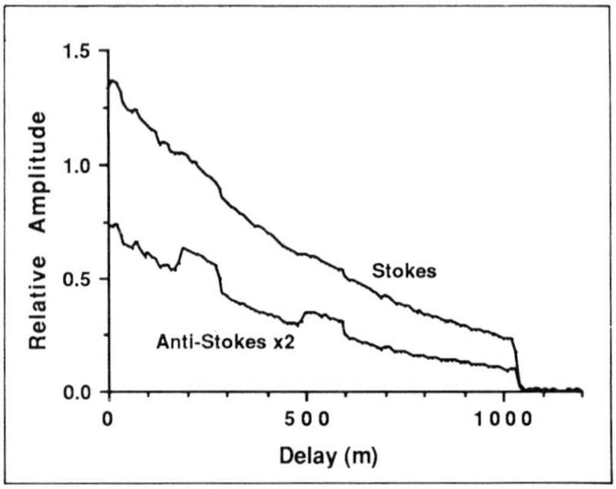

Fig. 2 Raman back-scatter

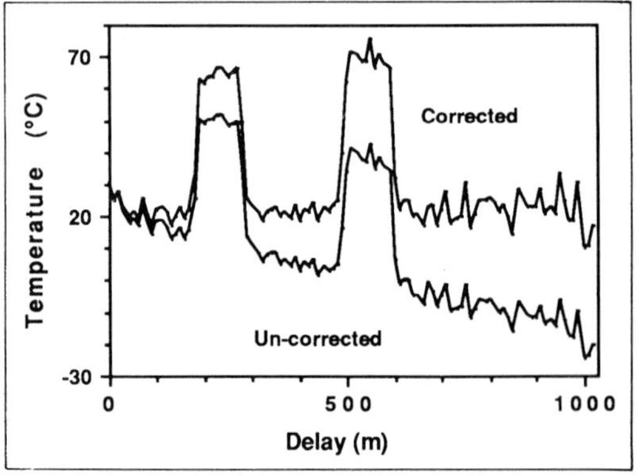

Fig. 3 Ratio of anti-Stokes to Stokes

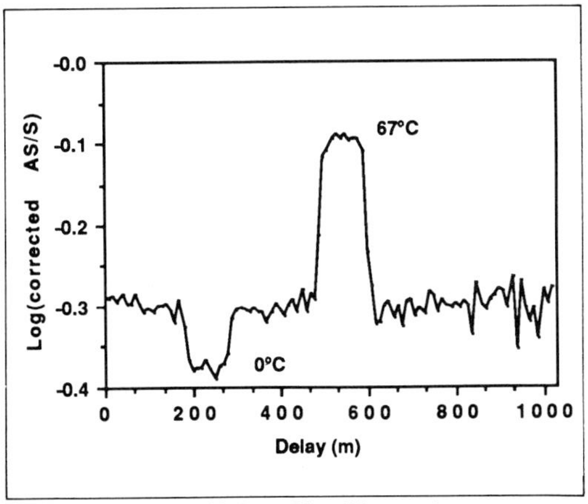

Fig. 4 Corrected AS/S for cold and hot sections

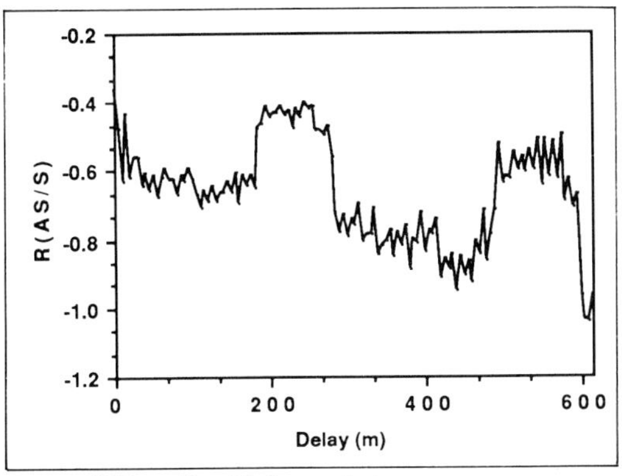

Fig. 5 AS/S profile with 5m resolution

These lines are temperature dependent and obey the Bose-Einstein thermal factor where:

$$I_{Stokes} = I\sigma_f\,(1 + N_{ft}) \qquad I_{anti\text{-}Stokes} = I\sigma_f(N_{ft})$$

where $N_{ft} = 1/(\exp(hf/kT) - 1)$. T is temperature in K, k is Boltzmann's constant, I is the incident optical intensity, h is Planck's constant, σ_f is the frequency dependent cross-section and f is the offset frequency.

Results

A plot of the Stokes and anti-Stokes lines measured over 1km of fibre with 10m resolution is shown in figure 2. Two sections of the fibre, between 200, and 300m and 500m and 600m were raised in temperature to 67°C. To obtain absolute measurements of temperature the ratio of the anti-Stokes to Stokes lines can be taken as shown in figures 2 and 4. In figure 4, the first section is now put in an ice bath and the second section is in an oven raised to 67°C.

However, the attenuation of the back-scatter is frequency dependent following a $1/\lambda^4$ characteristic and, therefore, the anti-Stokes/Stokes ratio changes with distance (lower curve in Fig. 3). A correction is therefore added to compensate for this (upper curve Figs. 3 and 4). Higher resolution measurements are easily achieved by doubling the clock frequency. A fibre anti-Stokes/Stokes profile with 5m resolution is shown in figure 5, which shows the first half of the fibre when the two sections are raised in temperature to 67°C. The difference in optical frequency causes a $1/\lambda^4$ variation in the optical cross-section which leads to a constant offset. However, the absolute value of the loss in the optical filters also causes a similar error and this also has to be taken into account. It is interesting to note that breaks in the fibre which cause Rayleigh back-scatter will not saturate the detector as this is filtered out optically.

Improvements

The present system has an average laser launch power of 5 milliwatts into the coupler. It is not unreasonable to expect that this could be increased by up to two orders of magnitude. The fibre is standard telecommunications graded index multimode silica fibre with a 50μm core and an NA of 0.2. By increasing the fibre diameter and NA, the power handling capability and capture fraction can be increased. The fibre can also be doped to increase the Raman back-scatter. All of these factors will enable a compact high-resolution temperature sensor to be produced with better than degree sensitivity, second time constants and metre/submetre resolution. It is interesting to note that for a scanning system the shot noise can be reduced by performing the correlation in the APD by switching the APD bias with the delayed PRBS, thus causing the APD to be off for half the time. It is also hoped to improve this by other techniques.[5]

Acknowledgements

We wish to thank the SERC (UK) for a grant which supported both this work and one of us (R.T.).

References

1. J.K.A. Everard, *Patent application GB 2190186A* 9 May 1986.
2. J.J. Bernard, J. Ducate, Y. Gauson, J. Guillon and G. Le Blevenec, 'Field portable reflectometer for single mode optical fibres' *Proceedings of the 10th ECOC*, mode optical fibres,' *Proceedings of the 10th ECOC*, Stuttgart, 3–6 September 1984, pp. 84–85.
3. J.K.A. Everard, 'Novel signal processing techniques for enhanced OTDR sensors', *Int. J. Opt. Sens.*, Vol 2, 1987, pp. 363–369.
4. J.P. Dakin, D.J. Pratt, G.W. Bibby and J.N. Ross, 'Distributed Optical fibre Raman temperature sensor using a semiconductor light source and detector', *Electron. Lett.*, Vol. 21, 1985, pp. 569–570.
5. J.K.A. Everard, *Patent application number 8816250*, 7 July 1988.

Progress in optical time domain reflectometry optical fibre sensor networks

F.X. Desforges, P. Graindorge, L.B. Jeunhomme and H.J. Arditty
Photonetics S.A., France

OPTICAL time domain reflectometry (OTDR) is a very useful tool and a well established technique in the telecommunications domain.[1] Opto-electronics components such as pulsed laser diodes, PIN or APD detectors are now well known and can be used in industrial applications. The use of OTDR in fibre optic sensor network could thus be considered;[2,3] In fact, the cost of an OTDR instrument prohibits its use in many fibre optic sensor applications. The actual number of sensors which can be practically inserted along a fibre line, monitored through classical back-scattered light analysis, is too small to afford the cost of an OTDR instrument. In this paper we introduce a big improvement in the number of multiplexed sensors through the use of an original retro-reflected light technique.

In the first section we describe briefly the OTDR technique. Important parameters such as dynamic range, spatial resolution and so on, will be discussed as well as their mutual interdependence.

A first design of an OTDR fibre optic sensor network is described in the second section. This network, based on back-scattered light analysis, works by the monitoring of optical losses induced by fibre bends. Such a fibre optic sensor network technique does not allow great performances and its most important drawbacks will be examined.

In the third section, a new design of an OTDR fibre optic sensor network will be described. This novel scheme allows the monitoring of 10 to 50 on/off microswitches along a single fibre under practical and realistic conditions. In order to obtain such a result, fundamental improvement of the sensor concept was performed as well as an original set up of the optical microswitch. Detection of retro-reflected rather than reflected or back-scattered light, makes much better use of the in-line loss introduced by each sensor yielding a large dynamic range which explains this very significant improvement in the number of multiplexed sensors.

OTDR technique

OTDR has been used for several years in telecommunications for cable testing. It allows the location of faults along the fibre together with the evaluation of connector losses. In this technique, a short pulse of light (about 10 to 100 nanoseconds) is launched into the fibre with a repetition rate of several kHz. The OTDR instrument records the light scattered back by the fibre toward the input end, versus time.

This light signal is composed of two parts (see Fig. 1). The first part corresponds to the pulses reflected each time the launched pulse passes across a fibre discontinuity, connector or break. This arises because of the Fresnel reflection at the glass-air interface. For a second part, a small fraction of the pulse energy is scattered by the core of the fibre, during its progress along the fibre. A part of this light is captured by the fibre and progresses backwards (back-scattered light). Thus, if the fibre is perfect, the back-scattered signal is a decreasing exponential whose exponent is twice the lineic loss of the fibre.

The most important parameters of the instrument are discussed below:

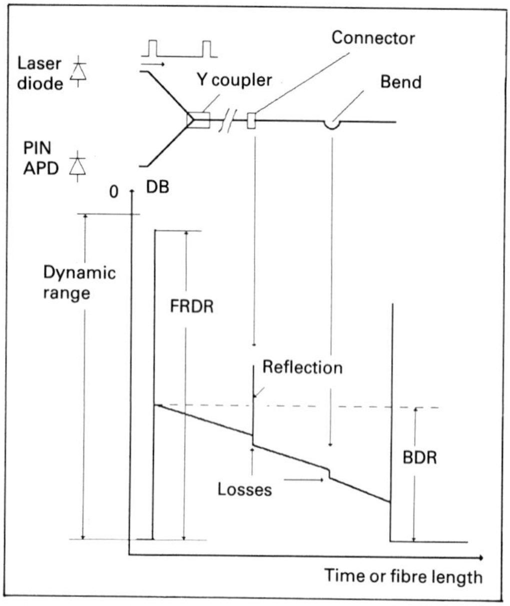

Fig. 1 Light Signal composed of two parts

- The 'dynamic range' is the level difference (in decibels) between the peak optical power launched into the fibre and the minimum optical power which can be detected by the photo-detector (PIN or APD).
- 'Maximum round trip range for fibre end detection' (assuming 4% Fresnel reflection) (Fresnel reflection dynamic range: FRDR). This parameter is 14dB less than the 'dynamic range' because of the 4% Fresnel reflection. This figure indicates the maximum round trip fibre loss which can be checked for breaks. In this case, the in strument detects the reflected signal.
- 'Maximum round trip range for back-scattered signal with SNR=1 (back-scattering dynamic range: BDR). The back-scattered signal can be written as:

$$P(z) = P0 \cdot 0 \cdot 5 \cdot T \cdot a \cdot S \cdot (c/N) \exp(-2az)$$

where:

P0 is the peak optical power launched into the fibre, T is the width of the optical pulse (FWHM), a is the attenuation of the fibre (in Np/m), S=(1/4). NA2/n(1)2 (for graded index fibre), S=(3/8). NA2/n(1)2 (for step index fibre), NA is the numerical aperture of the fibre, n(1) is the core index, c is the light velocity in vacuum, N is the group index of the silica (N=1.473 for λ =850nm and N=1.469 for λ =1330nm). With standard telecommunication fibres (50/125, NA = 0.2, 2.5dB/km) and T=100ns, this signal is 43 to 47dB less than the peak optical power which is launched into the fibre.

- 'Spatial resolution' is the parameter corresponding to the minimum separating between two faults, which can be observed separately by the instrument. The usual definition is given by:

$$D = cT/2N$$

where D is the spatial resolution T is the width of the light pulse.

There is a trade-off between the dynamic range for back-scattered light measurements and the spatial resolution. Thus, most instruments offer two selectable pulse widths. With a long pulse (100ns), the dynamic range for back-scattered light is important allowing the examination of long fibre lengths. On the other hand, a short pulse (20ns) allows a better spatial resolution for the location of connectors and breaks.

Fibre optic sensor network based on back-scattered light analysis

Summarised here is the most important results obtained with experimental sensor network, which has been described elsewhere.[3]

Sensor principle

It is now well known that a bend on the fibre induces local losses (see Fig. 2).[4] Multiplexing many sensors along the fibre requires minimisation of insertion losses. The use of this bend loss technique allows insertion of the sensor on the fibre without connectors, and insertion losses are thus minimised.

Experimental results

Figure 3 illustrates the feasibility of an on/off sensors network. The instrument used in this experi-ment is a commercially available Photodyne 5500 XFA OTDR. The graded index fibre has a 100μm core diameter and

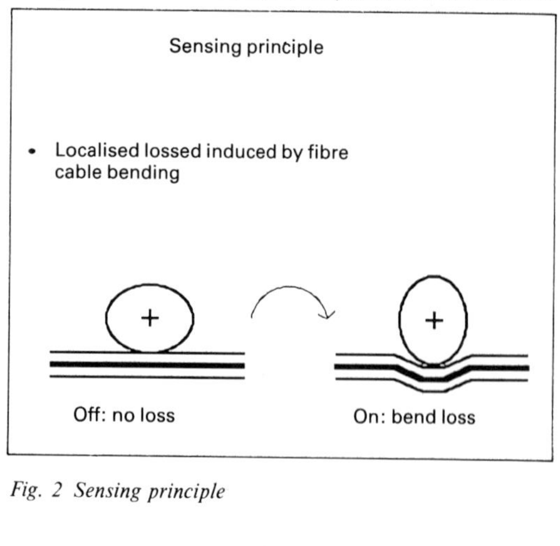

Fig. 2 Sensing principle

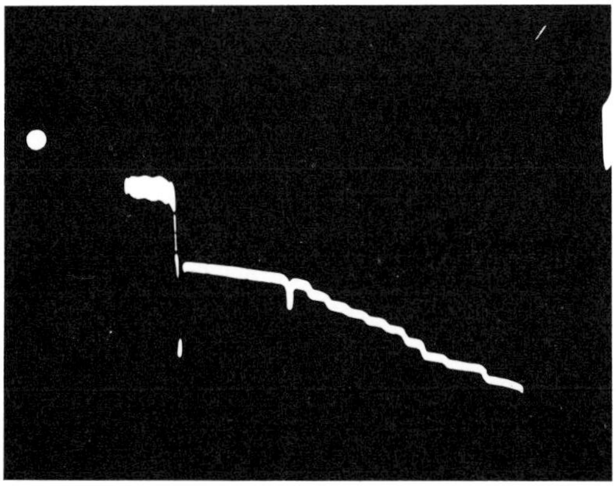

Fig. 3 BDR = 15.2dB. T = 100ns

its numerical aperture is 0.29. Nine bends were distributed along the fibre (ranging from 4.5mm to 9mm diameter) separated by 50 to 100 metres. A second fibre line is connected with a splice (and matching gel) to the bent fibre line. One can see the level difference between the amplitude of steps induced by fibre bending and the residual reflection of the splice (0.5% to 1%).

System evaluation

In order to evaluate the possibilities of such a system, we consider the following parameters:

- In order to minimise mutual influence between sensors, the loss of the first sensor is chosen to be 0.8dB while the losses of the others are chosen to be 0.3dB (single way).
- The minimum SNR for detecting 0.3dB local losses with a 1% error rate is 13dB.[5]
- We will consider two possibilities for the implementation of the system:
  —in an industrial plant, the average distance between sensors is 50 metres and fibre losses are 10dB/km, including connectors and splices;
  —in special environments, average distance between sensors is 20 metres and intrinsic fibre losses are 5dB/km.
- calculations will be made versus OTDR characteristics, in two cases:
  —a simple and cheap OTDR instrument having a 70dB dynamic range (without averaging);
  —a good performance OTDR instrument having a 105dB dynamic range (with one minute averaging).

The number of on/off sensors which can be multiplexed along the same fibre is:

	Simple OTDR	High performance OTDR
Industrial plant	11	27
Special environments	21	54

This shows that a simple commercially available OTDR instrument the allows monitoring of on or off position of more than 10 sensors multiplexed along the single fibre line. The use of a high performance OTDR increases the cost of the system but allows the multiplexing of more sensors. In fact, the additional cost of this high performance OTDR is shared between the additional sensors which can be multiplexed this way. However, such a network presents two disadvantages.

- Connectors should be avoided for the insertion of each sensor along the fibre line, because of their relatively high loss. This would reduce the dynamic range available for the sensing itself, thereby decreasing the number of sensors. Avoiding the use of connectors clearly complicates the installation of the system.
- The fibre is subjected to a sharp bend each time the on/off switch is actuated. This may cause fibre aging and makes this sensing principle questionable.

Fibre optical sensor network based on reflected light analysis

We present here a new sensor network which offers dramatic improvements over the previous concept described before. The new sensing principle is based on retro-reflected light instead of back-scattered light. This improves the dynamic range in such a way that connectors can now be used to insert the sensors along the cable, while still keeping the total number of sensors in the range of 10 to 50. Also the fibre is no longer subjected to varying bends, which improves the lifetime expectancy.

Reflected light analysis

The dynamic range of an OTDR instrument for reflected light analysis is greater (40 to 50dB depending on pulse width and fibre parameters), than the dynamic range for back-scattered light analysis. Moreover, this dynamic range does not depend on the pulse width and so, the trade-off between resolution and dynamic range encountered with back-scattered light disappears.

A fibre optic sensor network based on reflected light analysis is shown in figure 4. N commercially available T couplers are inserted along the same fibre, each one showing the same coupling ratio, C, and excess loss, P.

The peak optical power PR(N) which is reflected by

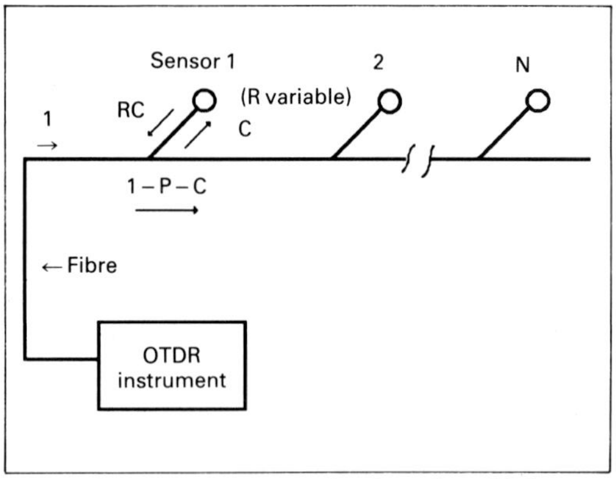

Fig. 4 Fibre optic sensor network using reflected light

the Nth sensor through the fibre line is given by:

$$PR(N) = (1-C-P)^{(N-1)} \cdot C^2 \cdot R \cdot (1-C-P)^{(N-1)} \cdot PO$$
$$PR(N) = (1-C-P)^{2(N-1)} \cdot C^2 \cdot PO \cdot R$$

where PO is the peak optical power into the fibre, and R is the reflection coefficient of a sensor.

For C = CO, this reflected power is maximum:

$$CO = (1-P)/N$$

For example, with N = 10 and P = 1dB (20%), which are usual values, we obtain:

$$CO = 0.08$$

In this case, the level difference between the peak optical power P0 and the reflected power PR(N) for the sensor number 10 is:

$$PR(10)/PO = -47\text{dB} \qquad \text{with } R = 100\%$$

Thus, there is not a great difference between reflected and back-scattered light analysis, although we did not take into account fibre losses, connector losses and so on. The use of commercially available T couplers with 1dB excess loss does thus seem to be incompatible with good performances of a fibre optic sensor network based on reflected light analysis.

Let us now consider the calculation of the reflected optical power PR(N) with a 0dB excess loss for the couplers.

In this case, for N = 10 we obtain:

$$CO = 0.1$$

$$PR(N)/PO = -28\text{dB}$$

This result is, of course, more interesting and there is a great interest to design a new transducer which minimises excess loss while using the reflected light. Such a transducer has been obtained by means of retro-reflected light.

Retro-reflected light analysis

Transducer principle. The transducer is based on permanent fibre bending. Near the bend, the light leaks in well-defined directions from the core of the fibre to the jacket.[6] By introducing a matching gel, we can extract the light from the jacket to the external medium. After inserting a retro-reflecting material in front of the leakage area, the light is retro-reflected into the fibre core, even if the retro-reflector is not perpendicular to the incident light (see Fig. 5).

Such a material is made of very small glass spheres which reflect light in its incident direction by a cube corner effect (see Fig. 6). The very low lateral offset between incident and retro-reflected rays, due to very small diameter glass spheres, allows the light to be launched back in the fibre core.

This arrangement allows the launch back into the fibre core of a variable amount of the light power lost at the bend. Thereafter, we shall call this relaunched light 'retro-reflected light'.

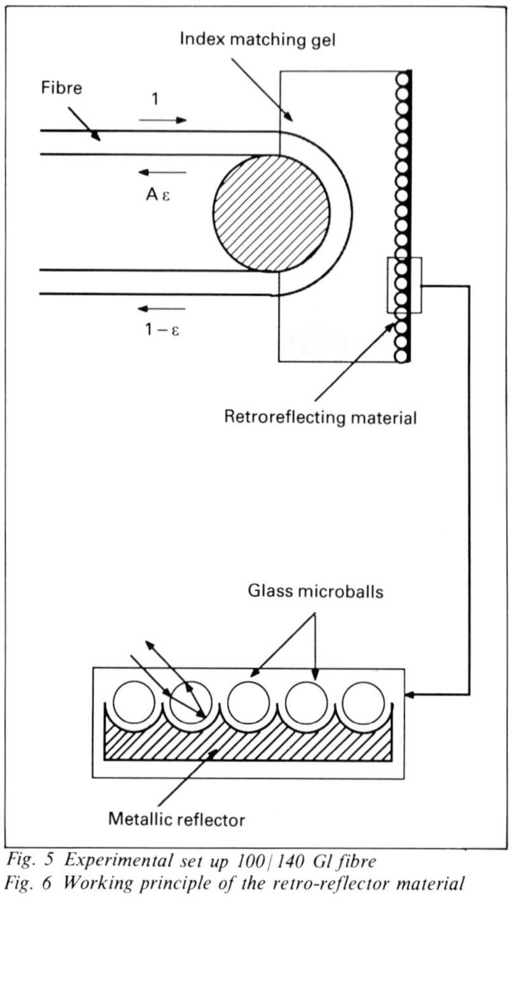

Fig. 5 Experimental set up 100/140 GI fibre
Fig. 6 Working principle of the retro-reflector material

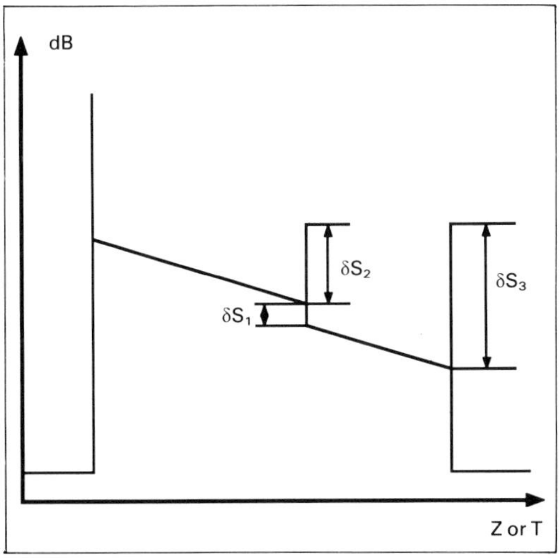

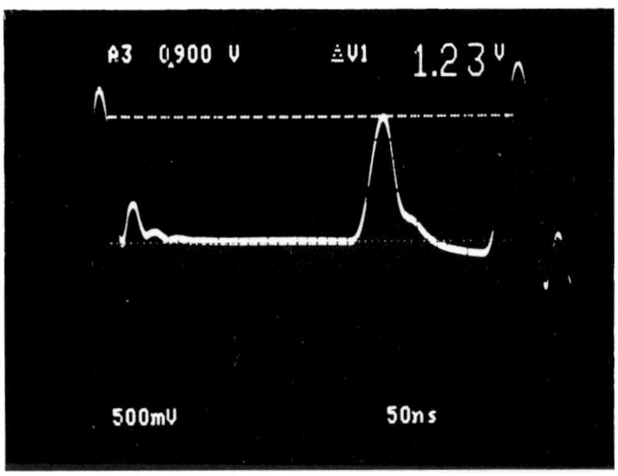

Fig. 7 1dB/div, T=20ns

Figure 7 shows the signal detected by the commercially available Photodyne 5500XFA OTDR instrument. It is clearly seen that the retro-reflected light signal is by far more intense than the back-scattered light signal variation at the bend. This illustrates, qualitatively, the increase in dynamic range allowed by this new sensor design. A 100/140/250 GI fibre is just bent around a steel cylinder without removing the mechanical jacket.

Let ϵ be the single way bend loss and A the 'reflection coefficient' of such a device. The optical power which is back-scattered just before the bend is:

$$Pb(N) = K \cdot T \cdot Pp(N)$$

where K is the back-scattered coefficient of the fibre ($K = 0 \cdot 5 \cdot S \cdot a \cdot c/N$) T is the pulse width (FWHM) and Pp (N) is the peak optical power just before the bend.

The optical power which is back-scattered just after the bend is detected as:

$$Pb(B) = K \times T \times (1 - \varepsilon)^2 \times Pp(N)$$

The back-scattered signal decrease at the bend is thus ($\epsilon \ll 1$):

$$\delta Pb(N) = 2 \times \varepsilon \times K \times T \times Pp(N)$$

On the other hand, the retro-reflected light signal is:

$$Prr(N) = A \times \varepsilon \times Pp(N)$$

Let us introduce the figure of merit between retro-reflected and back-scattered light as:

$$M = Prr(N)/\delta Pb(N)$$
$$M = A/2 \times K \times T$$

According to this formula, we observe that the detection of retro-reflected light is more advantageous for the following reasons:

- This figure of merit M is inversely proportional to the pulse width T. Thus, there is no more trade-off between dynamic range and spatial resolution.
- This figure of merit means that the same number of sensors can be monitored with an OTDR instrument having a dynamic range D and using the back-scattered light or with another OTDR instrument having a dynamic range D–10logM and using retro-reflected light analysis. Conversely, it can also be considered that 10logM is the two-way attenuation (in dB) which is now allowed for the insertion of connectors, while still maintaining the number of sensors equal to that achieved with the backscattering technique.
- As compared to back-scattered light analysis, analysis of retro-reflected light is easier, and we call the corresponding instrument an echometer as it has only to detect pulse echoes.

The results presented in figure 7 yield:

$$\partial S1 = 10 \log(1 - \epsilon) = 0.6 \text{ dB}$$

so, $\epsilon = 13\%$;

$$\partial S3 = 5 \log (0.04/KT) = 12.5 \text{ dB}$$

so, $KT = 1.265 \times 10^{-4}$

$$\partial S2 = 5 \log (A\epsilon/KT) = 6 \text{ dB}$$

so, $A = 1.5\%$. These early results yield a figure of merit $M = 60$.

Further improvements are expected to reach $A = 10\%$, but it is still clear that these promising results open new possibilities in fibre optic sensor networks. Furthermore, such a transducer allows various measurement possibilities such as position, level, temperature, pressure, refractive index and so on, and the performances of a network of such transducers are examined below.

Network capabilities. When considering *dynamic range* calculations will be for two echometer characteristics:

- A simple and cheap echometer with a PIN receiver and 70dB dynamic range.
- A high performance echometer with an APD receiver and 88dB dynamic range.

For *signal to noise ratio* considerations, with most applications, one can average the received optical power during one second. If the repetition rate of the laser is 1kHz, this averaging process increases the signal-to-noise ratio by about 15dB. It seems also reasonable to look for an error-free working period of 30 years. For this reason, the signal-to-noise ratio after averaging must be as high as 21dB.[5] So, the signal-to-noise ratio before averaging will be taken equal to:

$$(S/B)O = 6 \text{ dB}$$

With all these parameters, we can now compute the possibilities of the network.

Configuration of the network. Let us consider the tree configuration which is described in figure 8. The dashed lines surround the basic cell for the calculation of losses called 'Pe'.

If N1 is the number of sensors, one can write:

$$D-6 = 2(N1-1)[Pe - 10\log(1-\epsilon)] + 2Pe - 10\log A \times \epsilon$$

where D is the dynamic range of the echometer, ϵ is the in line loss induced by the transducer, $A\epsilon$ is the intensity reflected by the transducer, Pe is the loss of the basic cell as described previously. So,

$$N1 = \frac{[D-6+10\log A\epsilon - 20\log(1-\epsilon)]}{[2(Pe - 10\log(1-\epsilon)]}$$

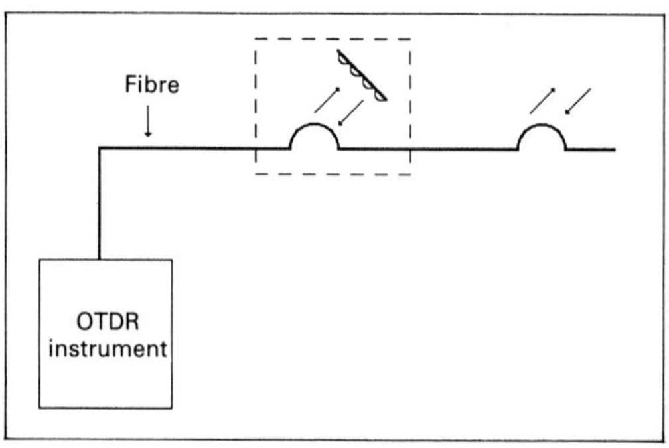

Fig. 8 Tree configuration

We will examine two possibilities for the evaluation of losses Pe as in the previous section:

- In an industrial environment, the transducer is connected to the fibre line by two connectors with 0.5dB insertion loss each. The average separation between two transducers is taken equal to 50 metres for which the added loss is 0.5dB (10dB/km including connectors and splices, 850nm, 100/140 Gl Fibre Optic Cable). So,
  Pe = 0.5 + 0.5 + 0.5 = 1.5dB
- In special environments, the transducer is installed with a splicing machine yielding to 0.2dB insertion losses (0.1dB for each splice). The average separation between two transducers is taken equal to 20 metres for which the added loss is 0.1dB (5dB/km, 850nm, 100/140 Gl Fibre Optic Cable). So, Pe = 0.2 + 0.1 = 0.3dB

Considering a typical value of $\epsilon = 10\%$, we obtain the following figures for two values of A:

Industrial environments Pe = 1.5dB:

	A = 1% (present results)	A = 10% (objective)
D = 70dB	N1 = 9	N1 = 11
D = 88dB	N1 = 13	N1 = 16

Special environments Pe = 0.3dB:

	A = 1% (present results)	A = 10% (objective)
D = 70dB	N1 = 23	N1 = 30
D = 88dB	N1 = 35	N1 = 41

According to these results, the present transducer technology allows the multiplexing of 10 to 20 sensors depending on echometer sensitivity and environment. In fact, these results are not very different from those obtained in the previous section. However, they are more encouraging for the following reasons:

- The fibre bend is permanent which is safer from the MTBF point of view than bending the fibre each time the sensor is actuated.
- The received optical signal is composed of a train of pulses. Thus, the preamplifier of the receiver module is simple to design. After this first amplifier stage, signal processing is easy for the pulse width currently used. For T = 20ns, a TTL circuit allows monitoring of the on/off position of the sensors (see Fig. 9).
- One can expect further improvements to yield A = 10% through an optimised transducer design. Noteworthy also is the fact that this type of transducing principle is no longer limited to position sensing but can be applied to other measurands.

In conclusion of this section, further improvements

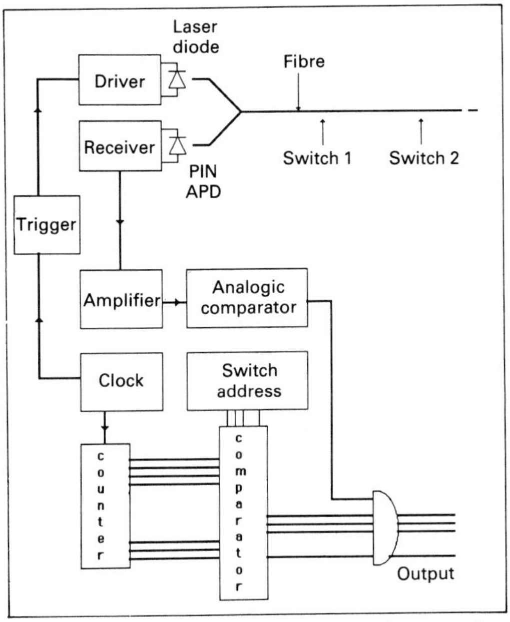

Fig. 9 Echometer scheme used for microswitches network

of the transducer performances are expected but present results show that practical feasibility was demonstrated as well as the possibility of industrial applications of such a device.

Concluding remarks

The introduction of retro-reflecting transducers brings the possibility of practically implementing OTDR sensor networks in industrial environments.

Furthermore, the use of retro-reflected light brings the following advantages:

- Low cost of the transducer due to the simplicity of the device.
- Detection of echoes rather than back-scattered light, suppressing the need for a trade-off between dynamic range and spatial resolution, and simplifying the signal processing inside the OTDR head.
- Various quantities can be measured by this transducer technique such as position, temperature, liquid level and so on.
- At last, such a device allows great improvements in terms of the number of sensors which can be practically multiplexed (up to 50 in some cases).

References

1. M.K. Barnoski, S.M. Jensen, 'Fibre waveguides: a novel technique for investigating attenuation characteristics' *Applied Optics, 15, No. 9* p. 2112.
2. S.A. Kingsley, 'Distributed fibre optic sensors: an overview', *Proceedings of SPIE, vol. 566 San Diego, 1985.*
3. P. Graindorge, L.B. Jeunhomme and H.J. Arditty, *Cannes 1985.* 'Simple OTDR network for on/off fibre optic sensing', *Proceedings of the SPIE no. 586.*
4. F.X. Desforges, 'Pertes par courbure pure dans une fibre à gradient d'indice fortement multimode'. *Septièmes Journées Nationales d'Optique Guidée, Nice,* April 1986.
5. S.D. Personick, 'Receiver design for optical fibre systems', *Proceedings of the IEEE, vol. 65, No. 12,* pp. 1670–1678.
6. L.B. Jeunhomme and J.P. Pocholle, 'T-coupler for multimode optical couplers' *AGARD Symposium on Optical Fibres, London,* May 1977.

Birefringent stress location sensor

R.B. Franks
University of Sheffield, UK

W. Torruellas
France

R.C. Youngquist
Stanford University, USA

DISTRIBUTED and multiplexed sensing of environmental parameters along an optical fibre have many advantages over conventional sensing methods. Systems for monitoring pressure, temperature or acoustic signals at multiple locations are of increasing interest, due to the possibilities of electrically passive measurements, low maintenance requirements, intrinsic safety and availability of data at sites remote from the sensor.

This paper presents a new type of sensor using two separate modes in an optical fibre. The light source is a frequency modulated semiconductor diode laser. Few other optical components, and only simple electronic processing are required. The system may be optimised for either mutiplexed sensing of known sensor sites or distributed location sensing over the length of the system.

Operating principles

In an optical fibre propagating two modes, each mode or light path may have a different group propagation constant. These modes may be polarisation modes in a birefringent fibre, for example, or the two sets of modes in a step index fibre which is used at a wavelength just below cut-off of the third order modes. In this system, a fibre of either type may be used.

In a fibre of total length l, a modal coupler[1,2] placed a length L from the output causes mixing of the modes, so that power initially propagating in just one mode, may propagate in two modes after coupling.

The time tl, taken for the light in path or mode a, to reach the end of the fibre, a distance L from the coupling point is given by:

$$tl = L/cl \qquad (1)$$

where cl is the group velocity of light in path a.

Similarly for path b:

$$t2 = L/c2 \qquad (2)$$

The time delay τ between light coupled between modes at the coupling point, on reaching the end of the fibre is given by:

$$\tau = tl - t2$$
$$\tau = L(1/cl - 1/c2) \qquad (3)$$

Therefore from equation (3), one can see that the time delay between modes, is directly proportional to the length between the coupling point and the end of the fibre.

The theory of the frequency modulated continuous wave (FMCW) technique is well covered.[3] In application to this system, the time delay necessary in the technique takes the form of the delay between two optical paths in an optical fibre.

Figure 1 shows a plot of frequency with respect to time for the two paths. The instantaneous frequency at the output of path a is calculated from the slope of trace a as:

$$\omega a = 2\pi \frac{\Delta f}{Ts} t + fl \qquad (4)$$

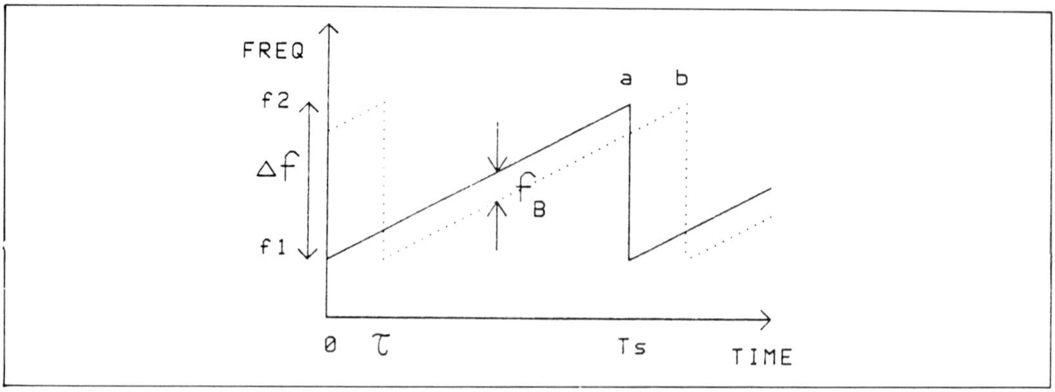

Fig. 1 Time delayed FMCW signals

within the region $0 < t < Ts$, where Ts is the frequency sweep, or ramp period.

The instantaneous frequency at the output of path b is similarly calculated from the slope of trace b as:

$$\omega b = 2\pi \frac{\Delta f}{Ts} (t-\tau) + f1 \tag{5}$$

within the region $\tau < t < Ts$.

The difference in frequencies $\omega a - \omega b$ is the angular beat frequency ωB, which is related to the time delay τ. By inspection:

$$\omega B = 2\pi \frac{\Delta f}{Ts} \tau \tag{6}$$

and

$$fB = \frac{\omega B}{2\pi} \tau \tag{7}$$

where fB is the beat frequency.
Thus:

$$fB = \frac{\Delta f}{Ts} \tau = \frac{\Delta f}{Ts} (1/c1 - 1/c2)\, L \tag{8}$$

It may be seen from equation (8) that the beat frequency of the output is directly proportional to τ, the time delay between the paths, due to the difference in group velocities between the two modes in the fibre. Thus the beat frequency is proportional to the distance L of the sensor location from the fibre output. Both Δf and Ts are constants of the system.

Fibre modes

In a practical system, the fibre modes may take two different forms; either the polarisation states in a birefringent fibre, or two of the lower order modes of a conventional step index fibre, such as the linearly polarised LP01 and LP11 modes of Gloge (5), which be obtained by a careful choice of laser wavelength and fibre core diameter. There is one pair of first order modes, which have a group propagation constant, and two pairs of second order modes, each pair having a slightly different group propagation constant.

Modal mixing

Coupling between the two modes of the fibre is necessary for the operation of the system, and the variety of parameters which are to be detected depends on which parameters the modal couplers can be made sensitive to.

In practice, most changes of index profile in the fibre cause some degree of modal coupling, and changes of stress along the fibre are accompanied by changes in index profile. Periodic stress devices[1,2] have been constructed which increase the fibre's sensitivity to stress by successively increasing the amount of coupled over power, over a spatially periodic region. Thus, the obvious choices of parameter to sense are those which may readily be converted to a stress by a ridge coupler device, e.g. strain, pressure displacement.

Examination of the beat signal

A comprehensive analysis for the output of an FMCW system has already been provided,[3] and the output from this system is of similar form, apart from an additional amplitude modulation produced by the diode laser. The spectrum consists of a series of discrete lines at frequencies:

1/Ts, 2/Ts, 3/Ts..........N/Ts

The spectrum features a series of lines of greatest amplitude at the lower harmonics, due to the undesired saw-tooth amplitude modulation of the laser. These lines decay in amplitude as the inverse of the harmonic number and may be reduced at the output by signal processing electronics. The beat frequency corresponds to the centre of a (sin x)/x envelope function which is imposed on the harmonic line spectrum. The width of the envelope and the ability to locate its centre determines the resolution and accuracy of the system, and techniques have been described to accomplish this.[4]

Operation of the system is possible in either distributed or multiplexed versions. In a distributed sensor, fB need not be an integer multiple of 1/Ts and determination of the centre of the (sin x)/x envelope yields the value of fB.

For a multiplexed system, the centre of the (sin x)x envelope is arranged to coincide with an integral multiple value of 1/Ts by placing the coupling sites at known locations, corresponding to known beat frequencies, so that fB=n/Ts for n an integer. This configuration is useful for sensing a coupling event, for instance determination of the state of any relay in an array along the fibre may be possible by arranging the relays to stress the fibre when in a certain state and, thus, giving a different frequency corresponding to each relay in that state. In this configuration, the maximum multiplexed dynamic sensitivity may be obtained, because the n/Ts frequency line of interest is at the peak of the (sin x)/x envelope, and the maximum coupling power is directed into one harmonic line. However, the minimum distance between sensors is limited by possible ambiguity due to sidebands of the neighbouring sensors.

If detection of only one sensor along the fibre is required, then the use of a period measurement technique of the beat signal yields greater accuracy. In this case, a triangle wave can be used, since the beat information is contained within the rising edge, and fast flyback is not necessary.

Examination of equation (8) shows that, in order to get the largest change in beat frequency for a given change in time delay between modes, then a large frequency excursion, Δf, is required. For semiconductor laser diodes the maximum realisable frequency excursions are of the order of 100GHz, however, to avoid mode hopping effects, a limit of 60GHz is practical.

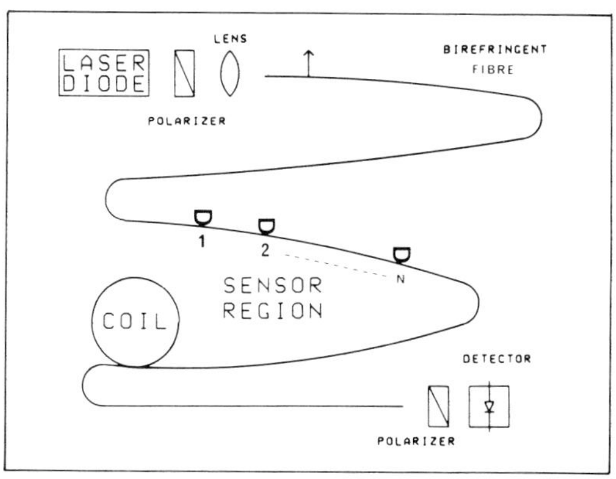

Fig. 2 Configuration for the birefringent fibre system

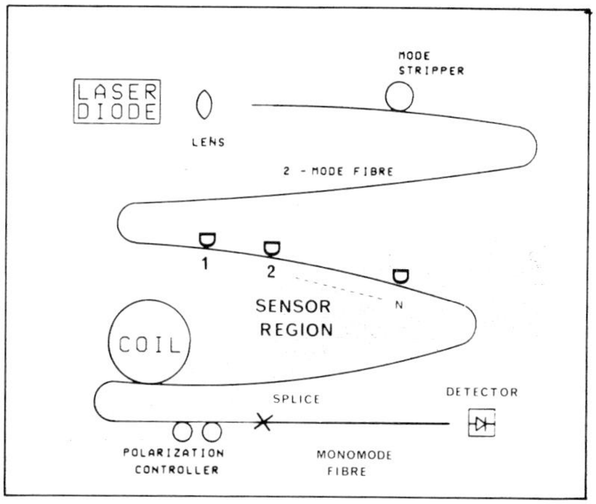

Fig. 3 Configuration for a two moded fibre system

System configurations

Figures 2 and 3 show two versions of the system using birefringent and two moded fibre.

In the first system, figure 2, frequency modulated light is launched down one polarisation of a birefringent fibre, any unwanted laser polarisations are rejected by the polariser. Rotation of the fibre about its axis enables the optimum launch orientation to be found.

Modal coupling is obtained at the first sensor location, which may be anywhere along the fibre, and, from this location, the light propagates in two optical paths to the end of the fibre. At the second and successive sensors, further power is cross-coupled until beyond the final sensor, after which no cross-coupling occurs. Phase information in the two modes is resolved into an amplitude modulation by a polariser orientated at 45 degrees to the fibre axes and then a detector. Simple processing of the detected signal is accomplished using a circuit which multiplies the inverted ramp intensity information with the detected signal, to reduce the effect of unwanted amplitude modulation of the laser diode. An additional length of fibre after the last sensor separates the received beat spectra to a frequency position above the spectrum of the laser amplitude modulation, by adding a fixed increase in fB due to the extra fibre length.

The second system, figure 3, using the two transversal mode fibre, operates similarly. Frequency modulated light is launched into the fibre, exciting two modes. Any higher order modes which may be present, are removed as bending loss, in a mode stripper, which is simply a loop in the fibre in this case.

Modal coupling occurs at the first sensor, and then at each succeeding sensor until the last sensor is passed. Modal mixing, converting the frequency chirped time delay information to amplitude modulation is achieved by use of a slightly offset splice to monomode fibre. At this splice spatial filtering occurs, with the modulation

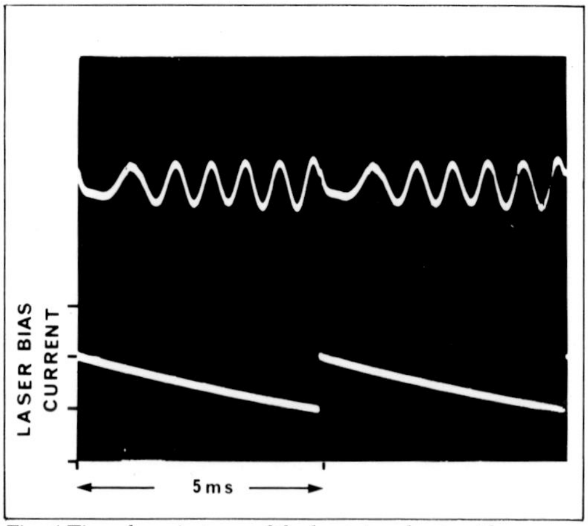

Fig. 4 Time domain trace of the beat signal output for one sensor located a distance L = 65m from the fibre output

depth depending on the geometry of the splice. A polarisation controller in the two-mode fibre section aids optimisation of the modulated power transmitted through the splice.

A detector at the output of the monomode fibre (which is lead insensitive to stress) is followed by simple processing and display.

A birefringent fibre experimental system comprised of an Hitachi HLP 1400 laser diode operating at 830nm, which was sawtooth current ramp modulated by 20mA (peak to peak) about a DC bias of 100mA. The laser threshold current was 70mA, and the frequency excursion produced was 60GHz. Sawtooth frequency, 1/Ts, was 200Hz. Operational temperature of the laser was controlled at 17°C to within 0.1°C. The light from this laser was passed through a beam-splitting polariser and the transmitted polarisation launched into one polarisation of a York polarisation maintaining fibre, of beat length 2mm, and total length 100m, by focusing with a microscope objective onto the end of the fibre. Figure 4 shows a time domain trace of the beat signal output for one sensor located a distance L=65m from the fibre output. The effect of the unwanted laser amplitude modulation has been reduced by the use of an analogue circuit which divides the beat signal by the sawtooth signal from the function generator. The technique leaves a slight residual modulation since attempts at complete division result in saturation of the circuit. The frequency of the beat signal is lower at the start of the ramp than at the finish, due to non-linearities of the frequency sweep caused by thermal effects in the laser.[7] The lower trace is the negative current ramp drive to the laser.

Figure 5 shows the frequency spectrum for two sensors placed at L=65m and 21m respectively, showing a multiplexed operation of two sites 44 metres apart. Ramp period is 100Hz, other laser parameters are as above. Total fibre length was 67 metres. Peaks

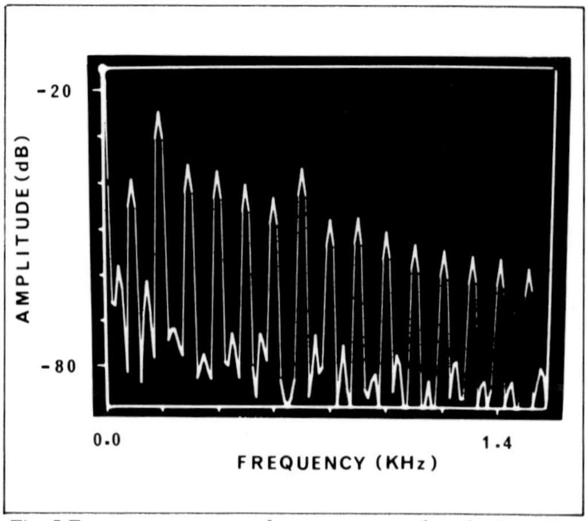

Fig. 5 Frequency spectrum for two sensors placed at L = 65m and 21m respectively

are at 12dB and 7dB above the adjacent frequency peaks. In this multiplexed configuration, the peaks of the (sin x)/x envelopes have been arranged to coincide with the harmonics of the ramp frequency. The modulator circuit has not completely removed the ramp modulation, as a decay in the frequency harmonics is still present.

Using a fine wire mesh as the coupling sensor, with a mesh spacing of 0.7mm, laid over the fibre, modulation depths of 13dB have been measured for a single sensor. For a practical system, the use of a wire mesh rather than a ridge coupler is desirable for increased flexibility, and in many applications the full dynamic range of the ridge device is unnecessary. The dynamic range of this system was limited by the sidelobe levels of the (sin x)/x spectrum, which is theoretically 13dB down on the peak. It may be possible to improve on the dynamic range by the use of amplitude tapers on the output signal.

Each measurement of inverse beat period for distances between 40 and 60 metres along the fibre, for a single coupling point indicated an accuracy of ±2 metres and was limited by the non-linearities of the

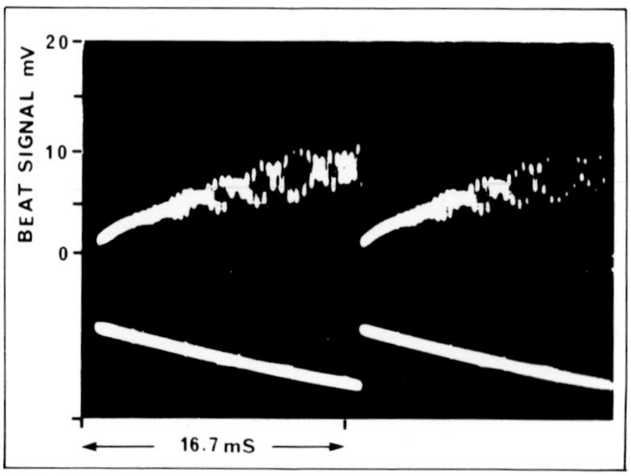

Fig. 6 An oscilloscope trace at the output of the fibre

frequency sweep. Improved accuracy should be possible by integration over a phase sweep of the beat frequency.

For the two mode system, initial investigation yielded an output of 24 oscillations in one sawtooth period for 145 metres of York Technologies step index fibre, bimodal at 830nm. By launching into both modes of the two moded fibre, the effect of having both second order mode sets can be seen. The launch end of the fibre was used as the coupling region.

Figure 6 shows an oscilloscope trace of the output of the fibre, with an analyser consisting of a splice to monomode fibre, of length 4 metres, preceded by a polarisation controller which is found to help align for the optimum modulation depth. The high frequency element at around 2.4 kHz is due to the beating of the first order mode pair with one of the second order mode pairs, and the lower frequency envelope is due to the introduction of the other second order mode pair, which produces beating between second order modes. The laser was a different HLP1400 device operated at a current of 72.0mA, with a 12.4mA sawtooth wave current modulation, of frequency 60Hz producing a ramp time of 16.7ms and giving a frequency excursion of 38GHz. Laser temperature was 12.7°C and was controlled to within 0.1°C.

As there are two sets of second order modes, each of which has a slightly different group velocity, then the higher frequency component of the beat signal is in fact composed of two beat frequency spectra adjacently spaced, with overlapping sidebands. The 3dB width of this spectrum can be up to 20% of the beat frequency and if two second order mode pairs are present, this effect limits the resolution. Suppression of one pair of second order modes is necessary to achieve resolution comparable with the birefringent fibre system, and for a single sensor location, may be achieved by either variation of the direct input launch, or orientation of an input polarisation controller on a monomode fibre lead-in to an input splice, hence launching only one second order mode set.

Limitations on resolution

The dominant limit on resolution arises in the case of a single sensor location in the birefringent system from the thermal effects of the laser, leading to a non-linearity in the frequency excursion. For this reason, the drive current used is a triangle waveform which allows one ramp time after a frequency sweep for the thermal transients of the laser to settle, before frequency ramping recommences. Approximately the first 10% of the frequency ramp is unusable at a ramp frequency of 1kHz, because of thermal effects, and around the first 5%, at a ramp frequency of 100Hz. Measurements should be taken in the last 80% of the ramp time.

The beat frequency is seen to change by about 10% of its value over the useful remainder of the ramp period, due to thermal time constants of the laser. This places a limitation on the resolution of the system until improved sources are available. For a period measurement technique used on the birefringent fibre system, using a 67 metre length of fibre, approximately 5.6 beats were seen under the envelope and so the error due to frequency non-linearities is approximately 1.8% of the fibre length, or around 1.2 metres. Calibration of the error should be possible. Previous workers[8] have reported optical path difference resolutions with a standard deviation of 2μm, obtained by averaging the non-linearity over a large number of samples. Translating this figure to the considered system suggests that resolutions of less than 5cm, limited by phase noise in the source, should be possible for a single sensor system.

Concluding remarks

Two versions of an optical fibre system which may be used either as a multiplexed or as a distributed sensor have been presented. Experimental results from the birefringent fibre version showed sensing of stress along a single fibre with a distance accuracy of about ±2 metres. Multiplexed sensing of two sensor positions has been achieved for two sensors spaced 44 metres apart, and this spacing may be improved upon. Initial experiments show that a similar system is possible using two moded fibre.

The systems are of potential value in the location of strain or stress in hazardous environments.

Acknowledgements

The authors would like to acknowledge the helpful discussions given by George Economou, Izumi Sakai and Darran Kreit in this work. This work is supported under the JOERS optoelectronic sensors consortium.

References

1. R.C. Youngquist, J.L. Brookes and H.J. Shaw, 'A birefringent-fiber polarisation coupler', *Opt. Lett.* Vol 8, 1983, pp. 656–658.
2. R.C. Youngquist, J.L Brookes and H.J. Shaw, 'Two-mode fiber modal coupler', *Opt. Lett.* Vol. 9, 1984, pp. 177–179.
3. A.J. Hymans and J. Lait, 'Analysis of a frequency-modulated continuous waveranging system', *Proc. IEEE, 107–B*, 1960, pp. 365–372.
4. D. Tuttam, I.P. Giles, B. Culshaw and D.E.N. Davies, 'Remote interferometric sensors using frequency modulated laser sources' *IEE Conference publication No. 221, OFS*, 1981, pp. 182–184.
5. S. Kobayashi, Y. Yamamoto, M. Ito and T. Kimura, 'Direct frequency modulation in A/GaAs semi-

conductor lasers', *J. of Quantum Electronics,* No. 18 1982, pp. 582–595.
6. D. Gloge, 'Weakly guiding fibers', *Appl. Opt. 10*, 1978, pp. 2252–2258.
7. M. Ito and T. Kimura, 'Stationary and transient thermal properties of semiconductor laser diodes', *IEEE J. of Quantum Electronics*, No. 17, pp. 787–795.
8. G. Beheim and K. Fritsch 'Remote displacement measurements using a laser diode', *Electron, Lett.* Vol. 21, 1985, pp. 93–94.

Distributed fibre-optic sensor using forward travelling light in polarisation maintaining fibre

K. Kurosawa and S. Hattori
Tokyo Electric Power Co., Japan

T. Yoshino
Gunma University, Japan

DISTRIBUTED sensing is one of the most important potentials of optical fibre sensors. Optical time domain reflectometry (OTDR) is at present the only available technique for distributed fibre sensing. This technique, however, relies on Rayleigh scattering which is so weak that a long time is needed for obtaining sufficiently high signal-to-noise ratios. Real time and high sensitivity detection are the most desirable features of distributed fibre sensors. In this paper we discuss a new approach to achieving such requirements by using polarisation maintaining fibres and frequency modulated laser diodes. The principle and demonstration experiments are presented.

Theory

The propagation mode of polarisation maintaining fibre (PMF) consists of two modes having orthogonal linear polarisations, denoted by 0 and 1. Each mode has a different phase velocity, β_0 and β_1 ($\beta_0 > \beta_1$). A linearly polarised laser light is launched into the fibre to excite one polarisation mode, say 0. The laser frequency is modulated as:

$$\omega = \omega_0 + \gamma t \tag{1}$$

where ω_0 is a constant frequency, γ is a modulation constant and t is time. We assume a mode conversion from 0 to 1 occurs at a distance L from the fibre end by an external perturbation. Then the 0 and 1 modes arrive at the fibre end with a time difference of:

$$\tau = \frac{(\beta_0 - \beta_1)L}{\omega_0} \tag{2}$$

or,

$$\gamma = \frac{\lambda_0 L}{c L_p} \tag{3}$$

in terms of a beat length:

$$L_p = \frac{2\pi}{\beta_0 - \beta_1} \tag{4}$$

where λ_0 and c are the light wavelength and velocity in vacuum, respectively. The output beams of 0 and 1 are interfered with each other by passing through an analyser, whose transmission axis is oriented at θ to the polarisation direction of the 0 mode. Then an optical beat signal:

$$V(t) = P_0\cos^2\theta + P_1\sin^2\theta + \sqrt{P_0 P_1}\,\sin 2\theta \times \cos(\Delta\omega t + \phi) \tag{5}$$

is generated. Here P_0 and P_1 are the light powers of the 0 and 1 modes:

$$\Delta\omega = \gamma\tau \tag{6}$$

which is the beat frequency, and:

$$\phi=(\beta_0-\beta_1)\left\{1-\frac{1}{2}(\beta_0+\beta_1)\frac{\gamma L}{\omega_0^{2}}\right\}L \qquad (7)$$

which is the phase difference between the 0 and 1 modes. From equations (3) and (6), one obtains:

$$L=\frac{cL_p}{\lambda_0\gamma}\Delta\omega \qquad (8)$$

Using this relationship, the location of mode conversion can be determined from the measurement of the beat frequency $\Delta\omega$. The resolving distance L_R is given from the condition that one cycle of the beat signal is involved within a measurement time T, i.e:

$$\Delta\omega T=2\pi \qquad (9)$$

Putting equation (9) into (8) leads to:

$$L=\frac{2\pi cL_p}{\lambda_0\gamma T} \qquad (10)$$

or,

$$L_R=\frac{2\pi cL_p}{\lambda_0\Omega} \qquad (11)$$

where:

$$\Omega=\gamma T \qquad (12)$$

is the amount of the frequency change for the measurement time of T. From equation (11) it follows that the resolving distance L_R is inversely proportional to Ω and proportional to the beat length L_p. The modulation depth of the beat signal is given, from equation (5), by:

$$m=\frac{\sqrt{P_0P_1}\,\sin 2\theta}{P_0\cos^2\theta+P_1\sin^2\theta} \qquad (13)$$

At $\theta=45°$ in particular:

$$m=\frac{2\sqrt{P_0P_1}}{P_0+P_1}\equiv M \qquad (14)$$

Obviously, at an angle of:

$$\theta_m=\tan^{-1}\sqrt{\frac{P_0}{P_1}} \qquad (15)$$

m becomes unity. The rate of mode conversion is defined by:

$$\eta=\frac{P_1}{P_0=P_1} \qquad (16)$$

which is related to M and θ_m as follows:

$$\eta=\frac{1\pm\sqrt{1-M^2}}{2} \qquad (17)$$

$$\eta=\cos^2\theta_m \qquad (18)$$

Experiment

In order to demonstrate the theoretical results, experiments were done using the scheme shown in figure 1. The light source is a single mode diode ($\lambda_0=780$nm) whose frequency depends on the injection current I as $\Delta\omega/\Delta I = 2\pi\cdot 4$GHz/mA. The diode current was modulated in a triangular wave form with a period of lms; then, referring to figure 2:

$$\gamma=5.3\times10^{14}\text{rad s}^{-2} \qquad (19)$$

A polarisation maintaining fibre used has a beat length of $L_p=2.7$mm, 125μm diameter cladding and 0.9mm diameter nylon jacket. Mode conversion was given by bending. At a distance of 100m from the fibre end, the PMF was bent at a radius of curvature of 1mm by means of the right-angle corner of a metal block; the polarisation axis of the PMF was inclined at 45° to the axis of bending. Figure 3(a) shows the oscilloscope signals of the injection current I and the photo-detector output V without mode conversion. Figures 3(b) and 3(c) represent the corresponding signals with mode conversion when $\theta=45°$ and $\theta=\theta_m$, respectively. The optical beat signals are observed. At $\theta=\theta_m$ in particular, 100% modulation is obtained, in accordance with the theoretical result. From figures 3(b) or (c), the beat frequency is obtained as:

$$\Delta\omega = 2\pi\cdot 8.0\text{kHz} \qquad (20)$$

Putting the data of equations (19) and (20) into equation (8) yields:

$$L=\frac{(3\times10^8)\times(2.7\times10^{-3})\times(8.0\times10^3)\times2\pi}{(780\times10^{-9})\times(5.3\times10^{14})}$$

$$=98\ [\text{m}] \qquad (21)$$

which agrees well with the real distance of 100m, within the measurement error. In the present case, from figure 3(a), $X=2\pi\times10.6\times4=2\pi\cdot 42$GHz so that the resolving distance is calculated from equation (11) to be:

$$L_R=\frac{2\pi\times(3\times10^8)\times(2.7\times10^{-3})}{(780\times10^{-9})\times2\pi\times(42.4\times10^9)}=24\ [\text{m}] \qquad (22)$$

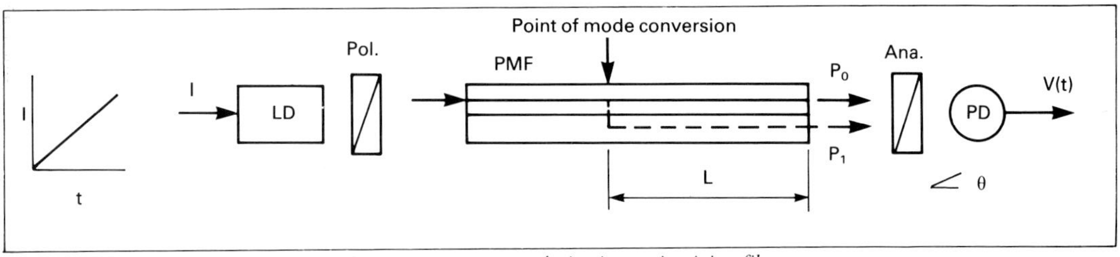

Fig. 1 Schematic diagram of distributed fibre sensor using polarisation maintaining fibre

Discussion

The measurement sensitivity depends on the efficiency of the polarisation mode conversion in PMF. High efficiency is achieved by using mechanical forces such as bending, tortion or pressure.

Bending

The fibre bending is the simplest but very efficient to cause polarisation mode conversion. The bending-induced birefringence is given by[1,2]

$$\Delta\beta_b = \frac{2\pi}{\lambda_0} \times 0.093 \times \left(\frac{2a}{R}\right)^2 \tag{23}$$

where 2a is the fibre diameter and R is the radius of curvature, respectively. The critical radius R_c of curvature under which the bending-induced birefringence compensates for the initial birefringence is given by:

$$\Delta\beta_b = \beta_0 - \beta_1 = \frac{2\pi}{L_p} \tag{24}$$

Thus, equating equations (23) and (24) with $\lambda_0 = 780$nm, $2a = 125\mu$m and $L_p = 2.7$mm:

$$R_c = 2.2\ [\text{mm}] \tag{25}$$

in reasonable accordance with the experimental result.

Magnetic field

When a uniform axial magnetic field H is applied to PMF over a length z, a mode conversion occurs by an amount of:

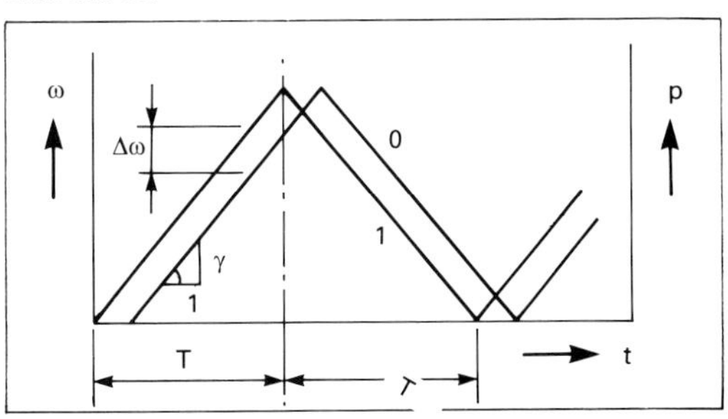

Fig. 2 Modulation characteristics of laser diode

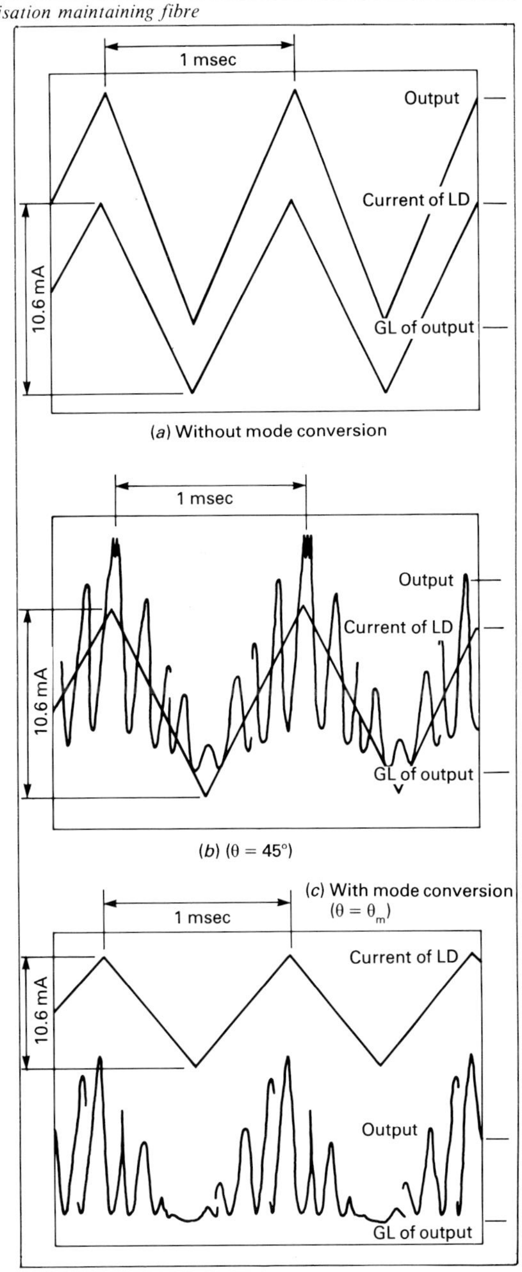

Fig. 3 Oscilloscope traces of output light signals

$$\eta = \frac{(VH)^2}{\frac{\Delta\beta^2}{4}+(VH)^2} \sin^2\left(\sqrt{\frac{\Delta\beta^2}{4}+(VH)^2}\, z\right) \quad (26)$$

and

where $\Delta\beta = \beta_0 - \beta_1$

and V is the Verdet constant of the fibre material. η is a periodical function of z. Since $\Delta\beta = 2\pi / L_p \gg VH$ in actual conditions:

$$\eta \approx \left(\frac{VHL_p}{\pi}\right)^2 \sin^2\left(\pi \frac{z}{L_p}\right) \quad (27)$$

which takes a maximum:

$$\eta_{max} = \left(\frac{VHL_p}{\pi}\right)^2 \quad (28)$$

when

$$z = \left(\frac{1}{2} + N\right) L_p \qquad (N = 0, 1, 2 \cdots) \quad (29)$$

In the silica fibre, as $V = 0.012 \text{min}/\text{Oe}\cdot\text{cm}$ or 4.4×10^{-6} rad/A, η_{max} is as small as 4×10^{-5} even for the saturation magnetisation of iron, i.e., 21KOe. A more efficient method is to use PMFs having longer beat lengths, and/or apply periodical magnetic fields in synchronism with the beat length. Longer beat lengths of PMF, however, increase the polarisation cross-talk and reduce the spatial resolution of sensing, thus leading to a compromise between measurement sensitivity and instability. The spatial resolution L_R is inversely proportional to the tunable range of the light source. In the present experiment, a conventional laser diode is used. Larger tunable range is obtained with the use of distributed feedback (DFB) or CW dye lasers. The use of such lasers allows the spatial resolution to be improved much more than in the present experiment. When mode conversion occurs at n different points at the same time, the output signal involves various beat frequencies, i.e. $\Delta\omega_i = L_i / \gamma$ and their cross-talks such as $\Delta\omega_i - \Delta\omega_j$, $\Delta\omega + \Delta\omega_j - \Delta\omega_k$, etc (i, j, k = 1, 2, $\cdots$, n); so that the interpretation of the output signal becomes complicated. But, if the mode conversion at each point is sufficiently small, to the first order approximation, the cross-talk components can be neglected; then, in principle, each point of mode conversion is discriminated by measuring the frequency components of $\Delta\omega_i$, involved in the beat signal.

Concluding remarks

The distributed sensing using forward travelling light in polarisation maintaining fibre has been proposed and its basic operation experimentally confirmed. The proposed scheme has the following advantageous features:

- Real time detection.
- High signal-to-noise ratio.
- Simplicity of the entire system.

References

1. Luc B. Jeunhomme, 'Single-mode fiber optics', Marcel Dekker 1983.

Further reading

After presenting this paper we discovered similar work by R.E. Franks, W. Torruellas and R.C. Youngquist, 'Birefringent stress locator sensor', *SPIE* (Cannes, 1985) Vol. 586.

Special shaped fibres for distributed sensors

A.C. Boucouvalas
Hewlett-Packard Laboratories, UK

N.S. Rayit and M. Greaves
GEC Research Ltd., UK

INTRINSIC optical fibre sensors have received much attention since high sensitivity has been demonstrated and due to the possibility of distributed sensing.

Although telecommunications-grade fibres are used for a great number of sensors, mainly due to availability and cost considerations, it is recognised that such fibres are far from optimum for most sensor applications.

As a result, special fibres for sensing applications, with geometrical and material variations, have emerged from communication fibre designs. Such fibres are low birefringence spun fibres, high birefringence, polarisation maintaining fibres, polarising fibres, twin core fibres, liquid core fibres, rare earth metal doped fibres and special coating fibres. The wide diversity of such fibres arises from the large number of possible applications of optical fibres in sensing.

In this work we examine the effect of a simple modification of the preform of a telecommunications grade fibre by forming a D shape. The resultant fibre is characterised and the usefulness of such fibre in sensing is examined.

Preform and fibre fabrication

A circularly cylindric fibre preform was fabricated by a standard MCVD process and, subsequently, a longitudinal slice was cut across the outer cladding, resulting in a preform with a D-shaped cross-section. The preform was subsequently pulled into fibre (see Fig. 1).

Clearly, the general features of the preform are maintained with rounding of the edges of the D. This indicates that the pulling temperature was higher than necessary. However, we have found experimentally that pulling at lower temperatures results in weaker fibres but maintains the preform shape. Long lengths of fibre were pulled and coated with silicone resin.

Measurements on D-shaped fibres

Higher order mode cut-off wavelength

This important measurement allows us to find the

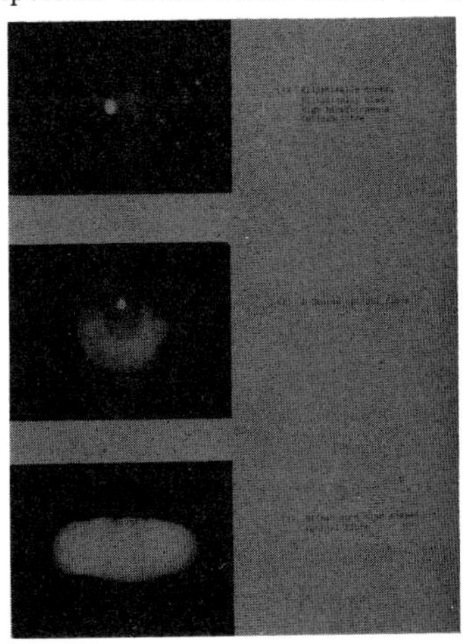

Fig. 1 The preforms from which special shape fibres were made. D-section preform (left); slab cross section preform (middle) circular preform (right)

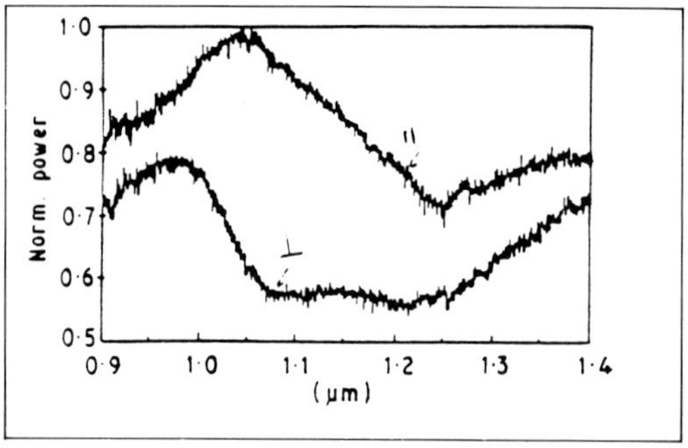

Fig. 2 Transmission for light polarised perpendicular and parallel to the flat at the D

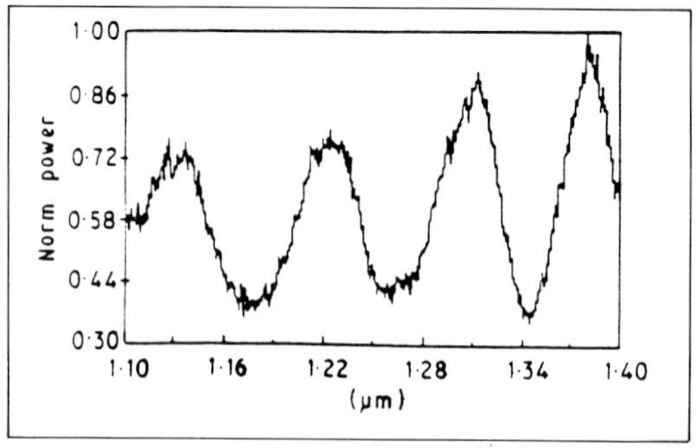

Fig. 3 Interference fringes at the D fibre

wavelength region for single mode operation of the fibre.

For this we have used a Rofin Sinar spectrum analyser and the transmission method was used for a 2m fibre. The fibre end was mounted on a rotation stage and the output was inserted into the spectrum analyser detection slit.

The spectrum analyser acts as a polariser at the output and the polarisation dependence of the cut-off wavelength can be measured by rotating the fibre using the rotating stage. A series of measurements revealed that the cut-off wavelength is dependent on the polarisation of the light.

Figure 2 shows the transmission for light polarised perpendicular and parallel to the flat of the D. The difference in the cut-off wavelength in this case is ~160nm. It is thought that the angular dependence is due to differential loss of the polarisations in the fibre rather than birefringence in the fibres.

Birefringence measurement

The large polarisation dependence of the cut-off wavelength suggested that the fibres might be birefringent. The measurement of birefringence was achieved using a white light source and the Rofin Sinar spectrum analyser. A polariser was used at the input of the fibre and adjusted to 45° to the birefringence axes of the fibre. The output from the fibre was directed to the Rofin monochromator (also acting as polarisation analyser). The principle of the measurement is well known,[1,2] and the measurement relies on detecting power oscillations as the wavelength is scanned. The method, although approximate, is sufficient for our purpose which is to give us a good indication of the birefringence rather than a precise measurement. Figure 3 shows the interference fringes of the D fibre. The beat length at 1300nm is ~ 1m which is large compared to typical polarisation maintaining fibres but low compared to ordinary circular core communication fibres.

Splicing

An important consideration, often conveniently bypassed, is the connection of sensing fibres with ordinary fibres. The successful jointing of circular core and cladding fibres to D fibres using a fusion splicing technique is reported. Typical splice loss of 0.5dB is obtained. The technique can be applied to fibres of other geometries as well.

Using the standard technique for splicing D-shaped fibres to circular fibres, losses of typically 10dB were

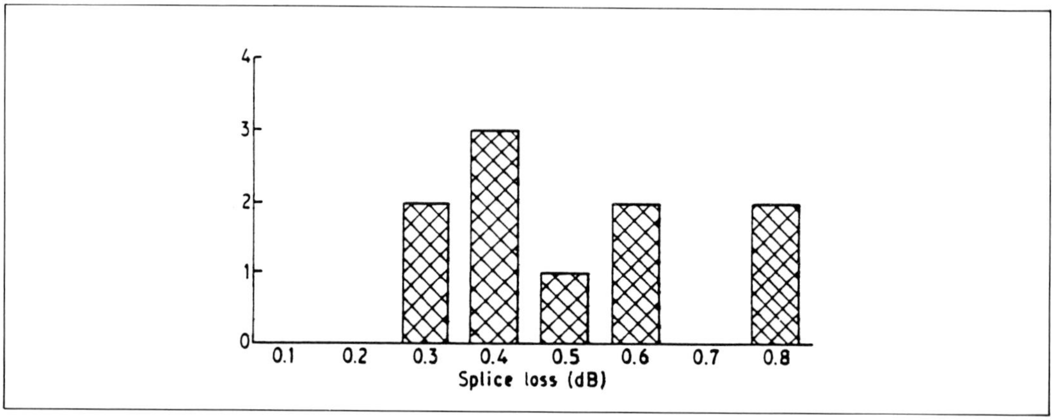

Fig. 4 Splice loss distribution for first ten splices

obtained. The method reported here is to use a short duration high temperature arc, to effectively 'spot weld' the fibres together. Minimum movement of the fibres upon splicing is hence obtained. The splice is subsequently reinforced with UV-curing epoxy, then encapsulated in a splice protector in the usual way. Figure 4 is a histogram of the first 10 splices using this technique. The average splice loss is just higher than 0.5dB, demonstrating low loss jointing. Further work in splicing can bring this loss to about <0.2dB.

OTDR sensing

The use of the D fibres as a pressure sensor has been demonstrated with the emphasis of our experiments in the comparison of sensitivities of the circular and D type fibres. We have found that the microbend sensitivity of the D fibre is superior to the circular cladding fibre. The bias pressure is at least five times smaller in the case of the D fibre, when the comparison is performed with rubber corrugated pressure pads of 4mm period 12mm long. The flat surface of the D was more sensitive than the curved surface.

References

1. K. Kikuchi and T. Oboshi, *Optics Letters*, Vol. 8, No. 2, February 1983, pp. 122–123.
2. S.C. Rashleigh, *Optics Letters*, Vol. 8, No. 6, June 1983, pp. 336–338.

Temperature sensing

THE papers in this chapter cover methods for distributed temperature sensing in optical fibres, an area where practical systems having real engineering applications are now becoming commercially available.

The first paper (Dakin *et al*) describes a method using Raman scattered light which, although it results in much weaker scattered signals, allows the use of conventional glass fibres due to the superior temperature coefficient of Raman scattered light in glass.

Historically, the first distributed temperature sensor was that of Hartog and Payne (*Proc. IEE Colloq. Optical Fibre Sensors, London, 1982*) This used the temperature-dependent Rayleigh scattering in liquid-filled fibres as the means of sensing temperature, using a conventional optical-time-domain reflectometer (OTDR) system. The second article in this chapter is a more recent paper by A.H. Hartog which gives a far more comprehensive coverage of the method.

The next two papers, by S.A. Kingley and V.D. McGinniss and by Dakin *et al*, provide reviews of temperature sensing in fibres.

The fifth paper (Lecoy *et al*) describes a novel temperature-dependent attenuation method using radiation losses from a spirally-wound fibre (and again using OTDR interrogation).

The sixth paper, by Farries *et al*, examines the use of rare-earth doped fibres, with their pronounced thermally-dependent absorption characteristics.

The last paper (Dakin *et al*) represents an attempt to utilise the optical Kerr effect in fibres as a temperature-dependent sensing mechanism (early work on the electrically-induced Kerr effect in glass showed a very strong temperature dependence). The paper showed, for the first time, that the *optical* Kerr effect is less temperature dependent than the electrical equivalent and, hence, the sensing method cannot be used in this simple configuration.

Temperature measurement using intrinsic optical fibre sensors

J.P. Dakin
York Ltd., UK

TEMPERATURE is one of the key variables to be measured in many physical and chemical processes. As the temperature of a material is a function of the population of its internal energy states, it is hardly surprising that it may be detected by optical means. However, the first optical fibre temperature probes were extrinsic* types based on optical radiation pyrometry, the earliest forms using optical fibres to allow the detector from a pyrometer head to be placed remotely from the objective lens. A specialised variation of this technique[1] involved an optical fibre detection head in which the fibre end was covered by an opaque end cover which radiated energy into the fibre when heated.

The possibility of using intrinsic optical fibre sensors to measure temperature or strain was first suggested in a paper by Kingsley and Davies,[2] in which highly sensitive interferometric techniques were proposed. Since that date, numerous intrinsic methods of temperature sensing have evolved, using both coherent and incoherent methods. The remainder of this review outlines the progress that has been made in intrinsic temperature sensors in the decade since that paper.

Coherent methods for the intrinsic detection of temperature

Basic principles

The coherent optical detection of temperature in conventional monomode fibres relies on a combination of the changes in refractive index (principally of the core medium) and fibre length which are induced in the fibre as the temperature changes. In classical solid state theory, these are both explained in terms of non-linearities in thermally-excited harmonic oscillations between atomic particles. In conventional monomode fibre, the phase shift, ϕ, at the output of a length of fibre, L, when subjected to a temperature change, $\Delta\theta$, is given by:

$$\Delta = \beta \times \Delta\theta\,(dL/d\theta) + L \times \Delta\theta\,(d\beta/d\theta)$$

β is the propagation constant of the fibre which describes the phase change per unit length. The first term describes the effect of thermal expansion, whereas the second term describes the effect on the fibre propagation constant due to temperature changes. The second term is complicated by the increase in diameter of the fibre which, along with the more significant effect of refractive index variations, causes modification of the fibre propagation constant with temperature.

Hocker[3] theoretically estimated the phase shift in a pure silica fibre with 633nm excitation as approximately 107 radians $\times$ $K^{-1} \times m^{-1}$, or 17 fringres, per degree, per metre. His measured value in doped-silica fibre was approximately 13 fringes, per degree, per metre, in reasonable agreement with the calculated value, considering the different materials involved.

The magnitude, and even the sign of the effect, may be modified by suitable choice of optical fibre material[3,4,5] and any sheathing materials[4,5] used to protect the fibre.

---

*An intrinsic sensor uses the fibre itself as the sensing element, whereas an extrinsic type uses the fibre only as a light-guiding medium to couple to the detector head.

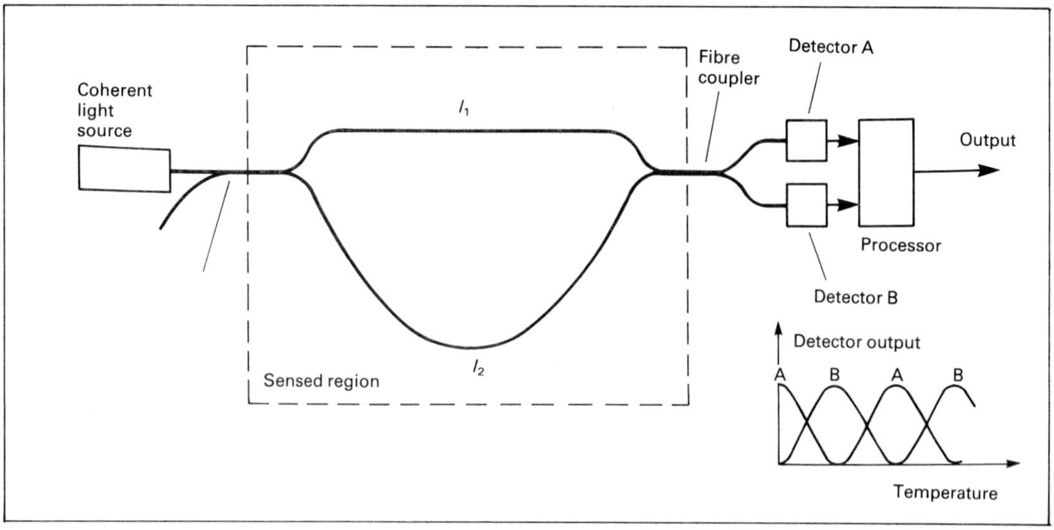

Fig. 1 Mach-Zehnder interferometer for temperature sensing

The overall phase shift is a function of the average temperature change in the length of fibre, and therefore the measurement is distributed in nature. However, it has not yet proved possible, using interferometric techniques, to determine the profile of temperature along optical fibres.

The only practical method of observing the optical phase shift in a single mode fibre sensor is to create an optical reference beam from the same source and arrange two-beam interference. For homodyne interference of equal intensity signals, a $(1+\cos\phi)$ variation in detected output with optical phase is observed. In order to detect the phase unambiguously, some form of optical intensity referencing must be employed, in order to avoid errors due to fluctuations in light source power or launching efficiency. In addition, if operating over more than a half fringe, an unambiguous fringe-counting method must be employed. In a homodyne system, this requires derivation of more than one interference output from the same interferometer, in order to detect the direction of fringe movement at the maximum and minimum positions.

Two major problems may occur in simple interferometric temperature sensors. The first is that the basic sensing method is incremental: i.e. it relies on a fringe counting method and will lose calibration if count is lost. The second problem is that, if conventional monomode fibre is used (as with virtually all sensor publications before 1981), serious errors can occur due to polarisation changes in the arms of the interferometer. The latter can lead to signal fading, non-linearity and serious calibration errors in homodyne systems (Polarisation fading is, however, less serious in heterodyne systems provided the intermediate frequency carrier does not fade to zero.)

Mach-Zehnder interferometer configuration

The most common configuration used for early systems was the well-known Mach-Zehnder interferometer shown in figure 1. Either the lengths l_1 and l_2 may both be heated in common mode as shown, necessitating a non-zero path imbalance in order to achieve a sensor, or alternatively the system may be operated in differential mode in a balanced interferometer ($l_1 = l_2$) configuration.

The unbalanced configuration has the advantage that the degree of imbalance could permit the possibility of path length calibration using frequency swept light sources by methods such as the FMCW technique reported in reference 6. The primary disadvantage of the unbalanced interferometer is, however, the related problem of sensitivity to random fluctuations in the phase or frequency of the light source, which can cause amplitude noise in the detected signal.

Michelson configuration

An alternative fibre interferometer is shown in figure 2. This is the fibre equivalent of the Michelson interferometer, and has the advantages of simplicity, the need for only one coupler, the convenience of source and detector at the same end as the sensor, and the sensitivity increase due to the double transit of the fibre. This system, reported by Corke *et al*,[7] again relied on incremental fringe counting and could lose calibration if count were lost. However, the temperature range and resolution achievable with the method was impressive, 0.1°C being resolved over a 300°C range in reference 7, and 0.005°C resolution being achieved with an improved system using a stabilised light source.[8]

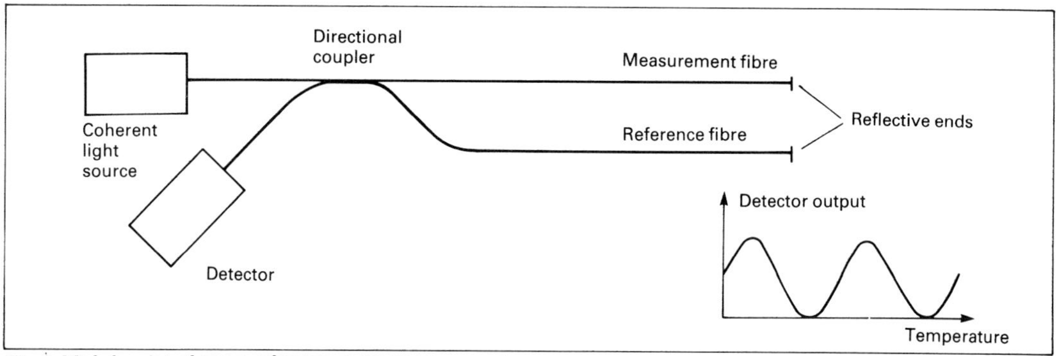

Fig. 2 Michelson interferometer for temperature measurement

Polarimetric sensor configuration

Both the interferometer arrangements discussed above use separate fibre paths for the detection and reference arms and attempt to arrange the interfering signals to be polarised in the same plane.

Polarimetric sensors use the two separate polarisation modes in a single polarisation-maintaining fibre as separate arms of a Mach-Zehnder type interferometer. The interference at the output now takes place between two orthogonally-polarised beams. The output is the vector run of the two fields at 90°, with their relative phase dependent on the differential mode delay. The output is read by analysing the resultant output polarisation state. The inherent simplicity of the conventional polarimetric sensor, first envisaged by Rashleigh and Ulrich[9] is shown in figure 3. In order to excite both modes equally, the input light is either polarised at 45° to the polarisation axis of the fibre, or is arranged to be circularly polarised. Eickhoff[10] was the first to recognise the potential of the technique as a temperature sensor.

Jackson *et al*[11] have operated the polarimetric sensor in a reflective (Michelson) mode, and have used a relatively insensitive interconnection lead of polarisation-maintaining fibre aligned at 45° to the sensing fibre, in order to allow remote operation (see Fig. 4). The same research group[12] has arranged an interferometer with greater unambiguous range by configuring both Michelson and polarimetric-Michelson interferometers within a single length of polarisation-maintaining fibre. In this manner, two sets of fringes may be obtained, one set from a Michelson interferometer operated in one of the fibre modes, the other a polarimetric interferometer, operated by analysing the resultant polarisation state from both polarisation modes. The full-wave temperature scan corresponding to the two sets of fringes differed by a factor of 20 (in the 'bow-tie' fibre used), thus enabling the output of the polarimetric sensor to be used to identify the fringe order of the more-sensitive Michelson interferometer.

The conventional polarimetric sensor has two main problems: first, a highly coherent light source is required in order to avoid phase noise at the output of the unbalanced fibre interferometer and second; the sensor is sensitive to ambient changes in both pressure and temperature. Both these problems have been solved using the compensated polarimetric technique,[13] which enables a differential sensor to be constructed by splicing two equal lengths of polarisation-preserving fibre, after a relative twisting of one fibre by 90° about its axis with respect to the other (see Fig. 5). In this manner a balanced interferometer is formed, thus in principle allowing the use of relatively-incoherent sources. In addition, the differential nature of the sensor renders it immune to common-mode changes of pressure and temperature.

In reference 13, the differential nature of the sensor and its common-mode rejection was demonstrated by using it as a temperature transducer. There has been experimental verification of the other expected advantage of the technique, the ability to operate it with LED light sources.[14]

The temperature resolution afforded by polarimetric

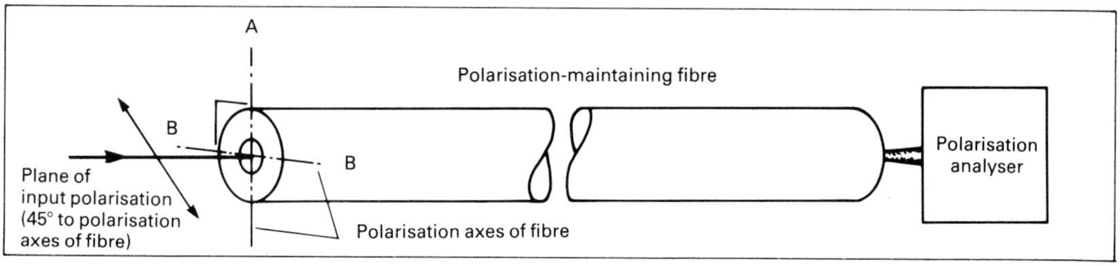

Fig. 3 Conventional polarimetric sensor

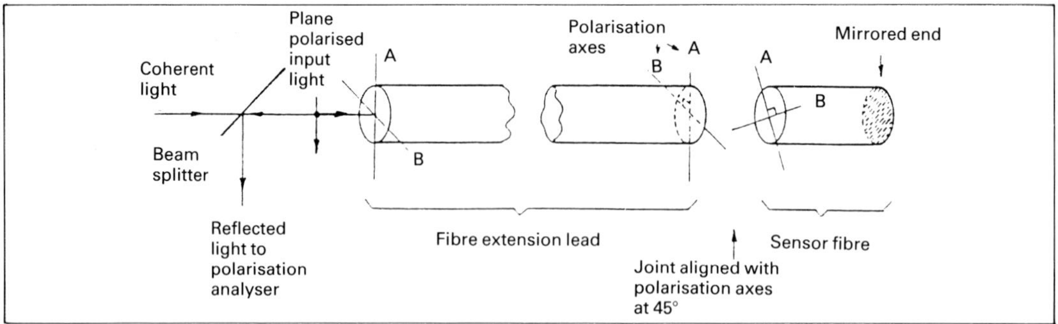

Fig. 4 Temperature sensor using polarimetric Michelson configuration (polarisation-maintaining fibre used for each section)

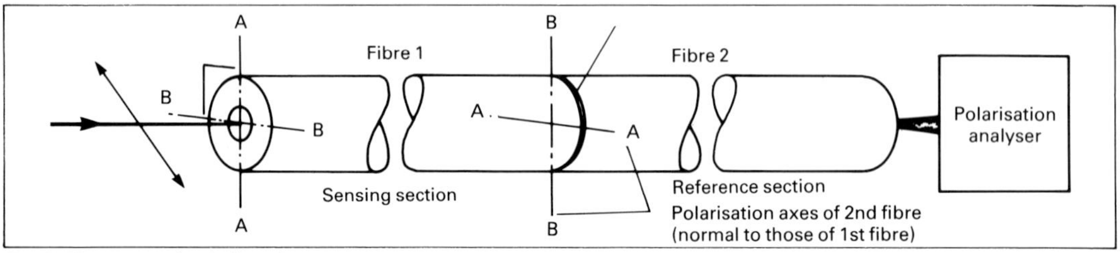

Fig. 5 Compensated polarimetric sensor

temperature sensors is more than adequate for most applications, and therefore the increased temperature range corresponding to each fringe spacing will normally be an advantage rather than a disadvantage. When used in the compensated mode as a differential sensor, the low drift and low phase noise, combined with insensitivity to light source fluctuations and external pressure changes, will generally permit operation close to photon noise limitations.

Fabry-Perot configuration

The fourth interferometer configuration which has been used for temperature sensing is that of the Fabry-Perot interferometer.[15] The basic method is to examine the spectral transmission or reflection of a fibre Fabry-Perot coupled to a fibre extension lead. The most convenient optical arrangement for most applications is the reflective one (see Fig. 6). In order to determine the temperature, free from any amplitude referencing problems, the light source may be frequency swept to determine the length of the remote cavity from the fringe spacing. Although an intrinsic sensor, as it uses fibre as the sensing element, the sensing head is a miniature fibre version of the two-mirror Fabry-Perot extrinsic sensor described in reference 16.

Incoherent temperature sensing techniques

The various classes of intrinsic, incoherent, temperature-sensing methods which have been reported are detailed below. In general, these methods do not have the extremely high temperature resolution offered by coherent types, but are relatively free of sensitivity to pressure changes, and are in general potentially simpler to engineer into practical systems. The variety of basic mechanisms of operation is greater, but there are fewer possible permutations of the optical configurations in which they may be operated.

Fibre with temperature-dependent attenuation

The absorption spectra of fibres doped with rare-earth ions show a significant temperature variation. Results using the transmissive arrangement shown in figure 7 were published by Snitzer, Morey and Glen[17] following earlier studies on sensing techniques for aerospace applications.[16] Temperatures in the range of 0–800°C were measured by comparing the transmission in a neodymium-doped, alumino-silicate glass fibre at two wavelengths. Problems were, however, reported due to drift in the LED light sources and variations in coupler

Fig. 6 Reflective Fabry-Perot temperature sensor

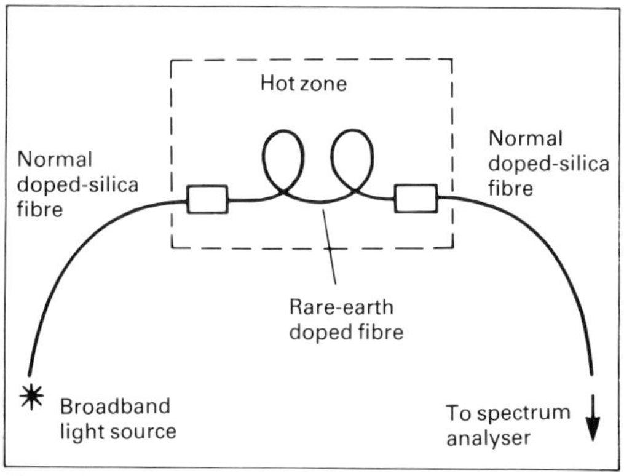

Fig. 7 Spectral filtering sensor using rare-earth-doped fibre

alignment.

Brandt and Gottlieb[18] have studied the possibility of constructing transmissive optical fibre sensors for detecting hot-spots in electrical equipment. They examined the possibility of constructing special optical fibres, with variable loss by virtue of a temperature-dependent variation in the core and cladding refractive indices. This should cause the evanescent field penetration into the cladding region to vary its penetration into lossy regions of the cladding or its external jacket. During experimental investigations of the method, however, the authors[19] discovered the possibility of using internaly-generated black-body radiation from the fibre to detect the presence of hot-spots. This method, as desired for their application, did not show a strong dependence on the length of the hot region.

More recently, the well-known problem of poor low-temperature transmission in plastic-clad silica fibres (due to refractive index increase of the polymer coating when cooled) has been turned to advantage in a patented method of detecting the temperature of such a fibre.[20] One variation of this[21] was for the detection of arbitrary-located cold spots due to leaks of cryogenic fluids from bulk containers. This latter technique has recently been applied commercially for monitoring bulk-liquified gas containers, probably the first industrial use of an intrinsic fibre sensor.

Distributed temperature profile measurement using optical time domain reflectometers

None of the temperature measuring methods discussed so far allows the temperature profile along the length of a fibre to be measured. In order to perform such a measurement, a means of monitoring the distance along the fibre is clearly a prerequisite. This is possible using the technique of optical time domain reflectometry (OTDR), where a short pulse of optical energy is launched into a fibre and the fraction of scattered energy guided back by the fibre is observed as a function of time (the latter, of course, being proportional to the distance along the fibre to the point from which the scattered light is received).

In vitreous fibres, the majority of back-scattered energy arises from Rayleigh or Rayleigh-Gans scattering from small refractive index variations which arise from high-temperature thermally-induced density changes which are 'frozen' into the glass structure as it is cooled from the melt. As a result, the elastically-scattered (i.e. scattered as the same wavelength as the incident light) light intensity varies very little with the temperature of the glass. In liquids, however, the scattering arises from real-time thermodynamic fluctuations in the refractive index and therefore the scattering shows a significant temperature coefficient. This effect has been made good use of in the system described by Hartog and Payne[22,23] using the apparatus shown in figure 8. The OTDR signals from the liquid-filled fibre are shown in figure 9. The temperature profile may be clearly seen from the measurements, which may be made with a somewhat greater signal/noise ratio than for conventional OTDR measurements on solid-cored fibre. However, the use of step-index liquid-filled fibres presents difficulties which, at best, complicate the method and, under some circumstances, restrict its use.

The following problems may arise:

- The use of liquid-filled fibres is inconvenient because of the need for expansion reservoirs.
- The choice of suitable low-loss liquids, compatible with low-loss silica cladding tubes, is limited and there are therefore temperature range restrictions imposed by the freezing and boiling points of the liquid.
- If the central region of a long length of such a fibre is 'rapidly' cooled, it is possible to create voids by rapid thermal contraction, before pressure differentials can fill the tube against the resistance of viscous forces.
- Impurities or dust particles could increase the scattering cross-section or increase the localised loss of the fibre giving the appearance of false temperature variations in the OTDR return.
- The numerical aperture of the liquid-filled fibre shows significant temperature dependence, generally reducing as the temperature is raised, allowing hot zones in particular, to affect the calibration of results further along the fibre (in the absence of mode-conversion, this effect may be reduced by the use of mode filters).
- The use of a step-index fibre may give rise to significant pulse dispersion, which will reduce the distance resolution as the measurement length is increased.
- Some liquids are prone to polymerisation.

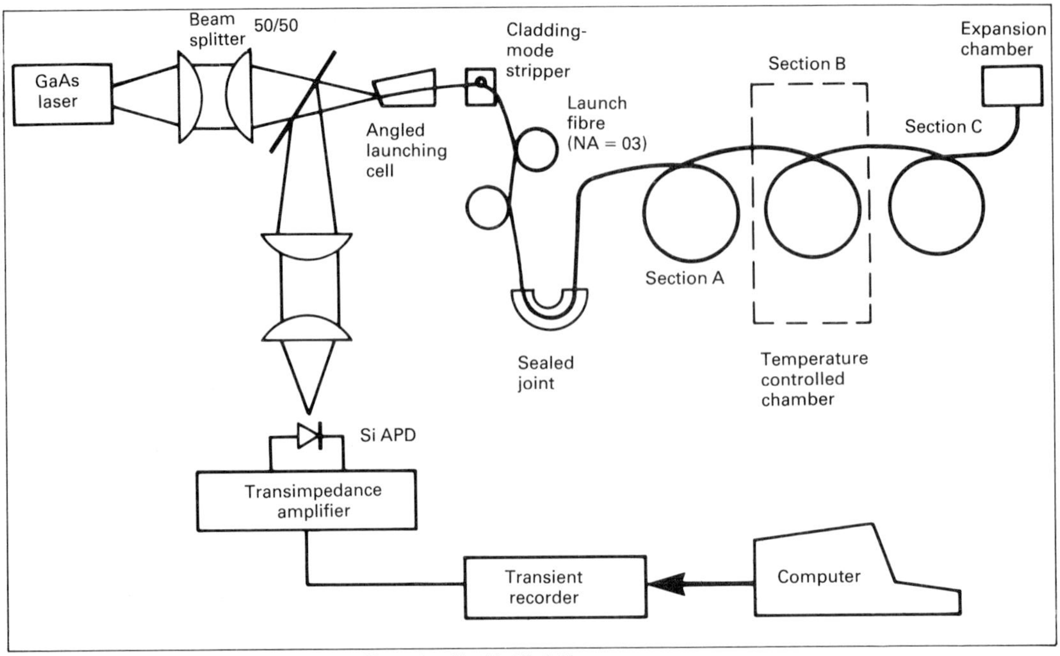

Fig. 8 Rayleigh-scattering temperature profiler using liquid-filled fibre

However, in spite of the above drawbacks, the system appears to work well over moderate temperature ranges, and is capable of monitoring the temperature distribution typically to ±0.2°C resolution, over several hundred metres of fibre, with a distance resolution of the order of 2 metres, a remarkable advance considering this was the first true temperature profile measuring method for use with optical fibres.

Although the Rayleigh scattering cross-section in solid-cored fibres does not change appreciably with temperature, Rayleigh OTDR returns can provide temperature profile information if temperature-dependent losses are present in the fibre. (Theocharous has briefly suggested this possibility in the introduction to a recent paper.[24]) Dividing the time differential of the back-scattered signal by the instantaneous value of this signal should yield a signal proportional to the loss of the fibre. The technique appears at first sight to be attractive, but a short semi-qualitative analysis suggests that the number of distance resolution elements would probably be low. To measure the average temperature within each distance resolution element, an accurate loss measurement is necessary, in order to detect any temperature-dependent changes in loss. A temperature dependent loss of at least 1dB in the resolution element (whatever its length) will probably be necessary in order to accurately detect changes in this loss in a relatively noisy OTDR return. Thus the round-trip attenuation in this example will be 2dB per resolution element, giving a likely limit to the number of measured resolution elements of perhaps 10, particularly bearing in mind the poorer signal/noise ratio from the later sections of fibre.

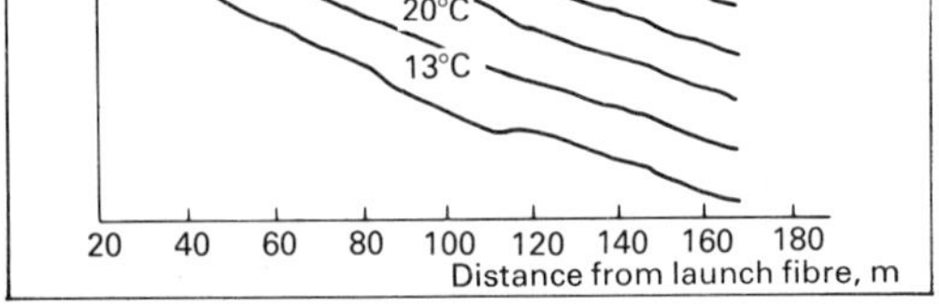

Fig. 9 Results from system shown in Fig. 10

If the spectral variation of back-scattering from a germania-doped-silica fibre is examined (see Fig. 10), it may be seen that there is a strong central line, primarily due to Rayleigh (or Rayleigh-Gans) scattering, but presumably also containing an unresolved contribution from Brillouin scattering. At each side of the central line, however, there are side-lobes due to Raman scattering, which may be used to detect temperature profiles in conventional communications fibres using modified OTDR techniques.[25] From standard texts on Raman scattering, the ratio R of anti-Stokes (higher frequency band) to Stokes (lower

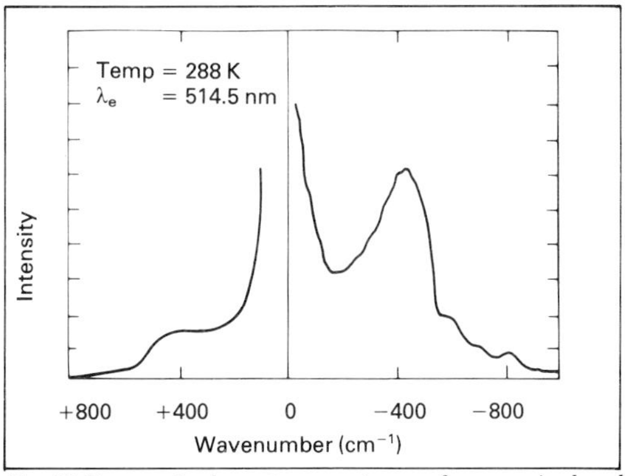

Fig. 10 Raman back-scatter spectrum of germania-doped silica fibre

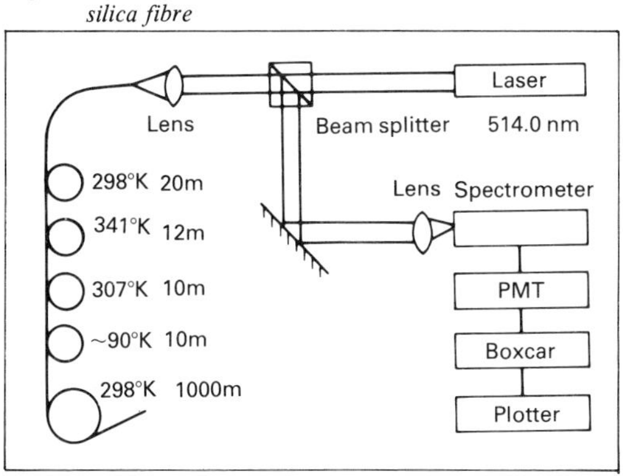

Fig. 11 Raman temperature profile equipment (first experimental layout)

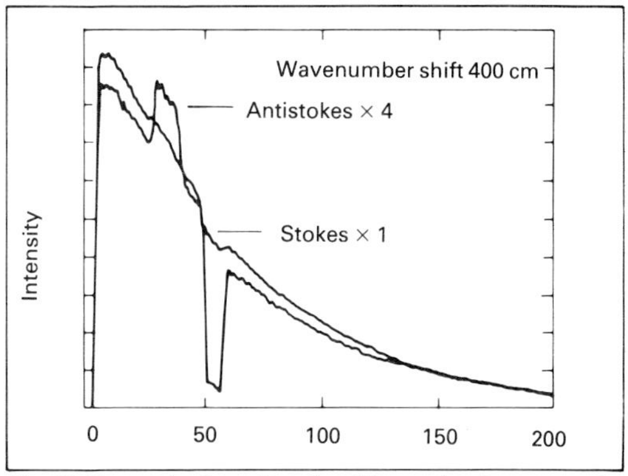

Fig. 12 Raman back-scatter signal (argon-ion laser source)

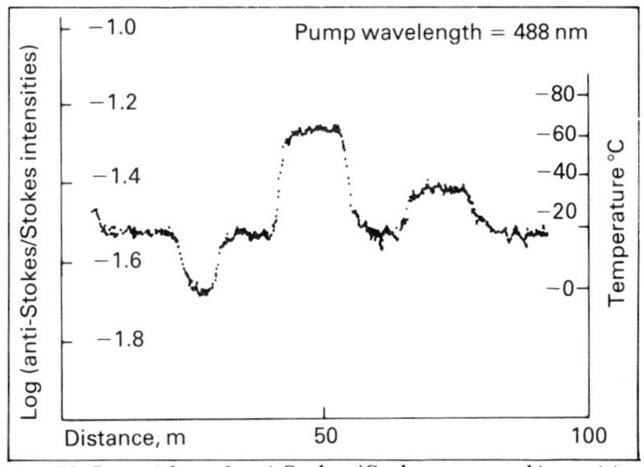

Fig. 13 Logarithm of anti-Stokes/Stokes scattered intensities

frequency band) intensity at wavelengths λ_a and λ_s, respectively at equal frequency separation from the central line is given by the relationship:

$$R(T) = (\lambda_s/\lambda_a)^4 \times \exp(-hc\gamma/KT)$$

where h is Planck's constant, c is the velocity of light in vacuo, K is Boltzmann's constant, T is the absolute temperature, and γ is the wave number of the incident light.

Therefore a measurement of the ratio of Stokes and anti-Stokes back-scattered light in a fibre should provide an absolute indication of the temperature of the medium, irrespective of the light intensity, the launch conditions, the fibre geometry and even the composition of the fibre. In practice, however, a small correction has to be made for the difference in fibre attenuation between the Stokes and anti-Stokes wavelengths.

The Raman technique appears to have only one significant practical drawback—that of a relatively weak return signal, the anti-Stokes Raman-scattered signal being between 20 and 30dB weaker than the Rayleigh signal. In order to avoid excessive signal averaging time, measurements have been taken using relatively high launched powers from pulsed lasers.[26,27] In reference 26, the first experimental demonstration of the method, a pulsed argon-ion laser was used in conjunction with Corning standard telecommunications-grade 50/125μm GRIN fibre (see Fig. 11). The OTDR results obtained are shown in figure 12. The anti-Stokes/Stokes intensity ratio is clearly a strong function of temperature. Figure 13 shows a curve from reference 27 where, with a different thermal profile along the fibre, the ratio of anti-Stokes and Stokes intensity has been computed to avoid the exponential decays due to fibre attenuation which are apparent in figure 12. Reference 27 also described preliminary results with a system based on a semiconductor laser, in the configuration shown in figure 14. The anti-Stokes return from this system (see Fig. 15) clearly shows the region of fibre which has been heated to 50°C above ambient. The method is believed, with further development, to be capable of measuring to 1k and 1 metre resolution of temperature and distance, respectively.

An alternative method of distributed temperature sensing is to utilise similar OTDR techniques with fluorescent fibres, taking advantage of the temperature dependence of fluorescent re-emission.[28] Such distributed sensors have yet to be demonstrated practically, but show promise of providing stronger signals than for Raman-based methods, with the disadvantage of requiring special fibres. Doped-silica fibres containing neodymium have already been reported with excellent transmission characteristics[29]

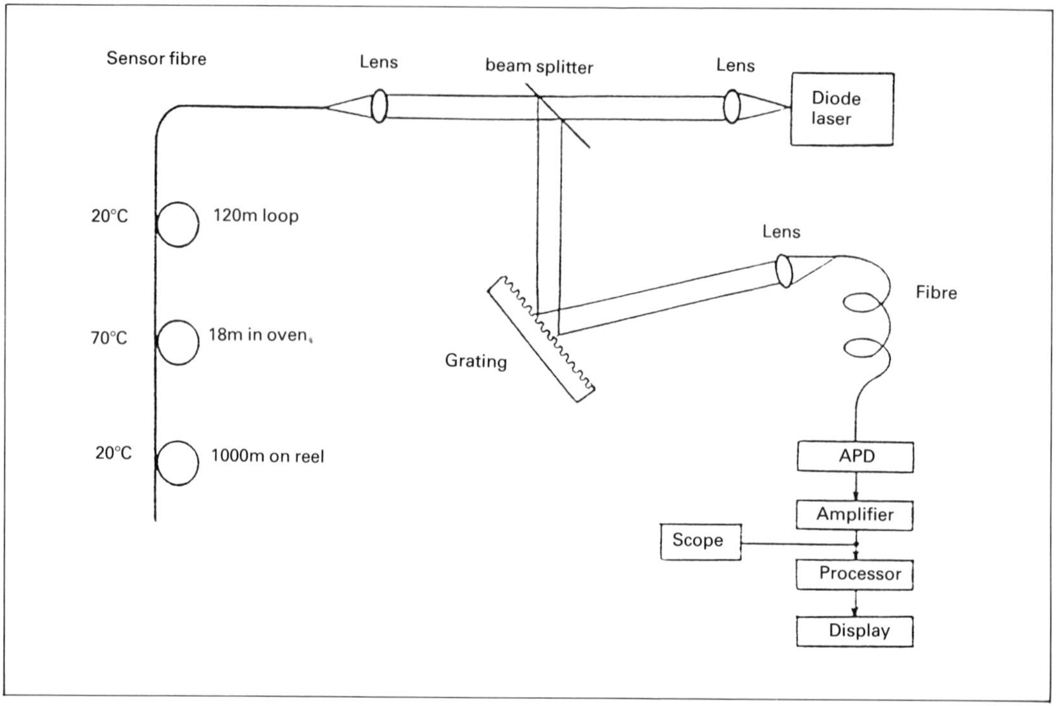

Fig. 14 Raman temperature profiler using a semiconductor laser source

but may have resolution limitations due to long fluorescent lifetimes. It is the view of the author that plastic fibres doped with organic fluorescent dyes may give excellent results for short distance systems in view of the high quantum efficiency and short fluorescent lifetime typical of such materials.

Concluding remarks

From the above review, it may be seen that significant progress has been made on methods for the measurement of temperature using intrinsic fibre sensors. The performance improvements in extrinsic temperature sensors have been no less dramatic. However, the two areas where intrinsic sensors have an advantage at present are: first, their potential for extremely high temperature resolution using interferometric techniques; and, second, their unique ability to perform distributed monitoring of either average temperature, the presence of hot or cold regions, or, most attractive of all, the temperature profile along a length of fibre. In terms of the highly important parameter of cost, the challenge is to match the price of extrinsic optical sensors or, perhaps more significantly for large scale application, that of the humble thermocouple.

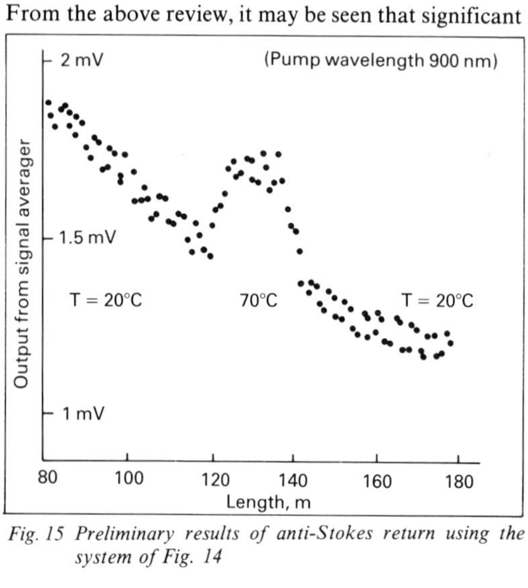

Fig. 15 Preliminary results of anti-Stokes return using the system of Fig. 14

References

1. J.P. Dakin and D.A. Kahn, 'A novel fibre optic temperature probe', *Optical and Quantum Electronics* 9, p. 540.
2. S.A. Kingsley and D.E.N. Davies, 'The use of optical fibres as instrumentation transducers', 1976, *Proc. Conference on Lasers and Electro-Optical Systems*, San Diego, pp. 24–25.
3. G.B. Hocker, 'Fibre-optic sensing of pressure and temperature', *Applied Optics* 18, 1979, pp. 445–448.
4. M. Tateda, S. Tanaka and Y. Sugaware, 'Thermal characteristics of phase shift in jacketed optical fibres', *Appl. Optics* 19, 1980, pp. 770–773.
5. N. Lagakos, J.A. Bucaro and J. Jarzynski, 'Temperature-

induced optical phase shifts in fibres', *Appl. Optics* 20, 1981, pp. 2305–2308.
6. D. Uttam, I.P. Giles and B. Culshaw, 'Closed loop FM interferometric remote optical sensor', *Proc. Conf. Fibre Optics and Laser Sensors*, Arlington, SPIE 412, 1983, pp. 256–260.
7. M. Corke, A.D. Kersey, D.A. Jackson and J.D.C. Jones, 'All fibre Michelson interferometer', *Electronics Letts* 19, 1983, pp. 471–473.
8. M. Corke, A.D. Kersey and D.A. Jackson, 'Temperature sensors with single mode fibres', *J. Phys. E. (Sci. Instr.)* 17, 1984, pp. 988–993.
9. S.C. Rashleigh and R. Ulrich, 'Magneto-optic current sensing with birefringent fibres', *Appl. Phys. Lett.* 34, 1979, p. 768.
10. W. Eickhoff, 'Temperature sensing by mode-mode interference in birefringent optical fibres', *Opt. Lett.* 6, 1981, p. 204.
11. M. Corke, A.D. Kersey, K. Liu and D.A. Jackson, 'Remote temperature sensing using polarization-preserving fibre', *Electron. Lett.* 20, 1984, pp. 67–69.
12. J.D.C. Jones, D.A. Jackson, M. Clrke and A.D. Kersey, 'Single mode fibre optic sensors: conventional interferometric, polarimetric and combined devices', *Proc. Fibre Optics 85 Int. Conf*, London, (to be published by SPIE Washington) 30 April–2 May 1985.
13. J.P. Dakin and C.A. Wade, 'Compensated polarimetric sensor using polarisation-maintaining fibre in a differential configuration', *Electron. Lett.* 20, 1984, pp. 51–53.
14. J.P. Dakin, S. Broderick, D.C. Carless and C.A. Wade, 'Opteration of a compensated polarimetric sensor with semiconductor light source', *Proc. 2nd Int. Conf. on Optical Fibre Sensors 'OFS 84'*, Stuttgart, September 1984, pp. 241–246.
15. R. Kist, S. Drope and H. Wolfelschneider, 'Fibre-Fabry-Perot thermometer for medical applications', *Proc. 2nd Int. Conf. on Optical Fibre Sensors 'OFS 84'*, Stuttgart, September 1984, pp. 165–170.
16. W.H. Glen, W. Morey and E.N. Snitzer, *NASA (USA) Report No. Cr 2159468.*
17. E.N. Snitzer, W. Morey and W.H. Glen, 'Fibre optic rare earth temperature sensors', *Proc. 1st Int. Conf. on Optical Fibre Sensors*, London, April 1983, pp. 79–82.
18. M. Gottlieb and G.B. Brandt, 'Temperature sensing in optical fibres using cladding and jacket loss effects', *Appl. Optics 20, pp. 3867*–3873.
19. M. Gottlieb and G.B. Brandt, 'Fibre optic temperature sensor based on internally generated thermal radition', Appl. Optics 20, 1981, pp. 3408–3414.
20. P.W. Black, 'Temperature sensitive optical fibure', *UK Patent No. 1582768.*
21. D. Pinchbeck, 'Optical fibres for cryogenic leak detection', *Proc. Conf. Electronics in Oil and Gas*, London, 1985, 15–17 January 1985.
22. A.H. Hartog, 'Remote measurement of temperature distribution using an optical fibre', *Proc. 8th Euro. Conf. on Optical Communication*, Cannes, 1981, pp. 215–222.
23. A.H. Hartog, 'A distributed temperature sensor based on liquid core optical fibres', *IEEE Journal Lightwave Technology*, Lt-1, 1983, pp. 498–509.
24. E. Theocharous, 'Differential absorption distributed thermometer', *Proc. 1st Int. Conf. on Optical Fibre Sensors*, London, April 1983, pp. 10–12.
25. J.P. Dakin, 'Temperature measuring arrangement', *UK Patent Appl. 2140554A.*
26. J.P. Dakin, D.J. Pratt, G.W. Bibby and J.N. Ross, 'Distributed anti-Stokes ratio thermometry', *Proc. 3rd Int. Conf. on Optical Fibre Sensors*, San Diego, Post-deadline session, February 1984.
27. J.P. Dakin, D.J. Pratt, G.W. Bibby and J.M. Ross, 'Distributed optical fibre raman temperature sensor using a semiconductor light source and detector', *Electron. Lett.* 21, 1985, pp. 569–570.
28. J.P. Dakin, *UK Patent Appl. GB 2156513A.* Priority date 28 March 1984. Published 9 October 1985.
29. S.B. Poole, D.N. Payne and M.E. Fermann, 'Fabrication of low loss optical fibres containing rare-earth ions', *Electron. Lett.* 21, 1985, pp. 737–738.

Distributed temperature sensor based on liquid core optical fibres

A.H. Hartog
York Ltd., UK

OPTICAL fibre sensors[1,2] are the subject of considerable interest and devices have been proposed for the measurement of a large number of physical parameters including temperature, [3,5], acoustic, electric and magnetic fields, pressure, strain rotation and displacement.[1,2] Optical fibre sensors are of advantage for remote measurements in environments advantage for remote measurements in environments which are hazardous or which suffer from severe electromagnetic interference or where space and weight are restricted. Sensors consisting entirely of optical fibres can be considerably more rugged than bulk transducers and although the sensitivity of glasses to external influences is usually lower than that of suitable crystals, this may often be compensated by the longer path lengths available in fibres.

Optical fibre sensors presented to date modify either the intensity, the phase, the state-of-polarisation or the time-of-flight of the light travelling in the fibre.[1,2] Elaborate techniques have been employed, especially in the case of the gyroscope,[1] to reduce the transducer's sensitivity to factors other than that to be measured. In most cases, the measured value refers either to a specific location in space or to some form of average of the sensed quantity over the entire length of fibre sensor.[3,4] It is then not possible to gain information about the spatial distribution of the measured value without resorting to multiple sensors.

In the present paper, a new class of fibre optic sensors is proposed which are able to monitor the physical quantity of interest at many points along the fibre simultaneously. The sensors are based on optical time domain reflectometry (OTDR) in specially designed fibres. OTDR is a well-known[6,7] technique for observing length-dependent changes in fibre parameters.[8–12] In order to use this technique in a fibre sensor, the fibre must be designed in such a way that its back-scatter signature is sensitive to the external physical parameter to be monitored.

The proposed technique may be seen as an extension of polarisation OTDR[13,14] where changes in the state of polarisation (which is affected by external factors) along a single-mode fibre are sensed using OTDR. In the present sensors, changes in scattering loss, numerical aperture (NA), or total attenuation may be used to provide the required sensitivity.

The sensors may be classed with other intensity-modulating transducers but they use timing information to multiplex the values sensed at many separate locations onto the one fibre.

In this paper, these ideas are applied to the development of a temperature-distribution sensor using liquid-core fibres which are known to exhibit a large sensitivity of Rayleigh scattering loss and NA to temperature changes.

Theory of operation

Review of OTDR

The basic arrangement of OTDR[7,8] is shown in figure 1. A laser is used to launch short optical pulses into the fibre via a directional coupler. As the pulse travels along

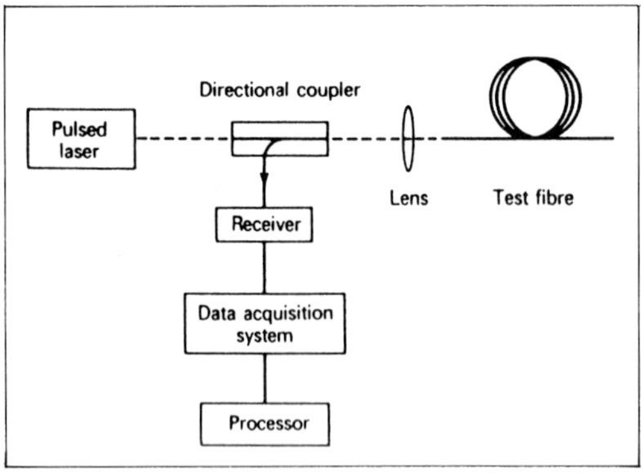

Fig. 1 Functional diagram of an optical time-domain reflectometer.

the fibre, it is attenuated via several mechanisms, one of which is Rayleigh scattering. Some of the light thus scattered is guided by the fibre back to the launching end where it is detected by the receiver. The resulting waveform takes the shape of a decaying pulse as the forward-travelling pulse and that portion of the light which is scattered backwards are both attenuated by propagation along the fibre.

The scattered power $p(t)$ returning to the launching end after time t is given by:

$$p(t) = P_0\ W\eta(z) \exp - \int_0^t \alpha(z)\, v_g\, dt \qquad (1)$$

where P_0 is the power launched into the fibre, W the pulse-width, v_g the group velocity, $\alpha(z)$ the local attenuation coefficient and $\eta(z)$ the back-scatter factor, i.e. the rati of the back-scatter power to the probe pulse energy. Distance z from the launching end is related to elapsed time t via $z = v_g\, t/2$. In the case of a multimode step-index fibre η may be expressed as.[15]

$$\eta = \frac{1}{2}\, \alpha_s\, v_g\, S = \frac{1}{2}\, \alpha_s\, v_g\, \frac{3}{8}\, \frac{n_1^2 - n_2^2}{n_1^2} \qquad (2)$$

where α_s is the Rayleigh scattering loss coefficient and S is the capture fraction, i.e. the proportion of the scattered light which is recaptured by the waveguide. In step-index multimode fibres, S is determined purely by the core and cladding refractive indexes (n_1 and n_2, respectively).

In a fibre which is uniform along its length, the back-scatter signal takes the form of a decaying exponential (see equation (1)), the rate of decay being determined by the attenuation of the fibre. In practice, back-scatter waveforms are found to depart from the expected behaviour owing to longitudinal variations in fibre parameters.[8-12,16] Figure 2 shows an example of a non-exponential back-scatter waveform; in this case, the increase in signal in the central section of the fibre is caused by a 12% reduction in fibre diameter. Fibre

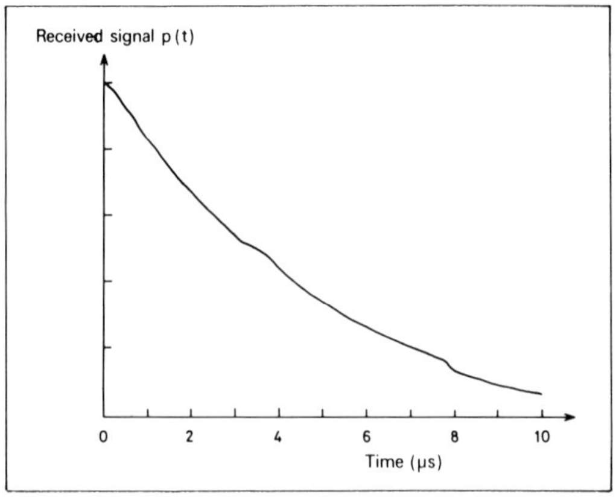

Fig. 2 Backscatter waveform obtained on a multimode fibre having a non-uniform diameter.

diameter variations are the most common cause for the features of back-scatter signatures in multimode fibres as they strongly alter the modal power distribution.[9,16] In addition, according to equations (1) and (2), a longitudinal change of the Rayleigh scatter coefficient or of the NA will have a direct influence on the back-scatter level, and the rate of decay is affected by variations in total attenuation.

Application to distributed sensors

If longitudinal variations of Rayleigh scattering, fibre diameter, NA, or total loss can be induced in a fibre controllably by external effects, then it is clear that OTDR can be used to provide spatially resolved sensing. The fibres used should be designed deliberately to enhance their sensitivity to the quantity of interest. Each fibre may be seen as a continuous set of sensors, receiving the light in turn.

The remainder of this paper deals specifically with a distributed sensor which uses liquid-core fibres to provide a high sensitivity to temperature and a good immunity to other external factors such as mechanical perturbations. Preliminary results have been presented.[17]

Of the fibre parameters which affect the back-scatter signature, total loss is probably the easiest to make sensitive to temperature using, for example, changes in the strength, width and position of absorption bands. A bulk device along these lines has been demonstrated recently[18] and, with suitable doping, fibres with similar properties could probably be developed. However, in a distributed sensor, modulation of the total loss is undesirable as it imposes an unwelcome trade-off between sensitivity and the amount of forward pulse energy available to probe the following sensing points. In addition, it is necessary then to differentiate the back-scatter waveform to retrieve the temperature distribution and it follows that the requirements on the signal-to-noise ratio are very stringent if a reasonable resolution (e.g. a few meters) is to be achieved.

Variations of NA result from the difference between the thermo-optic coefficients of the core and cladding materials. In glasses, these effects are normally so small as to be of no value; compositions specifically designed to enhance the sensitivity of the NA to temperature may become available in the future. In step-index multimode fibres a further, more fundamental, drawback exists, namely that a reduction in NA between the source and the sensing point will act as a mode filter and can completely mask the local NA change.[16] Thus sensing of the temperature at many positions in the fibre independently would be precluded. Note, however, that this is not the case in graded-index[10] or single-mode fibres[12] where NA variations may be detected all along the fibre using OTDR.

The remaining effect is that of changes in Rayleigh scattering. It shows some temperature sensitivity in glasses and is known to vary strongly in liquids.[19] Liquid-core fibres would thus appear to be an eminently suitable medium for a distributed temperature sensor. In addition, these fibres are likely to be largely immune to external influences other than temperature.

Temperature sensing in liquid-core fibres

Fabrication of liquid-core fibres

Liquid-core fibres[20,21] consist of a glass tube (the cladding) filled with an ultra-transparent liquid of higher refractive index which forms the core. For the present work, silica capillary tubing was obtained by drawing silica tubes on a fibre-pulling machine to an outer diameter of 300μm and to an inner diameter of 150–200μm. Prior to the drawing, high-purity silica was deposited by the MCVD process on the inside of the tubes to reduce the loss of the cladding region. The capillary tubing was coated on-line with polyimide, a material which can be operated up to temperatures above 400° C. Polyimide has the further advantage that it is optically absorbing and hence acts as an effective continuous cladding-mode stripper. Cladding modes are known[22] to give rise to spurious effects in back-scatter measurements and their total elimination is essential in this application.

The capillary tubing was filled with hexachlorobuta-1, 3-diene[20,21] using a high-pressure syringe system. Typically, 150m of 150μm bore capillary can be filled in 30 minutes. It is important to prevent the incorporation of particulate matter in the core. Dirt particles form troublesome scatter centres which move as the temperature varies and as the liquid expands and contracts. A 0.2μm pore filter in the syringe ensured that the fibres filled were free from scatter centres.

The fibres have an NA of 0.50 at λ=900nm (Δ=5.5%) and 0.54 at 589nm. Their loss is typically 13dB . km^{-1} at 900nm as measured by the back-scatter technique; lower losses have been reported with the same core liquid.[21]

Operation

The configuration of the liquid-core temperature sensor is shown in figure 3. The source is typically a GaAs single-heterostructure laser emitting short (~10ns) pulses of 1–2W peak power at a wavelength of 900nm. The laser output is focused through a beam-splitter onto the end of a launch fibre which is held in an angled index-matching cell to eliminate the front-face reflection. The remote end of the launch fibre is inserted into the core of the sensing fibre and the joint thus formed is sealed. For the experiments to be described, the sensing fibre is arranged into three coils, two of which (*A* and *C*) are held at room temperature whereas the central fibre section (*B*) can be temperature-cycled.

The scattered light returning to the launching end is

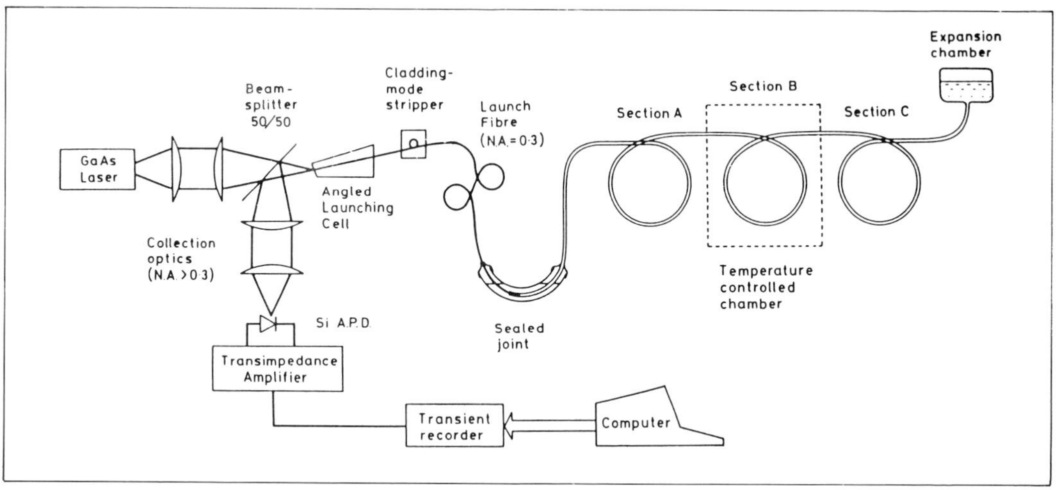

Fig. 3 Experimental arrangement for the distributed temperature sensor.

directed by the beam-slitter via the collection optics to the Si avalanche photo-diode detector. The photocurrent is fed to a 35MHz bandwidth transimpedance preamplifier. The amplified signal is sent to a high-speed transient recorder where the data are stored and averaged before being transmitted to the computer for further processing.

Two major effects occur in a liquid-core fibre as its temperature is increased. First, the core refractive index is found to decrease. The temperature dependence of the refractive index of hexachlorobutadiene at $\lambda = 569$nm is given by:[24]

$$n(\Theta) = 1.5660 - 5.23 \times 10^{-4} \times \Theta \qquad (3)$$

over the range $\Theta = 10$–$75°$C, where Θ is the temperature in Celsius. The fibre numerical aperture thus decreases strongly as the temperature is increased, which leads to a reduction of the back-scatter signal.

The second effect is the increasing scattering loss, resulting from the increasing thermal agitation, and given by: [19,23]

$$\alpha_s(T) = \frac{8\pi^3}{3\lambda^4}(n_1^8 p^2)\, kT\beta_T \qquad (4)$$

where p is the photo-elastic coefficient, k is Boltzmann's constant, T the absolute temperature, and β_T the isothermal compressibility. Since n_1, p, β_T all vary with temperature, α_s can have a large temperature sensitivity. In benzene, for example,[19] the temperature sensitivity $d\alpha_s R$ is 0.33 dB · K^{-1}. The measurements on hexachlorobuta-1, 3-diene (described in the section 'Sensitivity measurements in fibres having low-loss claddings') indicate a somewhat lower sensitivity of 0.022 dB · K^{-1} (to the author's knowledge, the temperature dependence of p and βT for this compound have not been published).

As the temperature sensitivity of the NA and of the scattering loss have opposing effects on the back-scatter signal, one of these must be eliminated if the influence of NA variations is eliminated from the back-scatter signal with the aid of a mode filter. These variations could not in any case be traced reliably along a step-index multimode fibre. The mode filter consists of a section of step-index multimode fibre, the NA of which is substantially lower than that of the liquid-core fibre. The filter, therefore, has the function of an angular stop and it transmits to the receiver back-scatter power carried only in the low-order modes of the sensing fibre. NA variations affect the power content in the high-order modes only and these modes are lost at the mode filter. The mode filter thus eliminates the sensitivity of the back-scatter signal to changes in NA.

A small residual sensitivity to temperature-induced longitudinal index changes exists, however, since ray bending will occur at the interface between fibre sections at different temperatures, as illustrated in figure 4. This ray bending causes light scattered, say, in a relatively hot section (see Fig. 4) to be converted from high-order to low-order modes and hence allows a larger proportion of the back-scatter power to pass

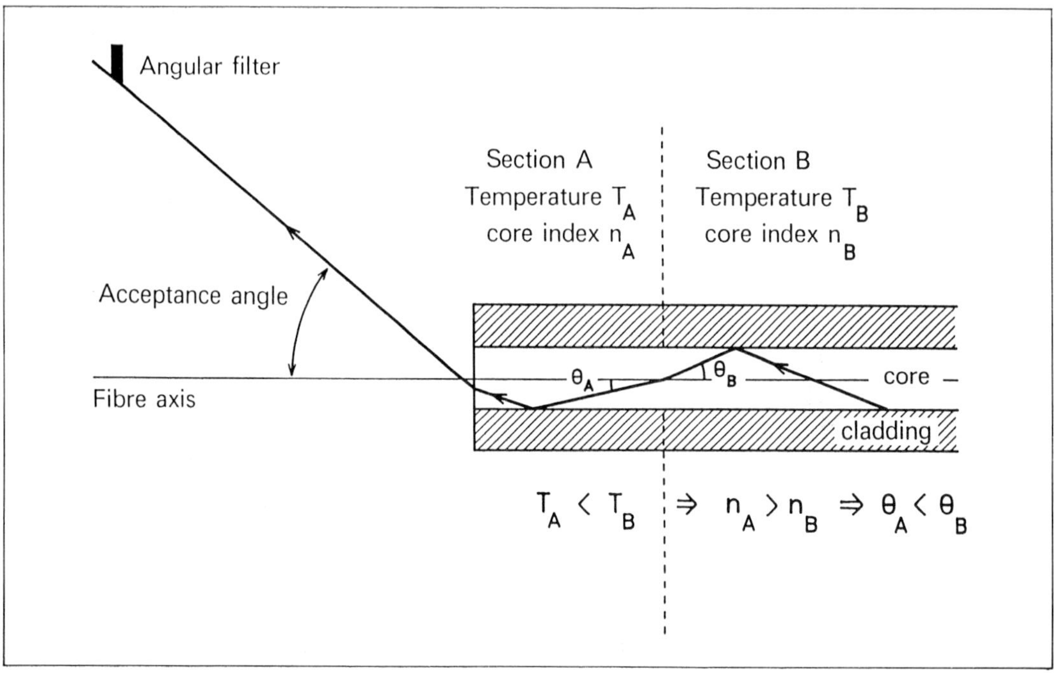

Fig. 4 Sensitivity of the mode-filter transmission to longitudinal variations of core index.

through the mode filter to the receiver. This residual effect fortunately has the same sense as the temperature dependence of scattering loss. It modifies the received back-scatter signal by a factor of $(n_1(0)/n_1(z))^2$ and adds typically 1.5×10^{-3} dB·K^{-1} to the sensitivity of the sensor.

Design study: measurement accuracy, spatial resolution and range

In this section, the potential performance of a liquid-core fibre sensor driven by a semiconductor laser is calculated. The numerical values of the important parameters are largely the same as those found in the experimental work (see section 'Results and discussion'). The main difference is that it is assumed that an A/D converter is available to acquire and average the entire back-scatter waveform at each laser pulse,[14,25] at a repetition rate of ~ 1kHz. Although flash converters operating at 75 Megasample/s with 8-bit resolution are available commercially, a transient recorder with a maximum averaging rate of 20 waveforms/s was used in the present work. In addition, the laser pulse power was somewhat lower than has been assumed in the calculations.

Signal-to-noise ratio. Table 1 summarises the properties of the components of the sensor system. The back-scatter factor η can be evaluated from the properties of the fibres, with the aid of equation (2). The numerical aperture NA_R of the launch fibre rather than that of the sensing fibre must be inserted in equation (2) since the former is the limiting NA in the receiving optics. The result is $\eta = 2.6 \times 10^3$ W/J at $\lambda = 904$nm. The laser output pulse energy is 20nJ.

Estimated losses L_T in the back-scatter system are given in Table 2 for the complete path from the laser to the start of the sensing fibre and back to the receiver. These amount to 19dB. The back-scatter signal $P_r(x)$ from a point at distance x along the sensing fibre is, therefore:

$$P_r(x) = P_0\, W \eta\, 10^{-(L_T + 2\alpha x)/10} \qquad (5)$$

Hence, for $x = 100$m, $P_r(x) = 0.36\mu$W. This optical signal is detected using an avalanche photo-diode and a transimpedance amplifier. At the preamplifier output, the signal-to-noise ratio is approximated by:[26]

$$S/N = \frac{(R P_r M)^2}{(2e(I_d + R P_r) M^2 + x_m + i_n^2) B} \qquad (6)$$

where R, M, I_d and x_m represent the responsitivity, avalanche gain, primary dark current, and excess noise factor of the photo-diode, respectively. i_n is the equivalent input noise current of the preamplifier and B, its bandwidth; e is the electronic charge. Values for

Table 1. Typical properties of the components used in distributed temperature sensor operating at a wavelength of 0.9μm

Component	Parameter	Symbol	Value
Laser	Peak power	P_o	2W
	Pulsewidth	W	10^{-8}s
	Wavelength	λ_o	904nm
	Pulse repetition rate		1kHz
Sensing fibre	Total attenuation	α	13dB/km
	Scattering loss	α_s	8.7dB/km @ 20° C (2.0×10^{-3} np/m)
	(temp. coeff.)	$\frac{1}{\alpha_s}\frac{d\alpha_s}{dT}$	21.8mdB°C
	Core diameter		150μm
	Outer diameter		300μm
	Core index	n_1	1.55 @ 20° C
	Group velocity	v_g	1.875×10^8 m/s @ 20° C
	Temperature sensitivity		23.3mdB/°C
Launch fibre	Length		10–50m
	Loss		<1dB
	Numerical aperture	NA_R	0.3
	Core diameter		100μm
	Outer diameter		140μm
Receiver	Detector responsivity	R	0.7A/W
	Unmultiplied dark current	I_d	~0.2nA
	Excess noise factor	x_m	0.35
	Input equivalent noise current	i_n	1.1pA/Hz$^{1/2}$
	Bandwidth	B	35MHz
	Transimpedance		10^5 Ω

Table 2. Loss budget for the optical path in the terminal equipment

Forward Path	
Beam-splitter	3dB
Reflections and aberrations	2dB
Launching efficiency	3dB
Loss of launch fibre	1dB
	9dB
Return Path	
Coupling loss	4dB
Launch fibre	1dB
Reflections and aberrations	2dB
Beam-splitter	3dB
	10dB
Total loss: 19dB	

these parameters are given in table 1 for the devices used in the section 'Results and discussion'.

The optimum avalanche gain is obtained by differentiation of equation (6):

$$M_{opt} = \left[\frac{i_n^2}{x_m e(I_d + R\,P_r)} \right]^{1/(2+x_m)} \tag{7}$$

Substitution of M_{opt} into equation (6) gives the signal-to-noise ratio at optimum gain which, expressed as a voltage ratio, takes the value of 100 for the parameters given above. Assuming that the signal is averaged for 1s (i.e. 1000 averages), the signal-to-noise ratio improves to $\sim 3 \times 10^3$. For a temperature sensitivity of 23×10^{-2} dB·K^{-1}, this signal-to-noise ratio leads to an uncertainty in the temperature of less than 0.1°C (at 100m).

If the temperature measurement is to be achieved to this degree of precision, attention must be given to the linearity of the receiver and data acquisition system, and to the processing of the back-scatter waveforms to extract the temperature information. The first point has been discussed in other literature[27,28] and the second will be discussed in the section 'Processing the back-scatter waveform'.

Spatial resolution. The spatial resolution of the sensor is directly related to the system impulse response, i.e. to the convolution of the pulse broadening in the fibre and the impulse response of the terminal equipment. Pulse propagation along a step-index fibre of length $2x$, results in a pulse broadening of:[29]

$$\Delta\tau = \frac{x}{c}\,\frac{NA_R^2}{n_1} \tag{8}$$

provided that mode coupling can be neglected (c is the speed of light in free space). Pulse broadening places a fundamental limit to the sensor resolution which degrades linearly with distance, but can be controlled by restricting the receiving NA. At a distance of $x = 100$m, receiving NAs of 0.1, 0.2 and 0.3 permit spatial resolutions of 0.2, 0.8 and 1.8m, respectively. Although liquid-core fibres generally have very low mode conversion,[30] the back-scattering process itself will result in an almost complete re-distribution of power amongst the mode groups. The figures given above are thus pessimistic, probably by as much as a factor of $\sqrt{2}$. The improvement in resolution achieved by restricting the receiving NA results in the signal-to-noise ratio being degraded in proportion to NA_R^4 since the launching efficiency into the launch fibre as well as the return coupling efficiency are reduced (owing to the mode conversion in the back-scattering process it is necessary to restrict both the launching and receiving NA).

A similar compromise between spatial resolution and measurement accuracy exists in the measurement system, as any reduction in the laser pulsewidth decreases the back-scatter signal, while increasing the bandwidth of the receiver will degrade its noise performance. Ideally, the impulse response of the fibre and of the receiver and the probe pulsewidth should all be of similar magnitude.

For the values given in table 1 the spatial resolution at a distance of 100m is approximately 2.5m. Suppose, however, that (at the same distance) a spatial resolution commensurate with the performance of a 75 Megasamples/s A/D converter, namely 1.25m, is required. A probe pulsewidth and a fibre impulse response of 7.7ns or less and a receiver bandwidth of 60MHz or more are needed to achieve it. It follows that, in each part of the system, parameter changes must be made, each of which results in a reduction to the signal-to-noise ratio. Thus it is estimated that the averaging time must be increased from 1 to ~ 16s to preserve the signal-to-noise ratio after averaging.

Dispersion in the launch fibre can be reduced to negligible proportions by use of a graded-index fibre, at the expense of additional launching and coupling losses.

Range. It is clear from the foregoing that the performance of the sensor is a complicated function of the spatial resolution, the temperature measurement accuracy, and the maximum sensing distance which are required. In figure 5, the signal-to-noise ratio expected after averaging 1000 waveforms has been plotted as a function of the maximum sensing distance (solid lines). Each line is calculated for a different value of the spatial resolution δx required at the most remote sensing point. In each case, it is assumed that the sensor system is fully optimised for the desired operating condition and the consequences for the performance of the various components (e.g. coupling losses or amplifier equivalent input noise) have been included in the calculation.

For comparison, the broken lines indicate the signal-to-noise ratio needed to achieve given levels of accuracy in the temperature determination. For example, if a spatial resolution of <3m at a distance of >100m is needed, together with a temperature measurement accuracy of <0.3°C, then the sensor can be designed to operate anywhere within the shaded area on the figure. Similarly, it is possible to sense temperature to within 1°C at 100m with a spatial resolution of 1m. However, should a precision of 0.1°C be required at the same distance and for the same spatial resolution, then it would be necessary to perform more averaging, to use a higher power laser, or to improve the noise performance of the receiver.

Processing the back-scatter waveform

The temperature sensitivity of the back-scatter signal

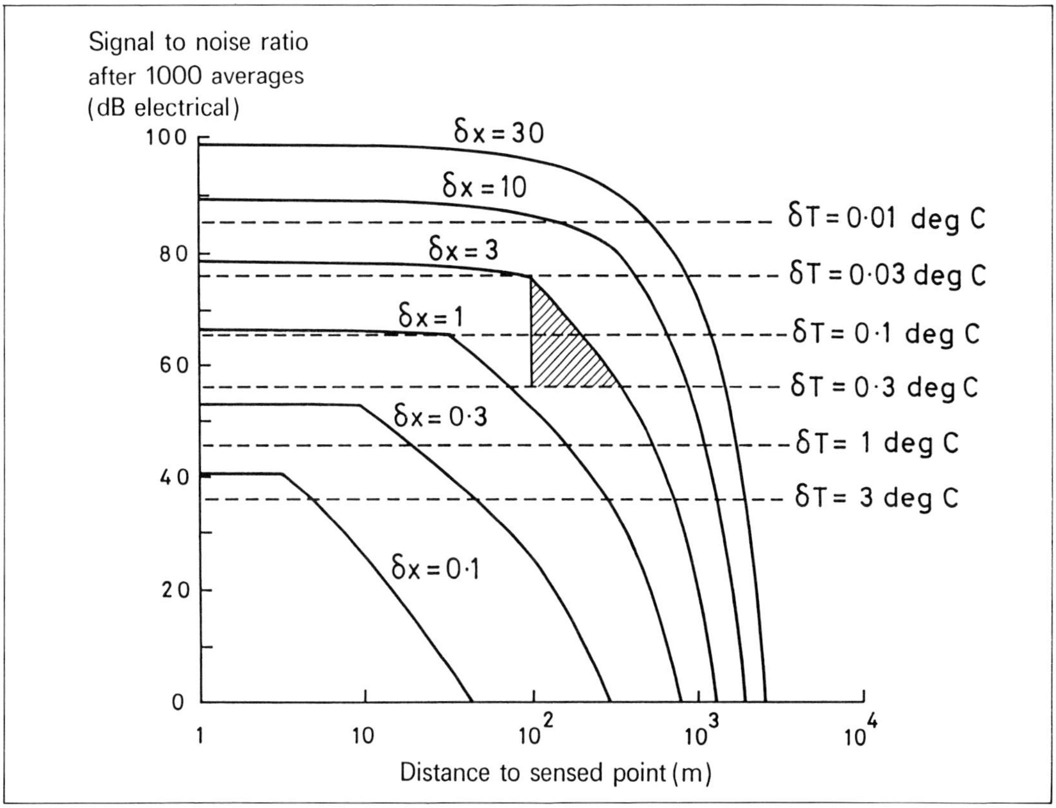

Fig. 5 Signal-to-noise ratio of the amplified back-scatter signal as a function of position. Solid lines are calculated for various values of the spatial resolution δx. Broken lines show the signal-to-noise ratio required in order to obtain the temperature accuracy δT indicated.

is, to first order, that of the scattering loss. In order to extract a valid temperature distribution from the measured back-scatter waveform, corrections must be made for several other effects which also have a bearing on the shape of the signal. First, the temperature is determined from a power ratio and it is, therefore, important to acquire the baseline accurately. This may be achieved by measuring the signal after the back-scatter has ceased. Similarly, in order to compensate for long-term drifts (e.g. of laser power) the entire waveform must be scaled to a power level external to the sensing fibre for example, to the back-scatter level in the solid-core fibre.

Core diameter variations produce fluctuations in η in proportion, for step-index fibres, to the change in core area.[9,16,22] In a fibre sensor such diameter variations will appear as spurious temperature non-uniformities. They can be corrected for, however, if the fibre is calibrated at a uniform temperature. Although fibre diameter control is now well understood,[31] the drawing of hollow tubes to a constant diameter is much more difficult and it may be that this calibration step cannot be avoided. Fortunately, the tubes used in the present study appeared (from their back-scatter traces) to have reasonably uniform bores.

Although temperature variations affect the received power $p(t)$ mainly via the coefficient α_S, the change in temperature also affects the overall loss of the fibre, which appears in the integrand in equation (1). Variations in the temperature distribution, integrated along the fibre, will result in a change in the loss to the sensing point. An error in the temperature reading is therefore produced for distant points. If the variation in fibre attenuation is due solely to a change in Rayleigh scattering, then a numerical integration of the temperature distribution up to the point of interest can be used to derive a correction factor. However, other contributions to loss, such as absorption are also likely to vary (perhaps unpredictably) with temperature, which would render the correction procedure inadequate. If back-scatter waveforms measured from each end of the fibre are available, then longitudinal variations of η and α can be separated unambiguously using a technique first proposed by Di Vita and Rossi.[8]

Finally, the group velocity in the sensing fibre is affected to a small extent by temperature changes and the locations corresponding to the time-samples will shift. This effect can be corrected completely adequately

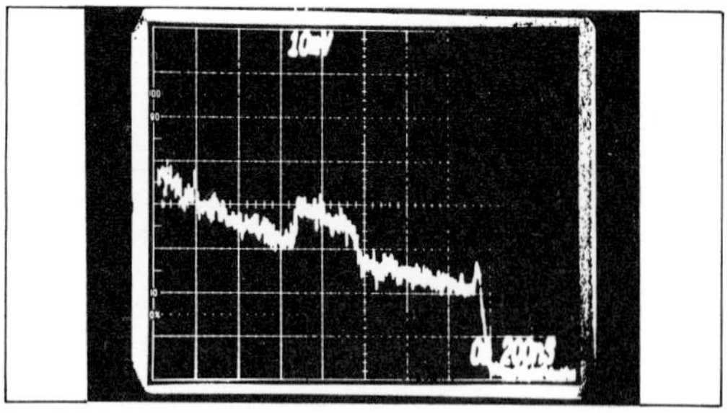

Fig. 6 Real-time back-scatter waveform measured in the distributed temperature sensor. Time scale is 200ns/div. The central section of the fibre is warmer by 33°C than the adjacent sections.

by computing the line integral of the temperature distribution between the launching end and the point of interest.

Networking of distributed sensors

The proposed sensors are intrinsically time-division multiplexed. In addition, the terminal equipment can be time-shared between a large number of fibre sensors using readily available coupling components. Several different configurations may be envisaged in which only one receiver is used. Multiple sources, independently triggered from the computer, are most appropriate as they ensure that no overlap between the various signals occurs. Alternatively, a single source could be used, in combination with power dividers and delay lines, at the expense of increased complexity and reduced signal-to-noise ratio.

The advantage of these multiplexing schemes is that the most expensive parts of the sensor system, namely the data acquisition and processing equipment, is shared between sensors and this reduces the cost per sensing point of the system.

Results and discussion

Back-scatter waveforms in liquid-core fibres

Back-scatter measurements were made on liquid-core fibres fabricated as described in the section 'Fabrication of liquid-core fibres' using the experimental arrangement of figure 3.

Figure 6 shows an oscilloscope trace of the back-scatter signal returning from the remote end of a 170m length of liquid-core fibre. The increase in signal which is clearly apparent betweeen 600 and 1000ns from the left-hand side is due to the corresponding fibre section being heated by ~33°C above that of the remaining fibre portions. The waveform shown was recorded without any signal averaging. Nevertheless, and in spite of the unusually high spatial resolution for OTDR, the signal-to-noise ratio is more than adequate to reveal the localised temperature increase, owing to the large back-scatter factor of liquid-core fibres. Clearly, with only a moderate amount of signal averaging, very small departures from temperature uniformity can be detected.

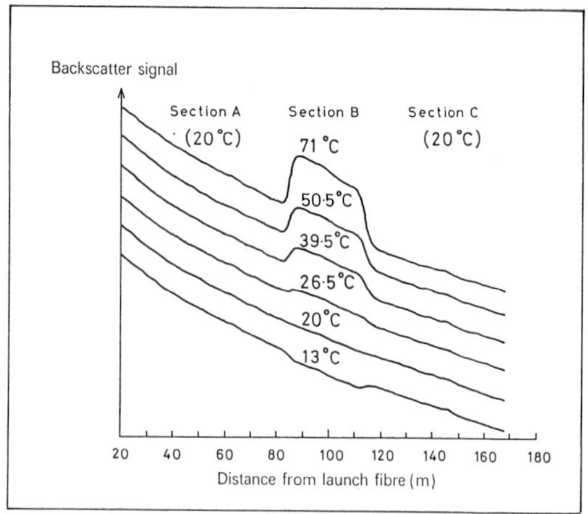

Fig. 7 Averaged back-scatter waveforms measured in the liquid-core fibre with the temperature distributions indicated.

A range of similar waveforms are shown in figure 7. In each case, the central fibre section, section B, was at the temperature indicated, whereas the remaining sections, A and C, were held at room temperature (20° C). It is clear from the figure that the back-scatter signal follows the local temperature variation, including when the temperature of section B falls below that of sections A and C.

In figure 8, the effect of a localised temperature increase on the back-scatter signal has been emphasised by dividing the waveform by one measured while the fibre had a uniform temperature of 20° C. This is a simple means of correcting the waveform for the exponential decay associated with fibre loss and for the effect of diameter changes. It also gives an indication of the magnitude of the noise remaining after averaging and of any long-term drift. The waveform of figure 8 is the result of averaging 2500 measurements;

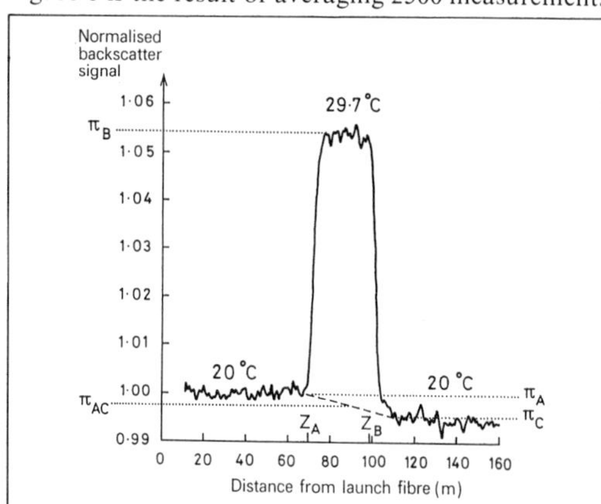

Fig. 8 Back-scatter signal measured in the liquid-core fibre whose three zones were held at the temperatures indicated. The waveform has been normalised to a back-scatter signal measured on the same fibre at a uniform temperature of 20° C.

it may be seen that the rms noise is less than 0.1% and that the normalised signal in section A is very close to 1, which indicates that the long-term drift is small.

Two effects are now apparent, namely, the increase in back-scatter from section B (which in this case is at a higher temperature than the rest of the fibre) and second, the accompanying increase in scattering loss which results in a reduced signal level in section C. The latter is one of the effects which would normally be corrected in a fully engineered system by the separation of the true local attenuation and the back-scatter factor.

π_A, π_B and π_C are defined as the mean back-scatter powers in regions A, B and C, respectively, measured (as in Fig. 8) relative to the signal level found at the same point in the fibre at a uniform temperature of 20°C. These quantities have been indicated in the figure, as has the geometric mean π_{AC} of π_A and π_C.

For the simplified temperature distribution used in these experiments, the sensitivity of the device may be deduced from measurements π_B/π_{AC} and the change in loss as a function of temperature is obtained from π_C/π_A.

Sensitivity measurements in fibres having low-loss claddings

The temperature sensitivity of a liquid-core fibre produced from a low-loss silica capillary tubing and filled with hexacholorobuta-1,3-diene is show in figure 9. The results are given in the form of the power ratio π_B/π_{AC} as a function of the temperature difference ΔT. The relationship is linear and has a slope of 23.3×10^{-3} dB/°C over the range $\Delta T=-4$ to 88°C. The device is not restricted, in theory, to this range and the limits to the operating temperature are discussed in the section 'Limits to the temperature range'.

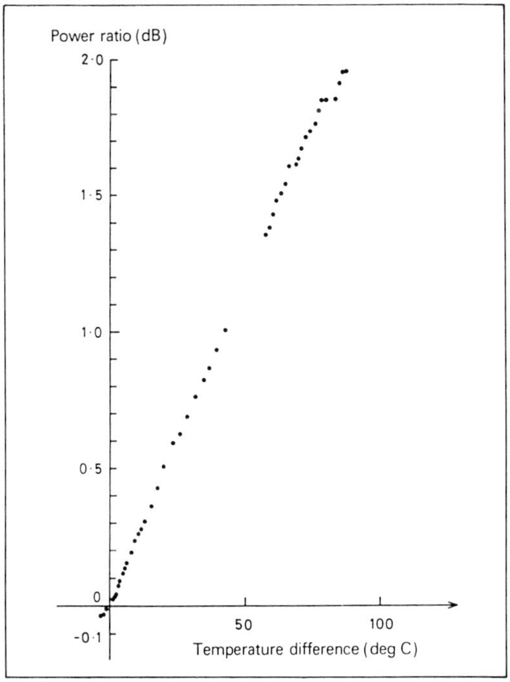

Fig. 9 Response of the sensor as a function of the departure of the temperature from 20°C.

The temperature sensitivity of Rayleigh scattering in hexachlorobuta-1,3-diene may be estimated by subtracting the (calculated) contribution of the longitudinal index changes from the measured sensitivity of the back-scatter signal. The result is 21.8×10^{-3} dB/°C. An independent estimate of the temperature coefficient of Rayleigh scattering is obtained from the change in loss of section B, π_C/π_A. For a direct comparison with the previously obtained temperature coefficient of α_s, π_C/π_A must be scaled to the room-temperature value of α_s and to the length of section B to give a relative temperature-dependent loss $\Lambda\,(T_B-T_0)$ defined by:

$$\Lambda\,(T_B - T_0) = 10\log_{10}\left[1+\frac{\pi_C\,(T_B-T_0)}{2\pi_A\,(T_B-T_0)\,\alpha_s\,(T_0)\,(Z_B-Z_A)}\right] \quad (9)$$

where T_B is the temperature of interest, T_0 is room temperature, and Z_B-Z_A is the length of section B. In

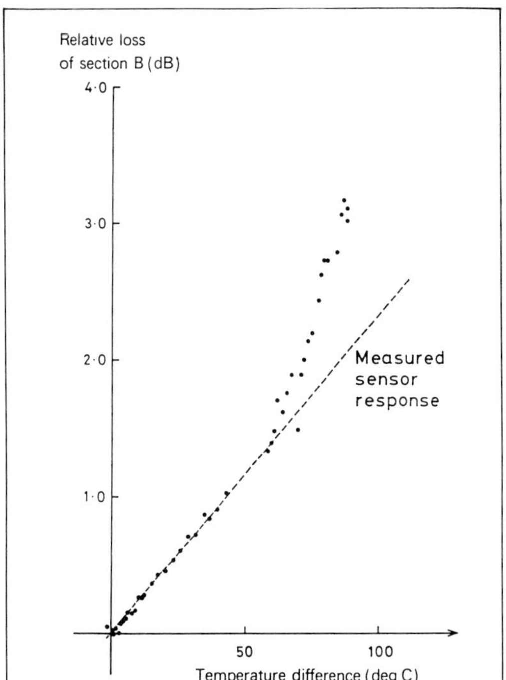

Fig. 10 Increase in loss of heated section, relative to its loss at 20°C, as a function of the imposed temperature change (dots). The broken line is a least squares fit to the data of Fig. 9.

figure 10, the quantity Λ is plotted (dots) as a function of the temperature difference $\Delta T = T_B - T_0$ (the ratio π_C/π_A is expressed in decibels). For comparison, the solid line is a least squares fit to the sensitivity data of figure 9. In the region $\Delta T = 0$–$60°C$, the excess loss closely resembles the temperature sensitivity curve. This demonstrates that the sensitivity of the device is largely due to the contribution of Rayleigh scattering. A straight-line fit to the data of figure 10 gives a slope of 24.2×10^{-3} dB/°C. This result is somewhat higher than that obtained for the Rayleigh scattering (21.8×10^{-3} dB/°C). The difference is possibly due to the temperature sensitivity of absorption loss in the core liquid. However, the scattering loss $\alpha_s(T_0)$ in hexachlorobuta-1,3-diene is not known to sufficient accuracy for the discrepancy to be considered definitely significant. In any case, comparison of the two curves shows that liquid-core temperature sensors can be built which operate very near to the theoretically predicted sensitivity. This ensures that the sensitivity is very reproducible from one sensor to the next as it is determined only by the basic physical properties of the materials used.

As the temperature difference between fibre sections increases, however, the back-scatter return from the hot zone carries an increasing proportion of power in high-order modes, since the NA of the hot zone is decreasing. In addition, the longitudinal refractive index gradient serves slightly to increase the receiving NA for the hot zone which results in a further increase in the proportion of power carried by high-order modes. The two effects combined lead to the much larger gradient of Λ ($T_B - T_0$) at temperature differences above $\Delta T = 60°C$. The temperature sensitivity of the device does not, however, appear to be affected until ΔT reaches 94°C, at least.

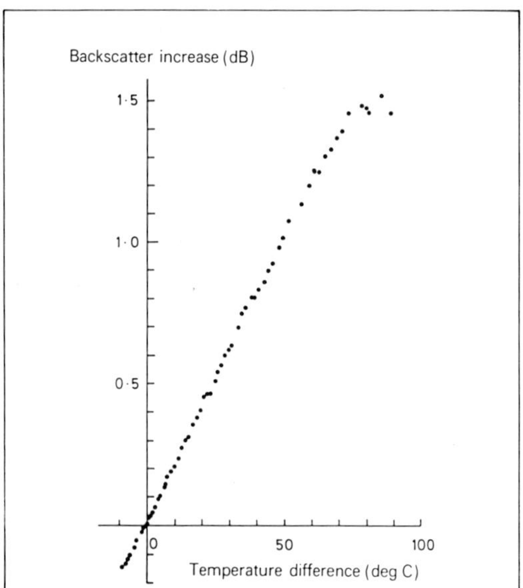

Fig. 11 Response of the temperature sensor when the launch fibre end is cut at an angle of 35°C from normal to its axis.

Effect of differential mode attenuation (DMA) on the performance of the sensor

Although the results presented so far indicate that the sensitivity of the sensor is not significantly affected by the attenuation of high-order modes, the loss of heated sections clearly is. Thus DMA could potentially become a source of error in the distributed temperature sensor and its effects deserve investigation.

The effect of DMA may be studied by enhancing the loss of high-order modes or by using the mode filter to select high-order modes in the back-scattered power.

In the latter approach, our objective may be achieved by cutting the end of the launch fibre at an angle markedly different from perpendicular to the axis, which causes ray bending at the liquid-glass interface. In our experiments, the fibre end was cut at an angle of 35° from the normal. Hence, the power travelling at ~1.4° from the axis in the liquid-core fibre couples to the lowest order mode of the launch fibre. The receiver thus detects power which is still restricted to a maximum numerical aperture, but which is now biased towards a higher order mode content. Figures 11 and 12 show, respectively, the sensitivity curve and

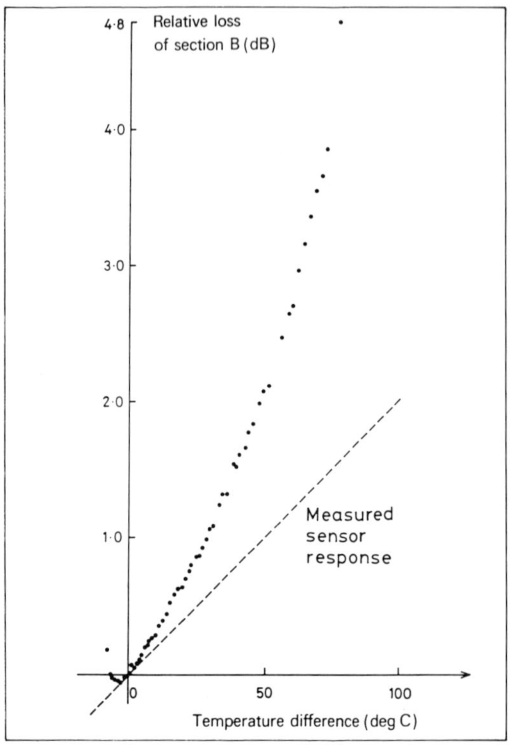

Fig. 12 Increase in loss of the heated section, relative to its loss at 20° C, as a function of the imposed increase in temperature (dots). The launch fibre is cut at an angle of 35° from normal to its axis. Broken line is fitted to the data of Fig. 11.

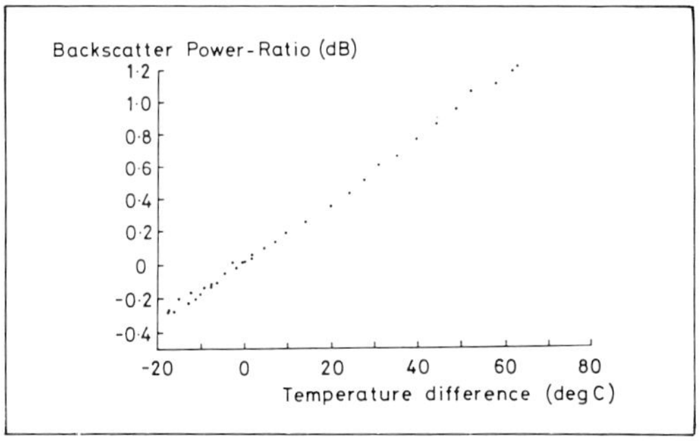

Fig. 13 Sensitivity curve of a liquid-core sensor having a lossy cladding.

Λ as functions of the temperature difference. In both cases, the effect of DMA on the sensor performance is obvious. Thus the sensitivity is now linear up to $\Delta T = 75°$C only, since the NA of section B approaches that of the mode filter at lower temperatures than in the case of an end cut normal to the fibre axis. It is calculated that the NA of the liquid-core falls to the effective NA of the mode filter at $\Delta T = 84°$C, allowing for the ray bending at the liquid-glass interface. The corresponding value is $\Delta T = 94°$C for a launch fibre cut at right angles. This is confirmed experimentally, as, in figure 11, the power ratio ceases to increase for $\Delta T > \sim 80°$C. In addition, the sensitivity of the device has fallen by some 15% to 20.2×10^{-3} dB/°C. The reason for the lower sensitivity is clear from figure 12 which shows that the excess loss of section B is now much more sensitive to temperature for all values of the temperature difference.

The effect of the cladding loss and the associated DMA can be studied by using capillary tubes in which no low-loss cladding layers have been deposited. This type of fibre was used in the preliminary work.[17] The sensitivity curve for such a fibre is shown in figure 13. Again the curve is linear (in decibels) over the region −15 to +60°C. Its slope, however, is reduced to 18×10^{-3} dB/°C in contrast to the 23.3×10^{-3} dB/°C measured on fibres with low-loss claddings. In addition, it is found that the back-scatter waveform is distorted by DMA at large temperature differences, as illustrated in figure 14. Trace (*a*) in the figure was obtained with $\Delta T = 27°$C. It is similar in appearance to waveforms obtained on fibres with low-loss claddings. For the second trace, trace (*b*), however, the temperature difference is 62°C and the back-scatter waveform is clearly distorted. At the beginning of section B, a region of rapid decay now exists which corresponds to the rapid removal of the high-order mode contribution to the back-scatter signal from this section. In addition, the apparent loss is reduced in the fibre section immediately following the heated part. DMA can, therefore, produce serious errors in the temperature reading.

In fibres with deposited low-loss claddings, errors

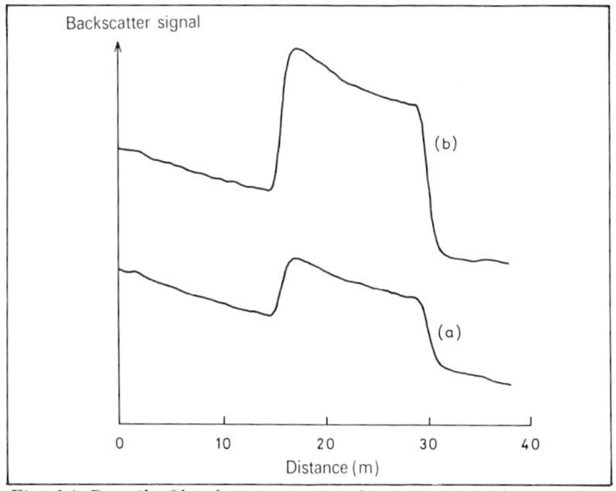

Fig. 14 Detail of back-scatter waveforms measured in a liquid-core fibre having a lossy cladding. For trace (a), the temperature difference between the sections is 27°C and is 62°C for trace (b).

resulting from DMA occur only within a few degrees of a 'cut-off temperature' at which the NA of the hottest fibre section falls to that of the launch fibre. In this case, DMA results from the received signal containing modes almost at cut-off. For example, figure 15 shows a series of back-scatter waveforms measured in a fibre with a low-loss cladding at various temperatures, all near the 'cut-off temperature'. In figure 15(a), the NA of the launch fibre is 0.3 and it may be seen that the shape of the waveforms becomes distorted for $\theta > 114°$C and that the sensor loses all sensitivity when θ reaches 131°C. In figure 15(b), however, the waveforms were measured with a launch fibre of NA = 0.2 and it is found that the temperature at which distortion to the back-scatter occurs is increased to ~140°C. The cut-off temperature may be calculated from the refractive index data for hexachlorobuta-1, 3-diene and for fused silica, and the results are given in table 3 for two values of an operating wavelength and of receiving numerical aperture. The values calculated for λ = 900nm agree well with the observed behaviour.

From the foregoing, DMA is clearly a potential source of inaccuracy in the temperature measurement. If, however, the CVD process is used to reduce the cladding loss, then the effect of DMA is largely eliminated over the entire operating range of the

Table 3. Cut-off temperatures at which the numerical aperture of the liquid-core fibre falls to that of the launch fibre

Receiving NA \ Operating wavelength	589nm	900nm
0.2	176°C	144°C
0.3	144°C	112°C

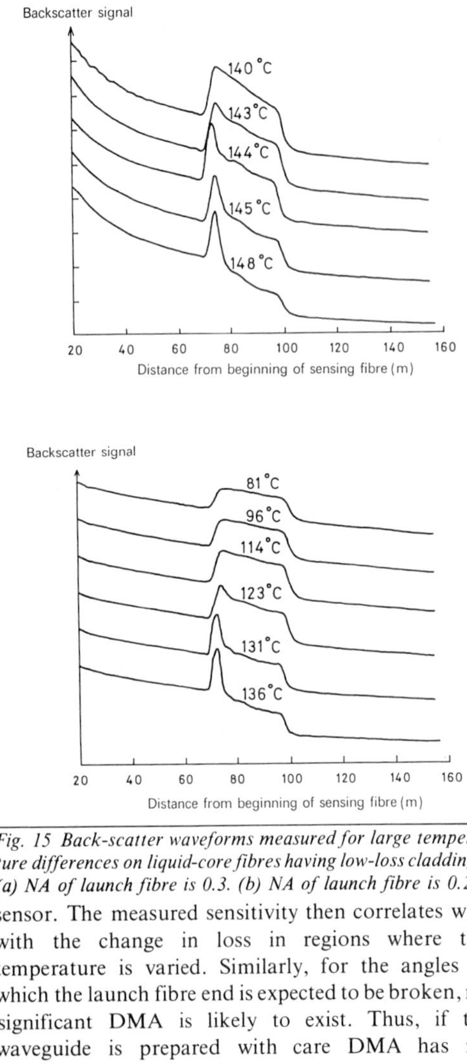

Fig. 15 Back-scatter waveforms measured for large temperature differences on liquid-core fibres having low-loss claddings. (a) NA of launch fibre is 0.3. (b) NA of launch fibre is 0.2.

sensor. The measured sensitivity then correlates well with the change in loss in regions where the temperature is varied. Similarly, for the angles at which the launch fibre end is expected to be broken, no significant DMA is likely to exist. Thus, if the waveguide is prepared with care DMA has no significant effect on the operation of the sensor over a wide temperature range.

Effect of cladding modes on the back-scatter waveform

Initial measurements in this study were made on fibres having a silicone primary coating. This type of structure naturally guides a large number of cladding modes which are able to reach the receiver and produce a spurious contribution to the back-scatter via a variety of mechanisms such as imperfect mode filtering, scattering in the core, or mode conversion at diameter changes or bends. As a result, anomalous effects were observed, such as an increase in the back-scatter signal at a variable-temperature section, regardless of whether the latter was heated or cooled. The experimental arrangement of figure 3, in which the launch fibre is inserted inside the core of the sensing fibre to form a joint, automatically selects modes bound to the core as power in cladding modes travelling through the core always falls outside the NA of the launch fibre. This mode-filtering arrangement, in combination with an absorbing primary coating (such as polyimide) completely eliminates the problem of cladding modes which can otherwise severely affect the shape of the back-scatter waveform.[22] Note that the scattering process itself sets a lower limit on the interchange of power between the modes of the core and cladding. Fortunately, this presents no problem for the fibre lengths and scattering loss envisaged here.

Limits to the temperature range

The ultimate limitations to the temperature range are the freezing and boiling points of the core liquid. For hexachlorobuta-1,3-diene, the resulting temperature range would be −19 to 212° C, i.e. a substantially wider range than is available from other suitable low-loss liquids. In practice, however, the range is limited by other effects.

The maximum operating temperature is that at which the NA of the liquid-core fibre falls to that of the launch fibre. This effect is discussed in the section 'Effect of differential mode attenuation (DMA) on the performance of the sensor', and table 3 lists the maximum temperatures for different NAs and operating wavelengths. To some extent, the temperature range can be modified by a suitable choice of operating wavelength, receiving NA, cladding index and ultimately of core material. Note that, at the maximum temperature range, the local attenuation of the fibre increases abruptly with increasing temperature. This effect can be used as an 'alarm' to indicate that, at a particular point along the fibre, the temperature has reached a dangerous level. The ability to vary the alarm temperature is thus of considerable advantage.

The operating temperature is further limited by damage, which occurs if the fibre is heated to about 160° C. The attenuation of the overheated section is found to suffer from an irreversible increase in absorption loss. Damage is also caused by exposure to ultra-violet light—an occurrence which is fortunately relatively easily prevented.

At low temperatures, the range is, surprisingly, limited by the fibre loss increasing with decreasing temperature. Although this mechanism is not at present understood, it appears to be associated with the heating damage already mentioned, since the temperature at which the excess loss becomes noticeable is increased after the fibre has been exposed to high temperatures.

In the present study, the range 5 to 110° C was covered. With further attention to eliminating DMA and to damage, a significant improvement in range is anticipated. An attractive possibility is to use a more robust liquid such as C_2Cl_4,[32] which is liquid over the range −22 to 121°C.

Concluding remarks

Optical fibre sensors of a new type have been proposed; they are able to monitor remotely the spatial distribution of the physical parameter of interest. The sensors use OTDR to detect externally induced changes in the back-scatter signature of specially designed optical fibres.

A design study and performance calculations have been presented for a distributed temperature sensor which uses a liquid-core fibre as the sensitive medium. The analysis shows that, for presently available components, the sensor is capable of a resolution of 1m over fibre lengths of more than 100m, with a temperature accuracy of 1°C.

Experimentally, fibres consisting of hexachlorobuta-1, 3-diene cores and silica cladding were found to have a sensitivity of 23.3×10^{-3} dB/°C (0.54% $°C^{-1}$) over the temperature range 5–110°C. The sensitivity measured results primarily from the increase in Rayleigh scattering coefficient with increasing temperature and this was confirmed by measurements of the temperature-induced excess loss.

The effect of DMA on sensor performance was investigated and it was found to be negligible provided that the fibres have low-loss claddings and that a suitable mode filter is placed in front of the receiver. Similarly, any detrimental effect of cladding modes can be eliminated very simply by use of a lossy primary coating and of a mode filter.

The device which has been described has a sufficient sensitivity and range to be useful in many situations in which the temperature must be monitored at many points simultaneously. Liquid-core fibres suffer from several drawbacks such as their susceptibility to damage and impurity ingression, which can be overcome by careful engineering. Owing to the simple construction of liquid-core fibres, the performance of the devices is expected to be extremely reproducible. Finally, it is hoped that, in future, solid core fibres with a usable sensitivity may become available.

Acknowledgements

The author is indebted to R.J. Mansfield for producing the capillary tubes used in this study, to E.J. Tarbox, S.R. Norman and M.P. Gold for many useful discussions, to D.N. Payne for advice and encouragement during the work, and to W.A. Gambling for his guidance.

This paper was financially supported by the UK Science and Engineering Research Council.

References

1. T.G. Giallorenzi, J.A. Bucaro, A. Dandridge, G.H. Sigel, J.H. Cole, S.C. Rashleigh and R.G. Priest, 'Optical fiber sensor technology' *IEEE J. Quantum Electron,* Vol. QE–18, 1982, pp. 626–665.
2. A.L. Harmer, 'Principles of optical fiber sensors and instrumentation,' *Meas. Contr.* Vol. 15, 1982. pp. 143–151.
3. M. Gottlieb and G.B. Brandt, 'Fiber-optic temperature sensor based on internally-generated thermal radiation', *Appl. Opt.,* Vol. 20, 1981, pp. 3408–3417.
4. —— 'Temperature sensing in optical fibers using cladding and jacket loss effects,' *Appl. Opt.,* Vol. 20, 1981, pp. 3867–3873.
5. J.P. Dakin and D.A. Kahn, 'A novel fiber-optic temperature probe,' *Opt. Quant Electron.,* Vol. 9, 1977, pp. 540–544.
6. M.K. Barnoski, and S.M. Jensen, 'Fiber-waveguides: A novel technique for investigating attenuation characteristics,' *Appl. Opt.,* Vol. 15, 1976, pp. 2112–2115.
7. S. D. Personick, 'Photon probe—An optical time-domain reflectometer,' *Bell Syst. Tech. J.,* Vol. 56, 1977, pp. 355–367.
8. P. Di Vita and U. Rossi, 'The backscattering technique: Its field of applicability in fiber diagnosis and attenuation measurements,' *Opt. Quant. Electron.,* Vol. 11, 1980, pp. 17–22.
9. A.J. Conduit, D.N. Payne, and A.H. Hartog, 'Optical-fiber backscatter-loss signature: Identification of features and correlation with known defects using the two-channel technique,' *Proc. 6th European Conf. on Optical Communication*, York, UK, 1980, pp. 152–155.
10. M. Eriksrud and A.R. Mickelson, 'Experimental investigation of variation of backscatter power with numerical aperture in multi-mode optical fibers', *Electron. Lett.*, Vol. 18, 1982, pp. 130–132.
11. N. Shibata, M. Tateda, S. Seikai and N. Uchida, 'Measurements of waveguide structure fluctuations in a multimode optical fiber by backscattering technique,' *IEEE, J. Quantum Electron.*, Vol. QE–17, 1981, pp. 39–43.
12. M.P. Gold, and A.H. Hartog, 'Determination of structural parameter variations in single-mode optical fibres by time-domain reflectometry,' *Electron. Lett.*, Vol. 18, 1982, pp. 489–490.
13 A.J. Rogers, 'Polarization optical time-domain reflectometry,' *Electron. Lett.*, Vol. 16, 1980, pp. 489–490.
14. A.H. Hartog, D.N. Payne and A.J. Conduit 'Polarization optical time-domain reflectometry: Experimental results and application to loss and birefringence measurements in single-mode optical fibres,' *Proc. 6th European Conf. on Optical Communication*,York, UK (post-deadline paper), 1980.
15. E.G. Neumann, 'Theory of the backscattering method for testing optical fiber cables,' *Arch. Elek. Übertragung.*, Vol. 34, 1980, pp. 157–160.
16. A.R. Mickelson and M. Eriksrud, 'Theory of the backscattering process in multimode optical fibers,' Appl. Opt., Vol. 21, 1982, pp. 1898–1909.
17. A.H. Hartog, and D.N. Payne, 'Remote measurement of temperature distribution using an optical fiber,' *Proc.*

8th European Conf. on Optical Communication, Cannes, France, 1982, pp. 215–220.
18. K. Kyuma, S. Tai, T. Sawada and M. Nunoshita, 'Fiber-optic instrument for temperature measurement,' *IEEE J. Quantum Electron.*, Vol. QE–18, 1982, pp. 676–679.
19. I.L. Fabelinskii, *Molecular Scattering of Light*, New York: Plenum, 1968.
20. D.N. Payne, and W.A. Gambling, 'New low-loss liquid-core fiber waveguide,' *Electron. Lett.*, Vol. 8, 1972, pp. 374–376, see also *US Patent 3 894 788*.
21. ——, 'The preparation of multimode glass and liquid-core optical fibers,' *Opto-Electron*, Vol. 5, 1973, pp. 297–307.
22. A.J. Conduit, D.N. Payne, A.H. Hartog, and M.P. Gold, 'Optical fiber diameter variations and their effect on backscatter loss measurements,' *Electron. Lett.*, Vol. 17, 1981, pp. 307–308.
23. D.A. Pinnow, T.C. Rich, F.W. Ostermayer, and M. DiDomenico, 'Fundamental optical attenuation limits in the liquid and glassy state with application to fiber optical waveguide materials,' *Appl. Phys. Lett.*, Vol. 22, 1973, pp. 527–529.
24. S.R. Norman, 1974, private communication.
25. R.D. Jeffery and J.L. Hullett, 'N-point pro-cessing of optical fiber backscatter signals,' *Electron. Lett.*, Vol. 16, 1980, pp. 822–823.
26. H. Melchior, M.B. Fisher, and F.R. Arams, 'Photodetectors for optical communications systems,' *Proc. IEEE*, Vol. 58, 1970, pp. 1466–1486.
27. A.J. Conduit, J.L. Hullett, A.H. Hartog and D.N. Payne, 'An optimised technique for backscatter attenuation measurements in optical fibers,' *Opt. Quant. Electron.*, Vol. 12, 1980, pp. 169–178.
28. J.L. Hullett and R.D. Jeffery, 'Long-range optical fiber back-scatter loss signatures using two-point processing,' *Opt. Quant. Electron.*, Vol. 14, 1982, pp. 41–49.
29. D. Gloge, 'Weakly-guiding waveguides,' *Appl. Opt.*, Vol. 10, 1971, pp. 2251–2258.
30 W.A. Gambling, D.N. Payne and H. Matsumura, 'Pulse dispersion for single-mode operation of multimode cladded optical fibers,' *Electron Lett.*, Vol. 10, 1974, pp. 148–149.
31. M.R. Hadley, D.N. Payne and R.J. Mansfield, 'Identification of sources of diameter fluctuations in smooth optical fibers by analysis of their spatial power spectrum,' *Proc. 6th European Conference on Optical Communication*, York, UK, 1980, pp. 53–56.
32. J. Stone, 'Optical transmission in liquid-core quartz fibers,' *Appl. Phys. Lett.*, Vol. 20, 1972, pp. 239–240.
33. G.J. Ogilvie, R.J. Esdaile and G.P. Kidd, 'Transmission loss of tetrachloroethylene-filled liquid-core-fiber light guide,' *Electron. Lett.*, Vol. 8, 1972, pp. 533–534.

Distributed optical fibre Raman temperature sensor using a semiconductor light source and detector

J.P. Dakin
York Ltd., UK

D.J. Pratt
Plessey Ltd., UK

G.W. Bibby
Central Electricity Generating Board, UK

J.N. Ross
University of Southampton, UK

THE use of Raman back-scattered light to measure the distribution of temperature along a conventional silica-based optical fibre has been recently reported.[3,4] The first experimental results with the technique[3] used a high-power pulsed argon ion laser as the source for optical time-domain measurements of the back-scattered anti-Stokes and Stokes Raman lines, the relatively weak returns being detected using a photo-multiplier with a single-channel boxcar integrator. We now report for the first time results taken using a compact semiconductor laser light source and avalanche photo-diode (APD) receiver followed by a multi-channel boxcar averaging system. In addition, a simpler diffraction grating system has been used to separate the Stokes and anti-Stokes Raman lines from the elastically-scattered Rayleigh line.

The overall result is a more compact system, which could form the basis of a practical laboratory or field instrument for monitoring the temperature distribution in doped-silica multi-mode fibre.

Principle of operation

A commercial optical time-domain reflectometer (OTDR) uses a semiconductor laser to launch a short pulse of light into a fibre and, by means of a directional coupler arrangement, the returning Rayleigh back-scattered light is monitored.

In liquid-filled fibres the back-scattered Rayleigh signal is temperature-dependent and has formed the basis of the first reported distributional temperature sensor.[3] However, liquid-filled fibres have potential problems with long-term durability, possible void formation under rapid cooling conditions, temperature dependence of their numerical aperture (NA) and the inability to distinguish between scattering changes due to contamination and temperature variations. Unfortunately, fibres with a vitreous core exhibit negligible change in the Rayleigh scattering cross-section with temperature and, therefore, cannot usefully be used in the same manner as liquid-filled types.

The Raman OTDR method relies on monitoring the anti-Stokes scattered signal as a function of distance along the fibre, which, unlike the Rayleigh line, does show a significant temperature dependence due to the Maxwell–Boltzmann energy partition between the various vibrational states of the vitreous macromolecule. In particular, if the ratio between the anti-Stokes and Stokes lines is measured, an absolute temperature measurement may be made (subject to the loss corrections discussed below). This ratio, $R(T)$, is given by:

$$R(T) = \left(\frac{\lambda_s}{\lambda_a}\right)^4 \exp\left(-\frac{hc\tilde{\nu}}{kT}\right)$$

where λ_a, λ_s are the measured anti-Stokes and Stokes wavelengths, $\tilde{\nu}$ their wave number separations from the pump wavelength, h is Planck's constant, c is the velocity of light, k is Boltzmann's constant and T is the absolute temperature of the core of the section of fibre from which scattered light is received.

This relationship is independent of the laser power,

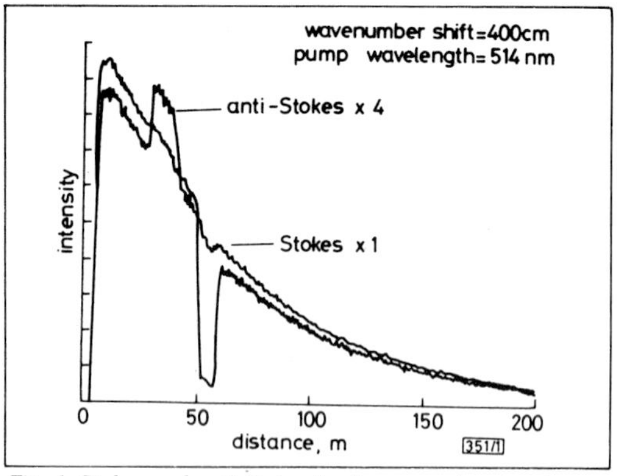

Fig. 1 Stokes and anti-Stokes back-scatter signals using an argon ion pump laser

the launch conditions and the composition of the fibre (provided no fluorescence occurs), and the only correction required is a small allowance for differences in the fibre attenuation at λ_a and λ_s when long lengths of fibre are used.

The earlier system of reference (2) used a high-power argon ion laser source (5W pulses at 514nm) and a photon-counting photo-multiplier system to obtain typical Raman OTDR curves, such as shown in figure 1, taken using 50μm/125μm 0.2NA graded-index germania-doped silica fibre (standard multimode telecommunications grade). The anti-Stokes 'peak' represented a section of fibre at 43°K above room temperature, and the 'dip' represented a section at approximately 90°K (205°K below room temperature). In this early result, the rapid reduction in intensity due to high fibre attenuation in the 500nm region is apparent. Recently, the Raman anti-Stokes/Stokes ratio has been computed and corrected for wavelength-dependent fibre attenuation, to obtain (typically) the curves shown in figure 2. (Note that, in this case, as shown in the diagram, the fibre has been subjected to a different thermal environment than that in figure 1.)

A schematic diagram of the solid-state experimental layout is given in figure 3. The pulsed laser diode used was a 900nm, single-heterostructure device (STC

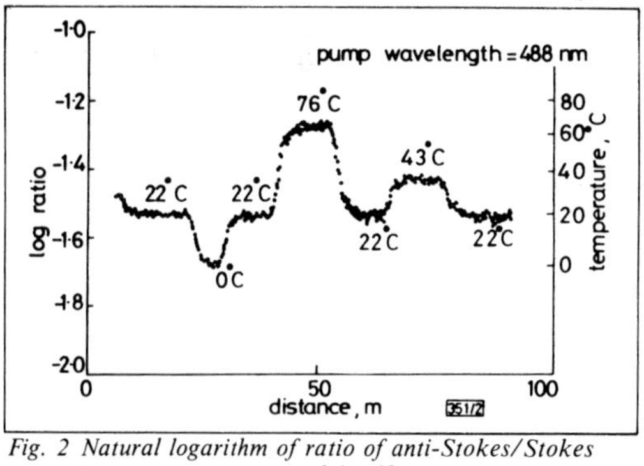

Fig. 2 Natural logarithm of ratio of anti-Stokes/Stokes scattering intensities corrected for fibre attenuation as a function of distance along fibre

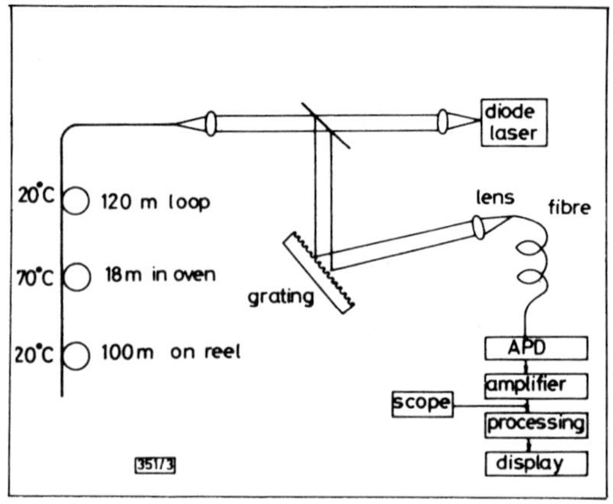

Fig. 3 Schematic layout of solid-state system

Components LA5-02). This produced an output of 3W of optical power in a 15ns pulse at a 10kHz pulse repetition frequency. The output from the diode laser was coupled into the sensor fibre using two back-to-back microscope objectives. Approximately 1W peak power was launched, representing a pulse energy of 15nJ. Back-scattered light returning from the sensor fibre was intercepted by a beamsplitter (nominally 50% reflectivity) and, after filtering by the simple grating spectrometer, coupled into the APD fibre tail using a third microscope objective. By careful positioning of this fibre tail, it was possible to select either the Stokes, anti-Stokes or Rayleigh wavelengths. The discrete-component spectrometer arrangement had a FWHM transmission linewidth of 10nm. The sensor fibre used in the work was Corning 100/140μm fibre, having the advantages of large core (easy launching), high NA (good capture of back-scattered light) and high germania doping of the core (a high Raman scattering cross-section). The APD signal was detected by a low-noise transimpedance amplifier connected to an oscilloscope and processing elec-

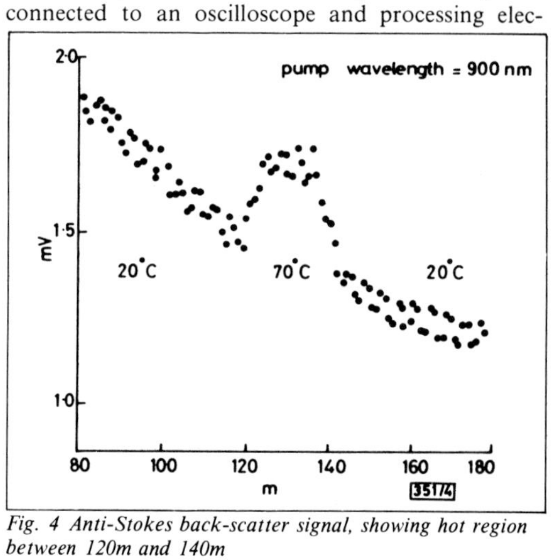

Fig. 4 Anti-Stokes back-scatter signal, showing hot region between 120m and 140m

tronics. The signal/noise ratio of the back-scattered signals was improved using a Le Croy 3500 SA multi-channel averaging system.

A typical anti-Stokes return, obtained using the solid-state system and 400 000 signal averaging cycles (equivalent to 40s integration time), is shown in figure 4. Easily visible is the peak from 120m to 140m, corresponding to the portion of fibre in the oven at 70°C. We estimate the resolution in this first measurement to be better than 10°C and 3m. From the figure it is clear that the remaining noise is dominated by instrumental noise from the signal averager, rather than the Gaussian noise content of the input signal. It is expected that further optimisation of the system could result in a ten-fold improvement in the temperature resolution for an equivalent averaging time. For clarity, figure 4 only shows the heated section of fibre and the regions around it. The anti-Stokes return from the entire fibre was actually observed and averaged, and a temperature variation at any point along the 1km length could have been detected.

Concluding remarks

We have presented the first experimental temperature distribution measurements in a solid-core optical fibre using a solid-state source and detector. This arrangement could form the basis of a more practical system than our earlier reported system using an argon ion laser source.

Acknowledgements

The authors wish to thank the Plessey Company and the UK CEGB for permission to publish this paper. The Plessey work has been carried out with the support of the Procurement Executive, UK Ministry of Defence.

References

1. Dakin, J.P. 'Temperature measuring arrangement'. UK Patent GB 2 140 554 A, May 1983.
2. Dakin, J.P., Pratt, D.J., Bibby, G.W. and Ross, J.N. 'Distributed anti-Stokes ratio thermometry'. 3rd international conference on optical fibre sensors, post-deadline session, San Diego, USA, February 1985.
3. Hartog, A.H. 'A distributed temperature sensor based on liquid-core optical fibres', *IEEE J. Lightwave Technol.*, 1983, LT1, pp. 498509.

Distributed fibre optic hot spot sensors

S.A. Kingsley
Fiberdyne Optoelectronics, USA

V.D. McGinniss
Battelle Columbus Division, USA

IN recent years, considerable interest has been shown in developing fibre optic sensors for the electrical power industry. These include voltage and current sensors, and temperature/hot-spot sensors. The electrical isolation properties and the 'distributed' nature of an optical fibre, offer unique solutions to measurement problems faced by the electrical power industry.[1-3]

In this paper we describe some aspects of the feasibility work done by Battelle on placing special polymeric coatings on fibres so as to sensitise them to temperature. In this initial investigation there was no requirement for much of a dynamic range at each sensing point, only that at a certain critical temperature, i.e., about 135°C, the losses of the fibre could be made to change drastically, and that several such sensing points could be time-multiplexed and resolved along a short length of fibre.

Other research groups have adopted similar and alternative approaches to distributed temperature sensing.[4-12] Some of these other approaches are discussed in the companion papers to be found elsewhere in this book.[16]

The other temperature-sensing techniques include:

- Rayleigh scattering in liquid-filled fibres.
- Raman scattering (anti-Stokes/Stokes ratio).
- Temperature-dependent absorption.
- Temperature-dependent fluorescence.

One particular criteria was important, and that was that the effect being measured should be reversible, so that the fibre could be used over and over again, and would not need to be replaced after one high temperature excursion. Because of proprietary aspects, certain information about the nature of the polymeric coatings and the systems in which they are employed, will not be disclosed at this point. However, we can say that the 'system' employs optical time domain reflectometry (OTDR) to determine the position of the hot-spot.[13-16]

The original application for this sensor was for hot-spot sensing in the stator-bars of large electrical generators, but obviously, other applications come to mind; such as hot-spot sensing in high-voltage transformers and electrical power cables, both overhead and buried.

The Battelle approach

Figure 1 shows a generalised multimode optical fibre with a part of the cladding removed. In this diagram we show the sensor in a 'quasi-distributed' form but, in practice, the sensor would be a continuously-coated section of fibre which may be interfaced to a conventional fibre for delay-line purposes. The

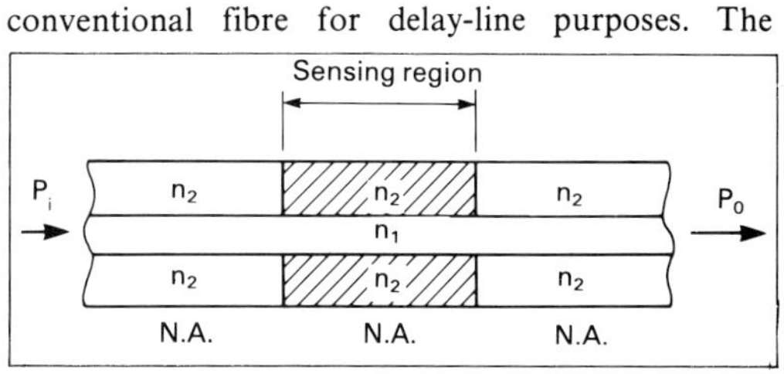

Fig. 1 Quasi-distributed fibre optic sensor

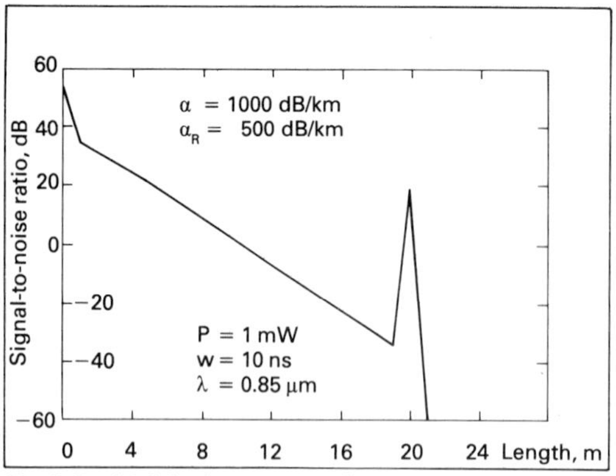

Fig. 2 Rayleigh back-scatter in a lossy sensing fibre

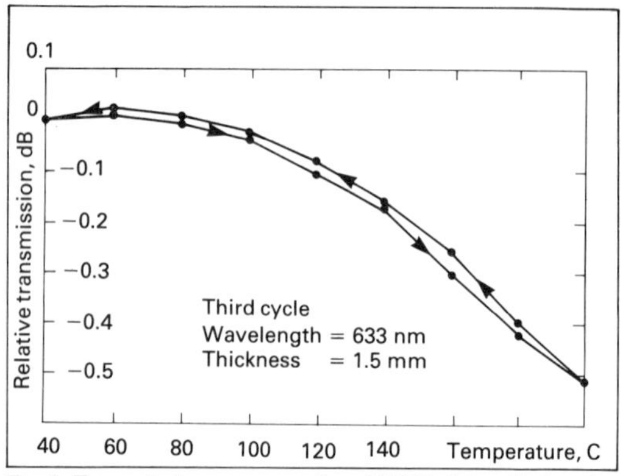

Fig. 3 Transmission response of an EPDM disk – third cycle

refractive index of the new cladding material is n_2'. The normal numerical aperture of the ordinary fibre is NA, while the numerical aperture of the modified fibre is NA′. By making the refractive index of this material dependent on temperature, the numerical aperture (NA) of the fibre can be made to vary with temperature. P_i is the input power and P_o is the output power.

There are basically two modes of operation of this sensor. First, if n_2' increases with temperature, then NA′ will decrease and the power in the high-order modes will be lost. Second, if the attenuation and scattering properties of n_2' change as a function of temperature, then the overall losses of the fibre will be changed.

In general, one or both processes may be operating at the same time. It can be shown, and easily seen, that a given change in n_2 will have a greater effect if the normal numerical aperture is small. This can be understood by considering that a greater percentage of the guided energy flows in the cladding when the index difference is small. There are, of course, certain trade-offs involved in making the numerical aperture very small. One of the major ones is that the fibre then becomes sensitive to any induced mechanical strains.

Figure 2 shows the theoretical Rayleigh back-scatter signal-to-noise ratio characteristics for a lossy sensing fibre. It is to be expected that custom fibres developed for their sensing properties may have very large losses compared with state-of-the-art communications-grade fibre. Even if the optical loss of the resulting sensing fibre should approach 1000db/km (the loss due to Rayleigh back-scatter is assumed to be 500dB/km), reasonable signal-to-noise ratios may be obtained over the short lengths of fibre required in the stator-bar sensor. Figure 2 assumes a 1mW pulse at a wavelength of 0.85μm, with a pulse-width of 10ns. The powerful Fresnel reflections from the ends of the fibre are indicated, assuming a perfect fibre-air interface. The far-end reflection may be exploited by using 'alarm-mode' sensing of this signal. The 'diagnostic-mode' of sensing would look at the Rayleigh back-scatter, and would yield position information of multiple hot-spots.

The signal-to-noise ratios (SNR) displayed in the graph assume a front-end bandwidth compatible with resolving the 10ns wide pulses, i.e. a bandwidth of approximately 100MHz. No signal integration is assumed. In practice, by integrating thousands of pulses over several seconds, SNR improvement ratios of about 45dB are available, i.e. a system conforming to the characteristic given in figure 2 is operable over the entire 20 metre length of the lossy fibre. Note that by increasing the optical power (P) above 1mW and/or by increasing the pulse width (w) above 10ns (at the expense of spatial resolution), signal-to-noise ratios can be improved further.

We investigated many different coating materials. Two of the most promising types of polymer are described below; one being more suitable for extrinsic sensing, while the other is more suitable for intrinsic sensing.

Ethylene Propylene Diene Monomer

A series of experimental tests were conducted on sheets of Ethylene Propylene Diene Monomer (EPDM) which is cured at 160°C for 30 minutes under pressure. Figure 3 shows the transmission characteristics of a thin sheet of EPDM, 1.5mm in thickness, and probed by a Helium-Neon laser beam at 633nm. There is a small amount of hysteresis which is probably due to the fact that the thermocouple was close to, but not in the same location as the beam, for obvious reasons. This graph is for the third cycle of temperature cycling. Similar characteristics were obtained for the first and second temperature cycles. These preliminary tests seem to indicate that over this limited temperature range, the material is perfectly reversible.

Because of the manner in which EPDM is fabricated, it was not found possible to coat an optical fibre with

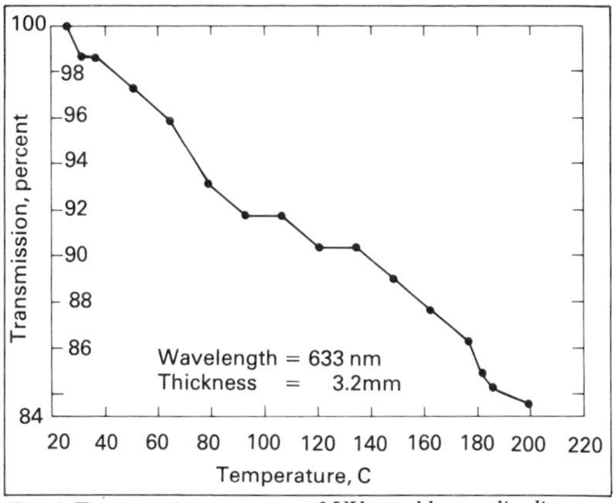

Fig. 4 Transmission response of UV-curable acrylic disc

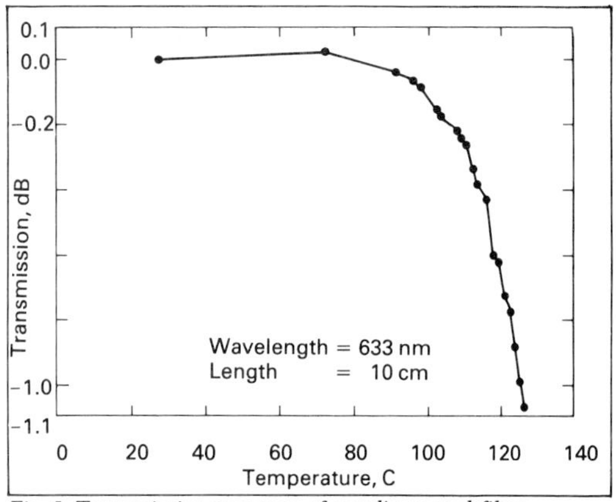

Fig. 5 Transmission response of acrylic-coated fibre

this material, though this is not to say that short lengths of fibre could not be coated by a sophisticated technique.

This material is more amenable for use in extrinsic sensors, where the material is either sandwiched between two lengths of fibre or placed at a fibre tip. A 'quasi-distributed' sensing system could be formed by having many such sensors arranged in tandem.

Ultra-violet – curable acrylic

This acrylic was cured with ultra-violet (UV) light into a ⅛ inch thick disc or sheet. The refractive index of the acrylic before curing was 1.4704 (measured with white light) while its refractive index after curing was 1.4945. At He-Ne wavelength (633nm), the cured acrylic was found to have a refractive index of 1.4937.

A thin disc of this material was prepared for bulk-optic throughput tests, in the same set-up as used to measure the EPDM characteristics. Figure 4 shows the transmission characteristics for rising temperature only. The falling temperature characteristic essentially retraced the rising characteristic except at a few points—which was thought to be due to our inability to simultaneously and accurately record both the throughput and temperature. As with the EPDM measurements, the sample was heated with a hot-air gun and allowed to cool under natural conditions.

A liquid photo-curable, thermally sensitive coating system was applied and cured onto a previously-etched glass optical fibre substrate. Because the refractive index of the acrylic was more than that of silica, which has an index of approximately 1.456 at 633nm, an 'old-fashioned' type of step-index fibre was chosen. Today, there is very little optical fibre on the market made from soda-lime-silicate glasses (borosilicates) which have refractive indices of about 1.5. Twenty years ago, virtually all optical fibres were made from such glasses.

At the time this feasibility project was undertaken, we had considerable difficulty locating such fibre (we have now located a source of low-loss borosilicate (BK) pre-form rod) but were eventually able to purchase a bundle of loose optical fibres (horses-hair), the type used for coherent imaging fibrescopes and endoscopes. This fibre was very fragile, since each of the thousands of 50μm diameter strands that made up the bundle were uncoated, i.e. they did not have any protective polymer coatings to prevent the ingress of water vapour and, hence, the propagation of micro-cracks.

A section of the fibre was etched in hydrofluoric acid to remove the thin glass cladding which had a thickness of about 10μm. The fibre was washed and then carefully inserted in a polymer tube which would be filled with the UV-curable acrylic. Attempts had been made to try to directly coat the fibre with the acrylic by dipping the fibre in the material and gradually pulling the fibre out. Unfortunately, surface tension effects were sufficient to prevent 'wetting' of the glass, and the material beaded. A more viscous acrylic and/or a larger diameter fibre should be coatable by this technique. (We understand that dip-coating or extruding polymer coatings on small-diameter fibres can be very difficult.)

Unfortunately, because of the diameter of the polymer tubing used to contain the fibre and acrylic during the curing process, the final rigidity of the system made the fibre very inflexible, and considerable difficulty was experienced in obtaining a sample completely coated and free from large flaws.

Upon addition of heat (100–130°C), the original light transmission dropped significantly, as shown in figure 5. Upon removal of the heat, the sample regained its original light power transmission value. The characteristic was found to be repeatable over many temperature cycles.

Because of flaws in the fibre, this was apparent from bright scattering centres that showed up very clearly in the darkened laboratory, there is a slight concern that the steep nature of the transmission characteristic may be connected with something occurring at one or more flaws. It was also observed that in some places along the fibre the acrylic had failed

Fig. 6 Reduction of transmission-line costs by distributed fibre optic temperature sensing

to wet the fibre, i.e. there were air voids. Thus, further tests on properly manufactured fibre are needed to remove these causes for concern.

Although the exact refractive index of the fibrescope fibre core is unknown, an estimate for the numerical aperture can be made. As previously stated, the refractive index (n_2) of the acrylic cladding was measured to be equal to 1.4937 at 633nm. Assuming that the refractive index (n_1) of the borosilicate core at 633nm was 1.5, we find that NA′=0.14. Observations of the far-field mode speckle pattern from the fibre would suggest that the NA′ was of this order.

Discussion

Besides the obvious benefit of monitoring stator bars and transformers, distributed hot-spot or temperature sensing is a powerful technique for measuring the temperature profile of high-voltage transmission lines. Because aluminium conductors anneal at elevated temperatures, power utilities normally under-load power lines to ensure that localised hot-spots do not develop. If a method could be found to monitor the distribution of temperature along a conductor, then conductors could be loaded higher, with substantial savings in transmission costs. For instance, figure 6 illustrates that by increasing the loading by 15% (100MW) on a 230kV system, a saving of $3.5 million per year can be obtained.

Presently, thermal mapping of the high-tension transmission system may be done with a thermal imaging camera mounted on a van or in a helicopter. A passive, continuously monitoring system would be very useful indeed as a means of providing data for thermal aging of the conductors.

For the past decade, many fibre optical systems have been installed throughout the world in the ground (shield) conductor at the top of the towers (the shield conductor protects the high tension conductors from lightning strikes), and inside high-voltage conductors or attached to the outside.[3]

At the moment, thermal mapping of the high-tension only for communications and data handling – not for sensing. It should be possible to use another fibre as a distributed temperature sensor, or even the same fibre for both communications and sensing, though the latter approach is likely to cause too many performance trade-offs.

Of all the sensing applications investigated for fibre optics over the past decade, the truly distributed sensing area is one which demonstrates some of the unique aspects of fibre optics, and where the fibre-sensor may not be in direct competition with alternative sensing technology. We believe that distributed hot-spot or temperature sensing will become one of the first applications for DFOSS – where better than in the electrical power industry.

Because the modified or on-fibre sensing technique employed here changes the local numerical aperture of the fibre as some function of environmental temperature, it is to be expected that in a system consisting of a number of such sensors in tandem which are activated simultaneously, crosstalk will occur between such sensors.[5] This multiplicative effect occurs because the effectiveness (modulation efficiency) of a second activated sensor depends on how well-filled the numerical aperture is prior to that sensor, which in turn depends on how a previous activated sensor has modified the guidance properties of the fibre. Of course, if effective mode-mixing occurs between successive activated sensors, then the crosstalk effect can be minimised.

As mentioned at the beginning of this paper, there are alternative techniques for the distributed measurement of temperature which have been reported in recent years. These include: Rayleigh scattering in liquid-filled fibres, Raman scattering (anti-Stokes/Stokes ratio), temperature-dependent absorption and temperature-dependent fluorescence; all in conjunction with OTDR to obtain the range information. Dakin and Pratt have compared the merits of each of these techniques.[10] They have concluded that the Raman method is the more likely technique to find practical application in the near future. One particular advantage of the Raman approach is that conventional communications-grade optical fibre can be used—a custom-made special sensing fibre is not required. On the other hand, a more powerful laser system is required. See the FODAR paper in chapter 1 of this book for further discussion on anti-Stokes/Stokes ratio thermometry.[16]

Concluding remarks

This preliminary work on the feasibility of sensing hot-spots with a distributed fibre-optic sensor indicates that short sensors may be constructed using fairly simple fibre structures. Longer sensors, say greater than 20 metres, might require considerably different system techniques, perhaps using non-linear phenomena in fibres and more powerful lasers. It should be noted that the approach adopted here to hot-spot sensing may be extendible to much greater lengths of fibre if losses in the acrylic can be kept low, below critical temperature, and if a source of low-loss borosilicate or heavily-doped silica pre-form rod can be acquired.

Incidental to the main aim of this project, which was to develop an intrinsic distributed fibre optic hot-spot sensor, we have also established materials that can be used for extrinsic hot-spot sensing and quasi-distributed hot-spot sensing.

An extrinsic sensing material based on a proprietary Battelle material was found to have the desired switching temperature and be reversible. Unfortunately, its method of manufacture, involving high temperatures and pressures, probably eliminates the possibility that this material could be coated onto an optical fibre.

Finally, we developed a UV-curable acrylic which exhibited the desired switching temperature, reversibility and could be coated onto a borosilicate fibre. The exact mechanism responsible for the sharp switching characteristic of the acrylic sensing fibre, it is thought that localised refractive index changes cause increased scatter and changes in the guiding properties of the optical fibre. The fact that the sensor fibre has a low numerical aperture (NA), so that a not-so-insignificant fraction of the guided power is transmitted through the cladding, probably assists in this respect. By changing the chemistry of the polymer and the refractive-index difference between core and cladding, it should be possible to produce a sensing fibre with a more gradual switching characteristic. A gradual characteristic would allow an increased dynamic range at each sensing position, thus producing a fibre that could be considered to be a distributed temperature sensor—not just a distributed hot-spot sensor. This would be particularly important for application in temperature profiling of buried and suspended high-voltage cables.

Acknowledgements

The authors wish to acknowledge that the distributed fibre optic hot-spot sensing work reported here was sponsored by the Electric Power Research Institute (EPRI), under contract No. RP 2308-6, and monitored by Jan Stein of the Plant Electrical Systems and Equipment Division.

References

1. B. Culshaw, *Optical Fibre Sensing and Signal Processing*, Peter Peregrinus Ltd., London. An IEE publication, 1984.
2. *International Journal of Optical Sensors*, Vol. 1, No. 1, January 1986.
3. *Optical Research Development and Applications Workshop* held at CERL, Leatherhead, England, November 1985.
4. P.M. Kopera, J. Melinger and V.J. Tekippe, 'Modified cladding wavelength dependent fiber optic temperature sensor'. *Proc. SPIE, Fiber Optic and Laser Sensors*, Arlington, Virginia, 5–7 April 1983, pp. 82–89.
5. A. Arie, M. Tur and S. Goldsmith, 'Measurement and analysis of a modified cladding optical fibre with various input illuminations'. *Proc. of SPIE's Fibre/Lase '86, Fiber Optic and Laser Sensors IV*, 22–24 September 1986, Vol. 718.
6. J.P. Legendre, 'A simple reflectometer configured to sense temperature', *Proc. SPIE's 29th Annual Technical Symposium, Fiber Optic and Laser Sensors III*, San Diego, California, 18–23 August 1985, pp. 321–324.
7. D.N. Payne and A.H. Hartog, 'Remote sensing of temperature distribution using an optical fiber', *Proc. ECOC 8*, Cannes, 1983, pp. 215–220.
8. E. Theocharous, 'Differential absorption, distributed thermometer', *Proc. 1st International Conference on Optical Fibre Sensors*, London, England, IEE Conference Publication No. 221, 26–28 April 1983, pp. 10–12.
9. J.P. Dakin, D.J. Pratt, G.W. Bibby and J.N. Ross, 'Distributed optical fibre raman temperature sensor using a semiconductor of light source and detector', *Electronics Letters*, Vol. 21, No. 13, 1985, pp. 569–570.
10. J.P. Dakin and D.J. Pratt, 'Fibre-optic distributed temperature measurement—a comparative study of techniques', *Colloquium on Distributed Optical Fibre Sensors*, IEE, Savoy Place, London, England, 12 May 1986.
11. A.H. Hartog and A.P. Leach, 'A practical optical-fibre distributed temperature sensor', *Colloquium on Distributed Optical Fibre Sensors*, IEE, Savoy Place, London, England, 12 May 1986.
12. E. Theocharous, 'Distributed sensors based on differential absorption', *Colloquium on Distributed Optical Fibre Sensors*, IEE, Savoy Place, London, England, 12 May 1986.
13. S.A. Kingsley, 'Fibre-optic sensors: opportunities for distributed measurement', *InTech*, Vol. 32, No. 8, 1984, pp. 44–48.
14. S.A. Kingsley, 'Distributed fibre-optic sensors: an overview', *Proc. of SPIE's 29th Annual Technical Symposium, Fiber Optic and Laser Sensors III*, San Diego California, USA, Vol. 566, 18–23 August 1985, pp. 28–36.
15. S.A. Kingsley, 'Distributed FODAR sensors', Presented at *The 3rd Annual Hostile Environments & High Temperature Measurements Conference*, Cincinnati, Ohio, USA, 25–26 March 1986. Also presented at *IEE Colloquium on Distributed Optical Fibre Sensors, Digest No. 1986/74*, London, England, 12 May 1986.
16. S.A. Kingsley, 'Advances in distributed FODAR (fiber optic detection and ranging)', *Proc. of SPIE's Fiber/Lase '86*, Vol. 718, 22–24 September 1986.

New fibre optic distributed temperature sensor

P. Lecoy, M. Groos and L. Guenadez
Ecôle Centrale de Paris, France

THE use of optical fibres as temperature sensors has been in development for several years. Many of these sensors, are, in fact, optical temperature measurement devices using absorption, scattering, fluorescence, index variations, thermal emission and so on, from a probe which is set at the fibre end, in the medium to be measured. The optical fibre has only a passive function, light guiding. In other sensors, the light propagation is modified by a temperature dependent parameter (phase delay, polarisation); very little temperature shifts can be detected by a very sensitive but sophisticated interferometric arrangement.[1] In all these examples the temperature is measured in a single (local or remote) point. A more specific interest of the optical fibre is to realise a distributed sensor measuring the continuous temperature profile along the fibre.

Review of distributed sensors

In any case, a spatial-resolved temperature profile measurement requires the optical time domain reflectometry (OTDR) technique. Previous works differ by the chosen parameter:

- Rayleigh temperature-dependent scattering.[2]
- Microbending losses caused by the thermal contraction of the jacket.[3]

(Both these methods require special fibres, are difficult to implement and are also sensitive to parameters other than the temperature.)

- Raman spontaneous scattering,[4] where the Stokes/anti-Stokes ratio is only temperature-dependent, is the best method but, despite sophisticated optical components and signal processing, is of limited resolution and sensitivity. It can be used in silica fibres or in plastic fibres,[5] in which the measure is slightly more sensitive but also more disturbed by propagation effects.

Principle of the new temperature-sensor

Our sensor uses the strong temperature dependence of the refractive index difference between the silica core (n_1) and the organic cladding (n_2) in the plastic-cladded-silica (PCS) fibres.[6] This normally unwanted effect disturbs the transmission when the temperature varies. To use it with efficiency, we induce bend-losses under a strong curvature; radiation losses are a sharp decreasing function of the index difference and then of the temperature at this point. As the value of n_1-n_2 increases from 0.027 at −20° C to 0.065 at 60° C (mainly due to n_2 variation, see figure 1) a good sensitivity over a wide range can be predicted.

This principle can be used as a very simple single-point temperature sensor, or as a distributed sensor by means of OTDR. In both cases, the fibre will be wound around a rod with a strictly constant radius of curvature R_o, which is generally not very large, versus the core diameter 2a (see figure 2); the actual radius of curvature of the bent fibre is then $R = R_o + b$, where 2b is the outer diameter of the fibre. This rod can be flexible and can follow any desired way along which the temperature is to be measured. The sensor is

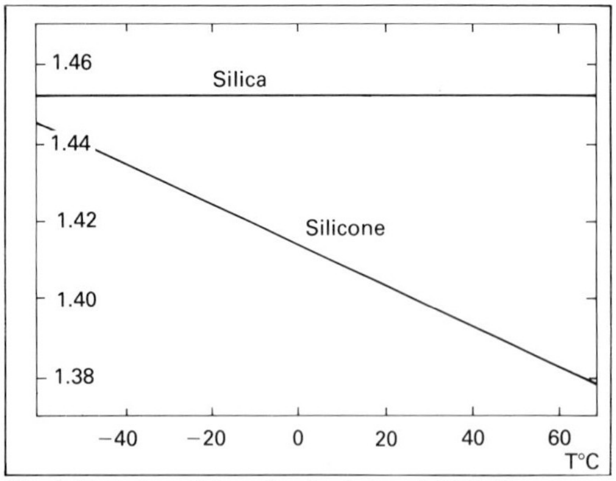

Fig. 1 Variation of the refractive index of PCS fibres versus temperature

calibrated by recording the overall loss or the OTDR curve before winding the fibre, and comparing these results with those of the wound fibre.

Theory

Losses along a curved fibre rapidly decrease when the core-cladding index difference $n_1 - n_2$ and, consequently, the temperature are increased. However, this dependence has several factors and is difficult to predict with accuracy. The three main factors are detailed below.

Spatial filtering Spatial filtering, of the high-order modes which are refracted in the cladding, is one factor. The relative number of modes guided by the curved fibre is given by Gloge:[7]

$$\frac{N_{eff}}{N_\infty} = 1 - \frac{g+2}{2g\Delta}\left[\frac{2a}{R} + \left(\frac{3}{2n_2kR}\right)^{3/2}\right]$$

where g = exponent of the index-profile law, $g = \infty$ for a step-index fibre or g = 2 for a parabolic graded-index fibre $\Delta = \frac{n_1 - n_2}{n_1}$ is the relative index difference; and k is the wave number in the vacuum.

We can write a more practical formula for standard step-index fibres. Assuming the power is equally distributed between all guided modes in the straight fibre, the input loss due to the curvature is:

$$\alpha(\mathrm{dB}) = 10\mathrm{Log}\left[1 - \frac{1}{2\Delta}\cdot\frac{2a}{R}\right] = 10\mathrm{Log}\left[1 - \frac{2a\,n_1^2}{R\cdot NA^2}\right]$$

where $NA = (n_1^2 - n_2^2)^{1/2}$ is the numerical aperture. Defining a critical radius of curvature as

$$R_c = \frac{2a\cdot n_1^2}{NA^2}$$

we have found, experimentally, a rather good accordance (except when R nears R_c) with this formula, when

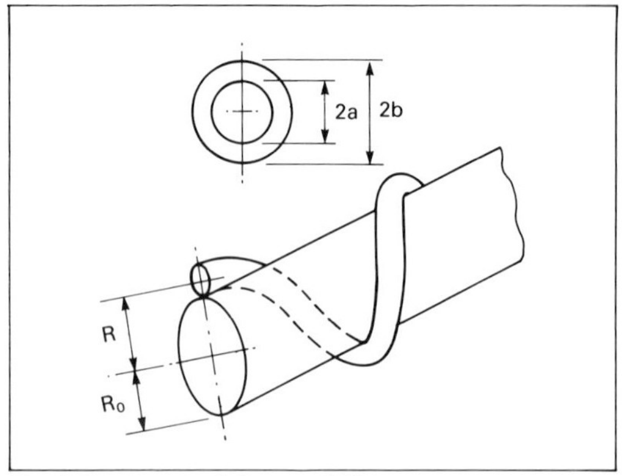

Fig. 2 Parameters of the wound fibre

all the modes are excited (for example by a He-Ne laser beam)—as good for step-index as for graded-index fibres; however, this value is dependent on light injection. Attenuation becomes lower but more stable, when a pig-tailed or collimated diode is used, exciting only the low-order modes. This term will decrease when the temperature (and then NA) is increasing.

Leakage The second dependence factor is leakage of the guided modes along the curved fibre, leading to a loss coefficient per length given by Marcuse for the mode with a constant β:

$$\alpha = K\cdot\exp\left[-\frac{2}{3}kn_2 R\left(\frac{\beta^2 - k^2n_2^2}{k^2n_2^2} - \frac{2a}{R}\right)^{3/2}\right],$$

a value which is negligible except when R nears R_c, at the mode cut-off.

Mode coupling Mode coupling between guided and refracted modes is the third factor in loss dependence. This effect is greater than mode leakage; it gives a non-negligible loss which is dependent on the light injection and on the mechanical stresses (and becomes larger when R decreases). We have found, experimentally, a rather larger dependence with graded-index than with the step-index PCS fibre, which is used in our sensor. Dependence of mode coupling on the temperature is due to thermal constraints, and can increase at extreme temperatures.

Temperature measurement at a single-point

We first used our sensor, in the configuration shown by figure 3, in order to measure the temperature at a single-point. The attenuation of the wound fibre was compared to the attenuation of the same fibre before winding. A calibration of the sensor was made (see figure 4) for different radii of curvature and numbers of turns. It has given the sharing of the attenuation between:

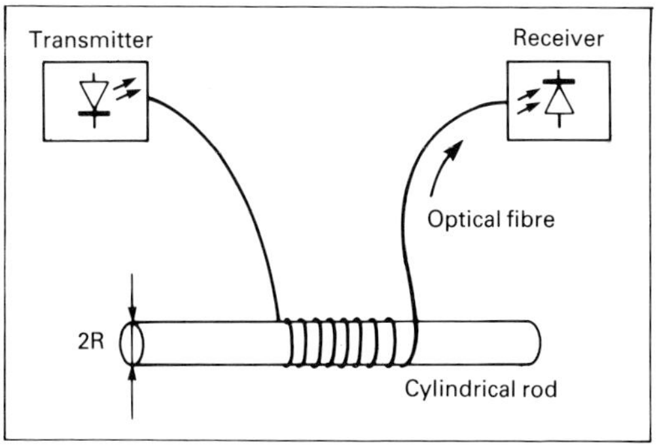

Fig. 3 Wound fibre with constant curvature

- The input attenuation due to the spatial filtering, of about 5dB for R = 5mm and decreasing by 0.1dB/°C, in good accordance with the theory.
- The continuous attenuation along the curved fibre, of about 0.05dB/turn, which is reached after a quite short transition zone and does not depend clearly on the temperature.

Measurements of the loss versus the temperature were performed with a He-Ne laser at 632.8nm (see figure 5) and with an LED transmitter at 850nm (see figure: 6). The first one is more sensitive, but the second is more fair. In any case, the weak difference between them is due more to the light injection than to the wavelength.

Interesting results were found:

- Measurements were made between -20°C and +60°C, which is at least the range of the sensor. The sensitivity decreases at 60°C (as theoretically predicted) but it would be possible to shift up (resp. down) the maximum sensitivity region by using a lower (respectively higher) radius of curvature.
- A sensitivity to a 0.2°C temperature variation was detected.

The basic idea is to measure the continuous attenuation coefficient along the wound fibre using

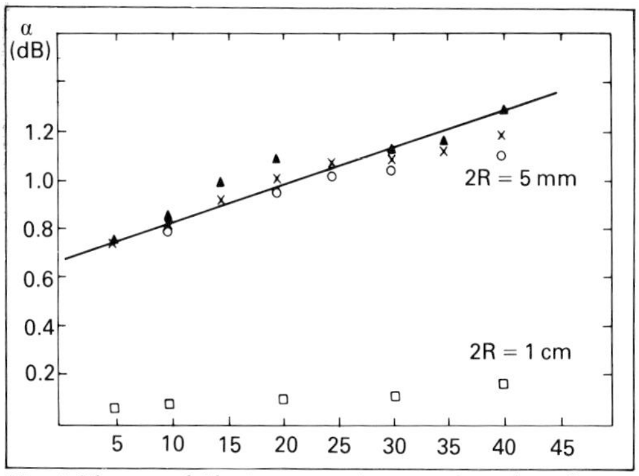

Fig. 4 The wound fibre's attenuation versus number of turns for different radii

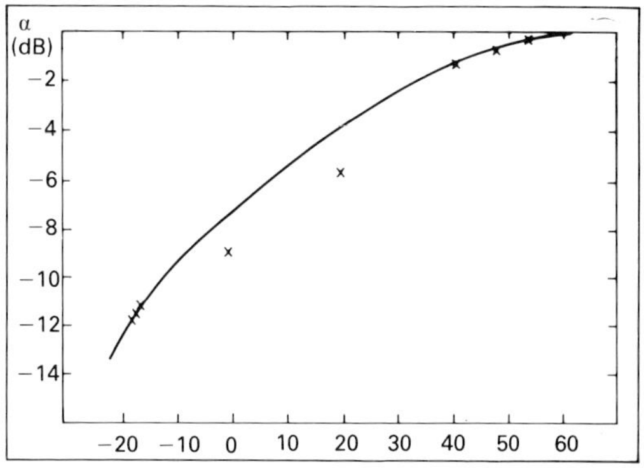

Fig. 5 The wound fibre's attenuation versus temperature with He-Ne laser transmitter

OTDR equipment, as shown in figure 7. Main advantages are:

- High spatial resolution, despite the relatively poor spatial resolution of classical OTDR (several metres), by winding a large number of turns on a short length. One can reach a few centimetres.
- As the refractive index and, then, the group propagation delay remain constant in the silica core, no location error is likely.
- High accuracy is expected after measurements above.
- Simple acquisition system, compared with the Raman OTDR technique; in any case, if more sophisticated signal processing is required, it can be time-shared between several sensors giving, for example, a multi-dimensional temperature chart.

An attractive application of this sensor is the accurate location of a temperature front. However, the reality is not as simple as a pure attenuation measurement along the fibre. In fact, instead of the back-scattered signal decrease, we observe (see figure 8), at first, a signal increase at the sensor place, followed by a shift down of the curve. The second effect corresponds to the overall insertion loss (including mode scrambling) of the sensor. More interesting is the first term, due (to our mind) to the scattering of the radiated modes by the cladding and the jacket

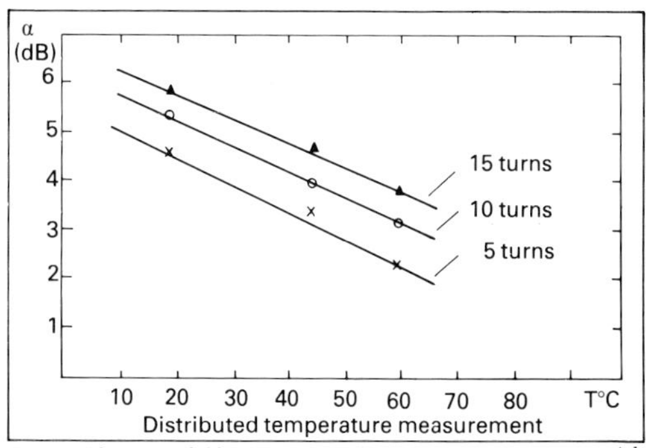

Fig. 6 The wound fibre's attenuation versus temperature with an LED transmitter

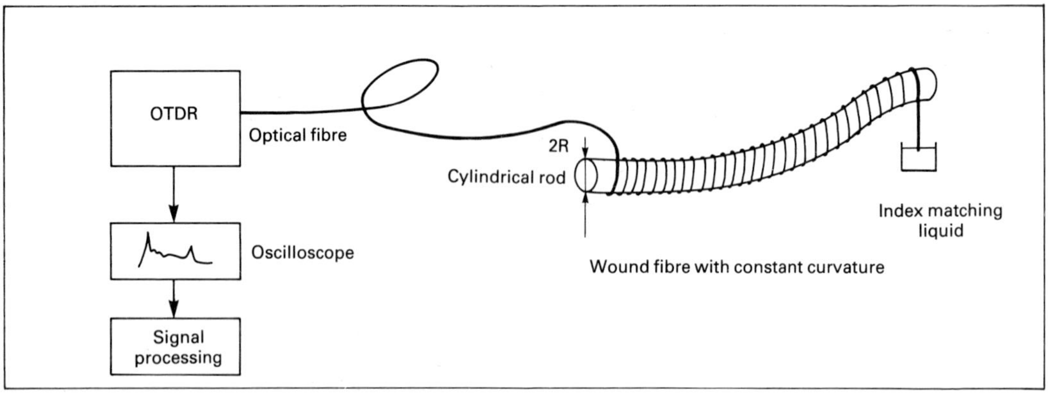

Fig. 7 Use as distributed sensor

materials, which is several orders of magnitude above the core scattering. It is even increased by unwanted effects induced by the strong bending, such as microbendings and flows. We obtain indeed an improvement of the back-scattered signal-to-noise ratio, but this signal becomes more difficult to interpret correctly. In another experiment on a force sensor based upon the microbending losses, we have also observed this enhanced back-scattering on the OTDR curve.

Nevertheless, our first results are showing (see figure 9) an effective temperature dependence of the curve. We will improve the sensitivity by signal processing, in order to reduce the effects of noise and pulse tail.

An alternative to this continuous sensor is a multi-point sensor composed of several single-point sensors, spaced with some metres of fibre. With this method, only the overall insertion loss of each sensor, which is mainly due to the temperature-dependent spatial filtering loss, is to be measured. However, the response of each sensor is disturbed by the mode filtering of the previous one, and no improved accuracy is achieved.

Concluding remarks

The principle of the sensor based upon the temperature-dependent bend losses in PCS fibres was demonstrated and used in single-point and multi-point sensors. The main part of the attenuation is due to the spatial filtering at the fibre input, which rapidly decreases when temperature increases.

However, the main problem is the reliability of the fibre under a very strong curvature. Because of the high Young modulus of the silica, very high constraints occur in the fibre and can break it. So, it would be necessary to use a less rigid material showing the same temperature-dependent index difference property, for example plastic fibres. Unfortunately, they are generally not available with small diameters. Another realisation of this sensor could use an integrated curved guide showing temperature-dependent bending losses but without stresses.

Acknowledgment

This work was supported by the D.R.E.T.

References

1. L. Jeunhomme, 'Single-mode fiber optics', Dekker,

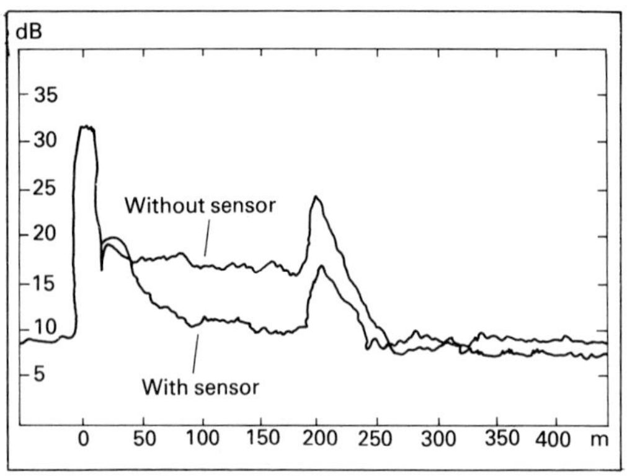

Fig. 8 Back-scattering curve with and without sensor

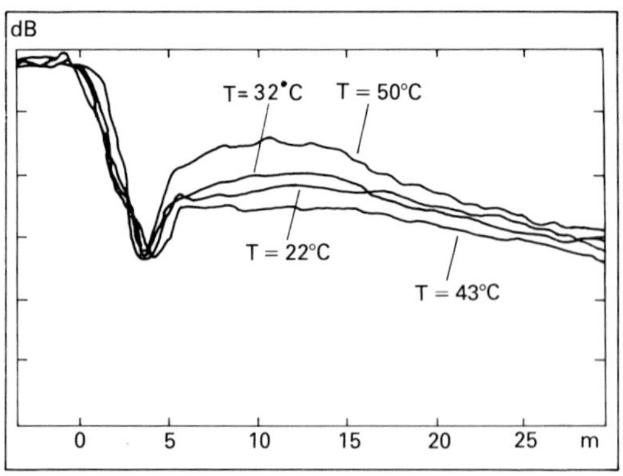

Fig. 9 Back-scattering curve (magnified around the sensor)

1984.
2. A Hartog and D. Payne, 'Remote measurement of temperature distribution using an optical fiber', *8th ECOC*, Cannes, 1982.
3. M. Gottlieb and G.B. Brandt, 'Temperature sensing in optical fibers using cladding and jacket loss effect', *Applied Optics*, Vol. 20, p. 3867.
4. J.P. Dakin, *et al.*, 'Distributed anti-Stokes ratio thermometry', *Conf. OFC-OFS*. 1985.
5. L. Guenadez, P. Dumas, P. Lecoy, 'Effet de la température sur la diffusion Raman spontanée dans les fibres optiques', *7èmes Journées Nationales d'Optique Guidée, Nice.*. 1986.
6. P. Heinzmann, R. Hofstetter, 'Temperature dependence P. Heinzmann, and R. Hofstetter, 'Temperature dependence of PCS fiber characteristics', *Conf. 584 of SPIE.*, Cannes, 1985.
7. D Marcuse, D. Gloge, and E. Marcatili, 'Guiding properties of fibers', *Optical Fiber Telecommunications*, Academic Press, 1979.

Temperature sensing by thermally-induced absorption in a neodymium-doped optical fibre

M.C. Farries
Plessey Research, UK

M.E. Fermann
University of Southampton, UK

SEVERAL temperature sensors based on optical fibres doped with rare-earth ions have been reported.[1,2,3] These are limited by two factors:

- The high absorption rapidly diminishes the useful signal.
- The absorption and fluorescence bands are considerably broadened by the host glass which dominates temperature dependent broadening.

We report initial studies of a novel temperature sensor using Nd^{3+} doped fibre, based on short wavelength fluorescence measurement first reported by Garlick *et al.*[4] The theory for this process is described in the next section. The experimental results show an absorption at 1.064μm which is strongly temperature dependent. Accompanying this absorption is fluorescent emission from 800nm to 1000nm. The absorption and, hence, the fluorescence are negligible at low temperatures, but increase with temperature following a Boltzmann distribution.

This process, which is strongly temperature dependent, is demonstrated in a distributed and a point temperature sensor.

Theory

The energy levels of the rare-earth ion, neodymium, with four-level lasing action is shown in figure 1. Normally, lasing action takes place between the levels $4_{F_{3/2}}$ and $4_{I_{11/2}}$, giving rise to emission at 1.064μm. Three level lasing action may also be observed between levels $4_{F_{3/2}}$ and $4_{I_{9/2}}$ at a wavelength of 900nm.

The energy difference E between the $4_{I_{11/2}}$ and the $4_{I_{9/2}}$ level is 3.97×10^{-20}J. The population N_2 of the $4_{I_{11/2}}$ level, with no optical pumping is due to thermally-excited electrons given by Boltzmann's equation:

$$N_2 = N_1 e^{-E/kT} \qquad (1)$$

At room temperature the ratio N_2/N_1 is 5.4×10^{-5}.

Absorption at 1.064μm between the $4_{I_{11/2}}$ level and the $4_{F_{3/2}}$ level is given by:

$$\alpha \propto N_1 e^{-E/kT} - N_3 \qquad (2)$$

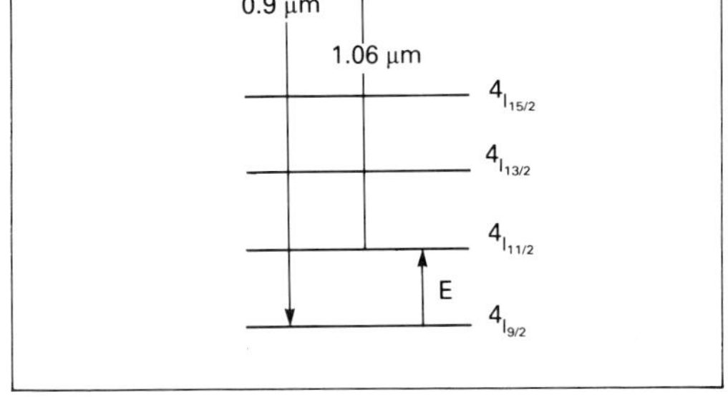

Fig. 1 Energy levels of neodymium ions

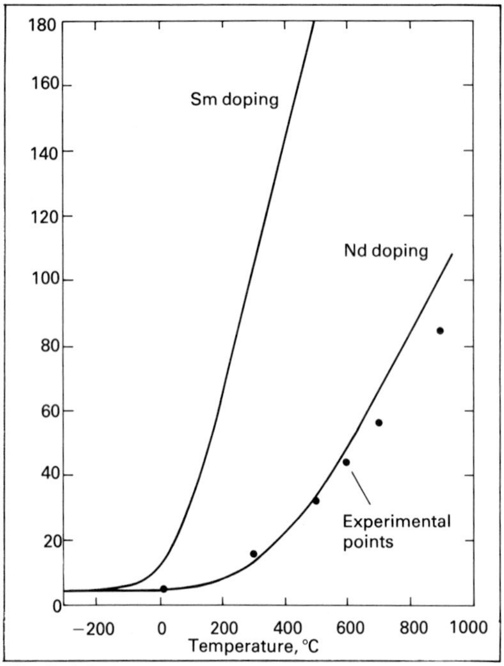

Fig. 2 Calculated thermally-induced absorption in neodymium and samarium doped fibre. Experimental points are from measurements of neodymium-doped fibre.

The magnitude of the ground state population N_1 may be ascertained from the measured absorption of the fibre at 900nm, which has been measured as 1100dB/km. The population of the $4_{F_{3/2}}$ level N_3 is small and may be neglected.

Fluorescence occurs from the $4_{F_{3/2}}$ level down to the $4_{I_{11/2}}$ and $4_{I_{9/2}}$ levels. The fluorescent intensity is also dependent on the Boltzmann relation of equation (1). The temperature sensitivity of a material depends on the energy of the first excited state above the ground level. Figure 2 shows the theoretical attenuation for the neodymium-doped fibre used here and also for a samarium-doped fibre with similar absorption.

The temperature sensitivity is given by:

$$\frac{d\alpha}{dT} = \frac{\alpha_0 E}{kT^2} e^{-E/kT} \qquad (3)$$

where α_0 is the absorption between the ground-state and the fluorescent level. The temperature sensitivity is maximised when the energy of the first excited state is given by:

$$E = 2kT \qquad (4)$$

At 20°C the ideal energy is 8×10^{-21}J which corresponds to an energy level of 400cm^{-1}.

The temperature sensitivity is also a function of absorption which, to maintain a reasonable signal-to-noise ratio in a point or a distributed sensor, should be 20dB total at 20°C. In an OTDR with 500m of fibre the

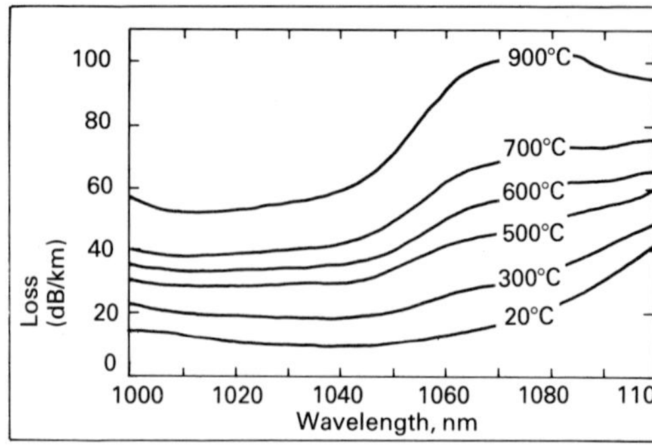

Fig. 3 Measured attenuation of neodymium-doped fibre

absorption in neodymium-doped fibre should ideally be 20dB/km. This corresponds to an absorption of 3.6×10^5 dB/km at 900nm which can be achieved in neodymium-doped fibres.

Experiment

To test this idea a point temperature sensor was constructed using 35m of single-mode optical fibre doped with 150ppm of Nd^{3+}. The fibre was inserted in a high-temperature oven and the absorption and fluorescence spectra were measured for temperatures up to 900°C.

The absorption spectra in figure 3 show a temperature-dependent peak at 1.08μm corresponding to the lasing wavelength of neodymium-doped glass. The change in absorption at 1μm is due to the temperature-dependent tail of the 900nm absorption. The thermally-induced absorption at 1.08μm is compared with the theoretical values in figure 2. There is good agreement between theoretical and experimental values allowing for the broad range of excited state energies in glass.

When the fibre was pumped at 1.064μm, fluorescence was observed over the wavelength range 800nm to 1000nm as shown in figure 4. The central peak at 910nm is due to fluorescence from the $4_{F_{3/2}}$ level down to the ground-state. The side peaks at 830nm and 970nm, which are significant at 800°C, are due to fluorescence from the $4_{F_{5/2}}$ level down to the ground

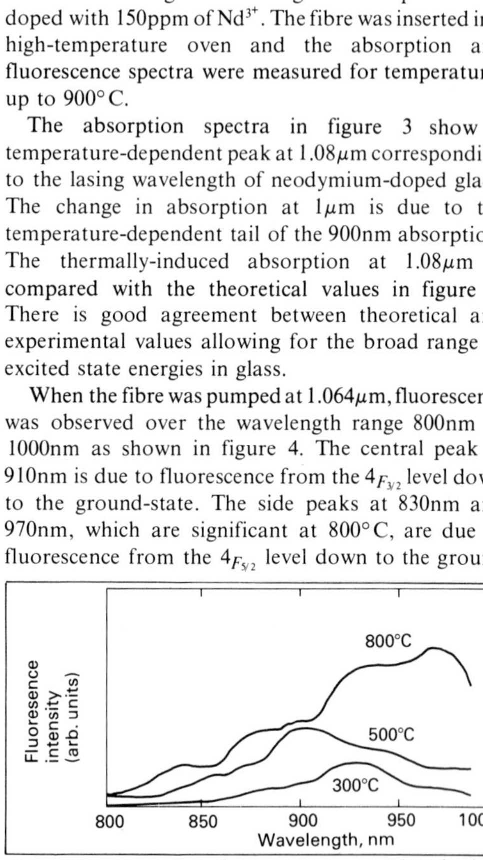

Fig. 4 Short wavelength fluorescence spectra of neodymium-doped fibre when pumped at 1.064μm

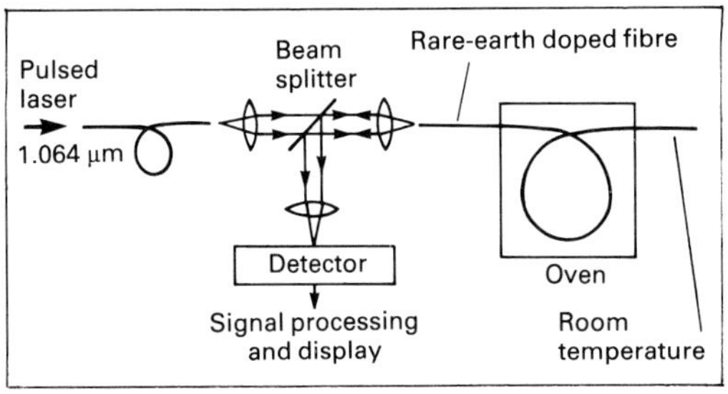

Fig. 5 Experimental OTDR used for distributed temperature sensing

level and the $4_{I_{11/2}}$ levels respectively. These emissions are strongly temperature-sensitive and are suitable for a dual wavelength temperature sensor. Unfortunately the fluorescence lifetime of 150μs prevents this emission being used in a distributed sensor.

The temperature-dependent absorption is, however, ideally suited to distributed sensing. An optical time domain reflectometer was constructed to demonstrate this application as illustrated in figure 5. 150ns pulses from a Q-switched Nd:YAG laser at 1.064μm were launched into the doped fibre via a launch fibre and a beam splitter. The launch fibre served to reduce front face reflections from the doped fibre by spatially filtering the mode pattern from the laser. The back-scatter traces were processed to obtain a measurement of local attenuation against position as shown in figure 6. The attenuation is clearly seen to increase in the hot fibre and to remain constant in the rest of the fibre at room temperature. This experiment demonstrates the fibre's application as a hot-spot detector with low loss at room temperature. The poor spatial resolution seen in figure 6 is due to the long laser pulses available in this particular experiment.

Discussion

The large energy of the first excited state in neodymium gives a weak absorption at 20°C and therefore this fibre is limited to applications requiring measurements of 500°C or more. However, other rare-earths such as samarium and europium have excited-state energies of 1.98×10^{-20}J and 7.15×10^{-21}J respectively. These would lead to absorption changes at room temperature of 1.7%/°C and 0.6%/°C respectively in fibres with similar doping levels to the neodymium-doped fibre used here. An increase in the doping level by two orders of magnitude for neodymium-doped fibre or a factor of 2 for samarium-doped fibre would optimise this sensor for room temperature.

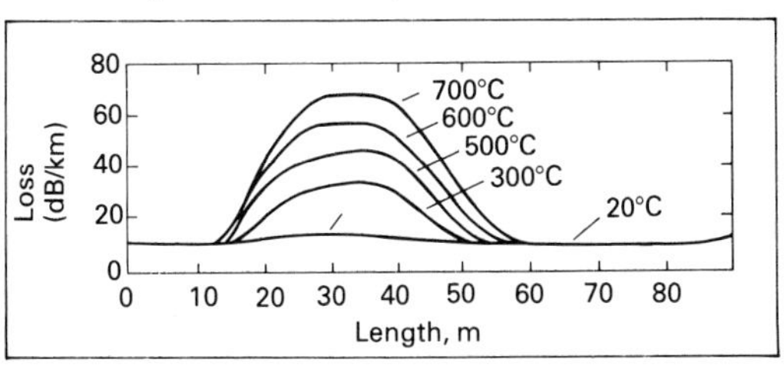

Fig. 6 Measurement of a hot-spot using an OTDR at 1.064μm in neodymium-doped optical fibre.

The attraction of this technique is the low intrinsic absorption at the pump wavelength which enables long range OTDR to be used. The temperature sensitivity is only dependent on the energy of the first excited-state of the rare-earth dopant. Monitoring of fluorescence for point sensing or low-resolution distributed sensors enables a dual-wavelength technique to be used. This will eliminate detector saturation due to end reflections and reduce errors due to bending losses.

Concluding remarks

We have demonstrated a novel temperature sensing scheme using thermally induced absorption at 1.064μm in Nd^{3+}-doped optical fibre. A basic OTDR was used to measure temperatures up to 700°C with a 10°C accuracy. We have also shown how fluorescence at a wavelength shorter than the pump may be used for temperature sensing. Temperature resolution and range may be improved by doping fibres with other rare earths and improving the measurement apparatus.

Acknowledgements

The authors thank Dr. A. Smithson for providing the optical fibre. M.C. Farries is funded by the SERC/DTI under the JOERS Optoelectronic Sensors scheme. M.E. Fermann is funded by the SERC/DTI under the JOERS Advanced Fibre Measurements scheme.

References

1. E. Snitzer, W.W. Morey and W.H. Glen, 'Fibre optic rare earth temperature sensors', *Proc. 1st International Conference on Optical Fibre Sensors*, IEE Conf. Publication 221, 1983.
2. M.C. Farries, M.E. Fermann, R.I. Laming, S.B. Poole, D.N. Payne and A.P. Leach, 'Distributed temperature sensor using Nd^3 -doped optical fibre', *Electron. Lett*, 22, 1986, pp. 418–419.
3. M.C. Farries, M.E. Fermann, S.B. Poole and J.E. Toownsend, 'Distributed temperature sensor using holmium-doped optical fibre', *Proc. 6th International Conference on Integrated Optics and Optical Fibre Communications*, Reno, Nevada, 1987.
4. G.F.J. Garlick and R.E. Nwokedi, 'Anti-Stokes effects for Nd^3 ions in solids', *Phys. Lett.*, 67A, 1978, pp. 205–206.

Distributed fibre temperature sensor using the optical Kerr effect

J.P. Dakin
York Ltd., UK

D.J. Pratt, C. Edge, M.J. Goodwin and I. Bennion
Plessey Research, UK

NUMEROUS methods have been proposed to measure the distribution of temperature along an optical fibre.[1–4] The first suggestion by Rogers[1] was to monitor the temperature dependence of the birefringence in optical fibres by means of polarisation optical time domain reflectometry. The first method, to experimentally measure temperature distribution, was the conventional OTDR method of Hartog and Payne.[2] This method, however, required the use of liquid-filled fibres in order to ensure a significant change of Rayleigh scattering with temperature. Theocharous[3] has also suggested conventional Rayleigh OTDR methods but, in this case, with a temperature dependent absorber incorporated in the fibre. This method leads to high optical losses which can limit the measurement capability. The most practical method of monitoring temperature distribution in conventional vitreous-core optical fibres appears to be the Raman method[4] where the well-defined temperature dependence of Raman scattering levels is used in an OTDR equipped with a suitable filter to extract the Raman light.

All the above methods suffer from the problem of a relatively weak signal return and considerable signal averaging is necessary to resolve the temperature profile to high accuracy. The methods are therefore unsuitable for high-speed temperature profiling, even if relatively powerful lasers are used. (A recently-proposed method for temperature profiling using fluorescent fibre[5] could provide stronger signals, but has yet to be realised in a practical implementation as it requires the development of special fibre). A fuller comparison of the above methods for intrinsic temperature measurement is given in reference.[6] The present paper will examine whether the sensitivity performance may possibly be increased using an interferometric method based on the optical Kerr effect.

The optical Kerr effect for distributed sensing of temperature

The optical Kerr effect results from the electric field component of an optical wave. This field component can influence the refractive index of a medium in a similar manner to that due to the conventional Kerr effect, which involves direct application of electric field. It is well known in the fibre gyroscope field that each of the two counter-propagating optical signals may influence the effective refractive index seen by the other, leading to possible errors if the intensities are not balanced.

In order to utilise the effect as a distributed sensor, it is proposed to perturb one arm of an optical fibre interferometer by launching an intense 'pump' pulse in the opposite direction to a continuous light beam (the 'probe' beam) used to interrogate the path imbalance in the interferometer (see Fig. 1). Injection of the pump signal causes a significant phase perturbation in the excited arm, an effect which continues until the pulse leaves the interferometer. The temporal variation of the detected signal used to monitor the phase condition of the interferometer, gives a measure of the magnitude and spatial distribution of the optical-Kerr-induced phase shift which occurred in the fibre.

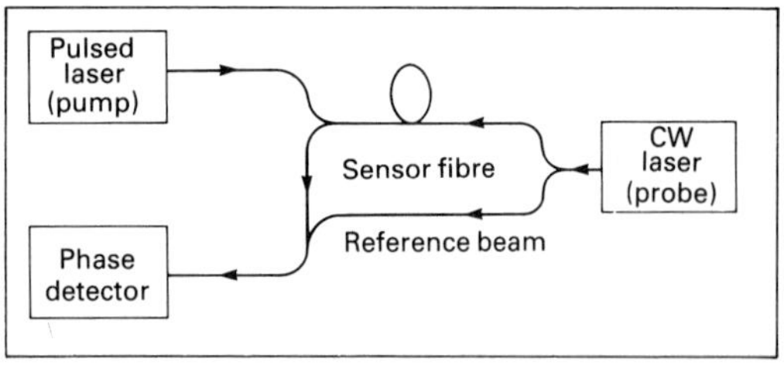

Fig. 1 Fundamental arrangement to determine the magnitude of the optical Kerr effect as a function of distance along the fibre

If the attenuation of the fibre is low, the pump pulse will cause a flat-topped pulse response on the output detector provided the optical Kerr effect is constant. In the presence of attenuation, however, it will have the more general form shown in figure 2.

It has been reported that the Kerr effect in doped-silica fibres shows a significant temperature dependence of the order of 0.56% per degree centigrade,[8] even when precautions have been taken to eliminate electrostriction effects. Thus the expected signal at the detector would be expected to show an amplitude modulation, the degree of modulation varying with time across the detected pulse in a manner corresponding to the temperature distribution in the fibre. This is provided, of course, that the optically-induced phase shift shows a similar temperature dependence, and that the time constants associated with the physical processes occurring are sufficiently rapid to respond at the rates required to provide the necessary measurement resolution.

In a later section, we shall calculate the anticipated magnitudes of the Kerr-induced phase shifts and deduce the expected variations, making assumptions of the expected temperature dependence from the previously published electrical values. Firstly, however, it is convenient to examine the interferometer arrangements which may be used in practice.

Interferometer arrangements for distributed measurement of the optical Kerr coefficient

The basic interferometer may take several forms, the only requirement being that the polarisation of the

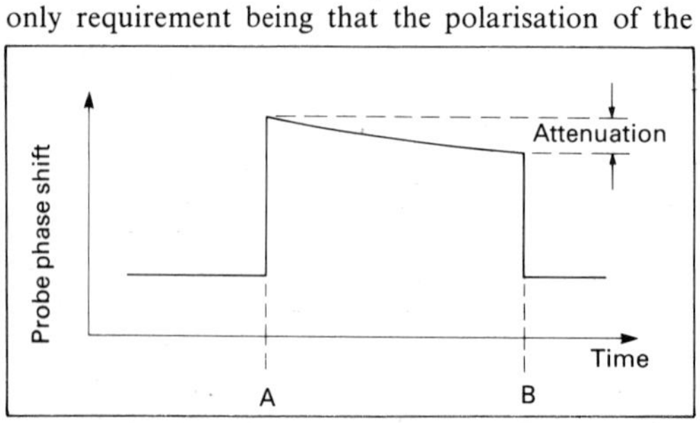

Fig. 2 Probe phase shift when the pump is subject to attenuation. Point A corresponds to the pump being launched, point B to the pump leaving the far end of the fibre

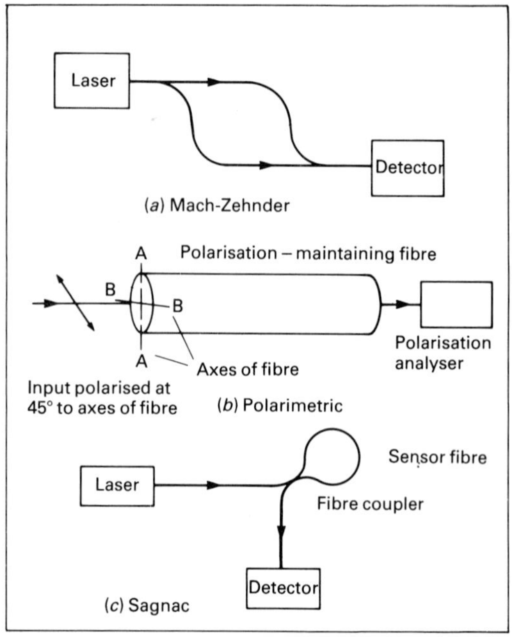

Fig. 3 Possible interferometer configurations

pump and probe beam must be held in alignment to avoid instrumental uncertainties. Three simple configurations are shown in figure 3: the well-known Mach-Zehnder (a); a polarisation-maintaining fibre polarimetric arrangement (b); and a Sagnac arrangement (c).

Arrangements (a) and (b) will have a slowly drifting quiescent phase relationship between the two arms, whereas the Sagnac will normally, without deliberate non-reciprocity being introduced, have a zero phase difference between counter-propagating beams.

In the case of the interferometer configurations (a) and (b), the quiescent condition of the interferometer will, in general, be subject to drift and, hence, both the magnitude and the sign of the detected pulse is expected to vary cyclically according to the condition of the interferometer at the time the pump pulse is launched. In practice, however, the quiescent state does not vary significantly during the short propagation time of the pump pulse along the fibre, particularly if care is taken to balance both interferometer arms and to eliminate any rapid, differential, externally-induced changes in temperature or strain. Such effects might otherwise cause differential interferometer transients, which could interfere with the desired temperature profile measurement.

In a practical system, using either of the interferometer arrangements (a) or (b), a closed-loop system to control the quiescent state of the interferometer will probably be necessary in order to avoid scale factor variations. With the Sagnac configuration, however, the inherent reciprocity of the system will normally ensure a zero phase bias between counter-propagating beams in the absence of rotation, magnetic field and fast transient changes in temperature or strain. Thus, in the latter arrangement using polarisation-maintaining (or polarising) fibre, the constancy of the response to a given pump energy is ensured. The sensitivity of the technique is optimised if the system is arranged to be close to quadrature when the pump pulse is propagating. This may be achieved by use of an appropriate power level for the pump, or by one of the phase-biasing methods conventionally used in fibre gyroscope signal processing schemes.

Calculation of the optical-Kerr-induced phase shift

We shall assume, as an example, a polarimetric interferometer using a York 'bow-tie' fibre. A sample available for our measurements had a birefringence of 2.5×10^{-4} (beat length = 2.5mm). The change, Δn, of refractive index due to an optical power, P, acting over a core area, A, is given by:

$$\Delta n = K \cdot \frac{P}{A}$$

where K is an optical Kerr coefficient, defined for convenience in terms of the refractive index change per unit optical intensity.

Thus, a phase shift, $\Delta\phi$, results, where:

$$\Delta\phi = \frac{2\pi}{\lambda} \Delta n \cdot \delta L$$

where λ is the wavelength of the probe beam, and δL is the interaction length, given by the product of duration and guide group velocity of the pump pulse.

Using a pump pulse power of 400W, phase shifts of the order of radians are anticipated, giving a very high signal-to-noise ratio in the detected output. However, such high power will lead to a non-linear response due to the sinusoidal response of the detected output to phase change, and could also lead to interference effects due to stimulated Raman and Brillouin processes. If powers of the order of 1W are used, good linearity of response is assured and the level is more consistent with the subsequent use of practical lower power pump sources (e.g. switched fibre lasers). Such pump powers, if of a few tens of nanoseconds duration, will give rise to a phase shift of the order of a few tens of milliradians at the interferometer output.

The temperature dependence of the electrical Kerr effect from reference (8) is 0.56% per degree centigrade. Therefore, a 70°C change in temperature would cause a 40% change in the phase shift if similar mechanisms are also associated with the optical Kerr effect at the rate required to respond to the fast optical pulse.

Experimental measurements

The arrangement used to conduct initial feasibility tests on the proposed method is shown in figure 4. We chose a polarimetric interferometer arrangement for its greater convenience for initial testing purposes, light being launched equally into both modes of a York 'bow-tie' polarisation-maintaining fibre from a 633nm He-Ne probe laser.

At the far end of the sensor fibre, a polariser was used to analyse the transmitted light, the emerging optical signal being transmitted to a remote receiver via an optical fibre delay section. (This delay conveniently avoided problems due to electrical interference from the pump laser supply.) The pump pulse from a Nd:YAG laser at 1.06μm was injected into the

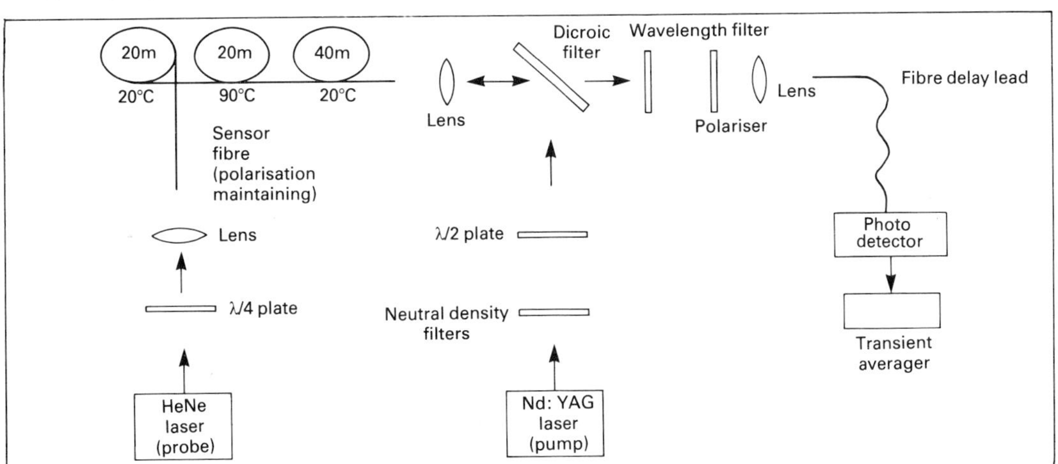

Fig. 4 The experimental arrangement to evaluate temperature distribution measurement using the optical Kerr effect

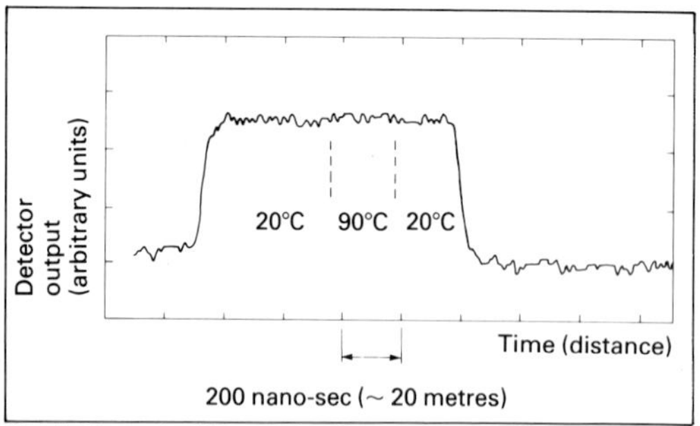

Fig. 5 The optical Kerr induced response on the detector output signal

sensor fibre in the reverse direction via an inclined dichroic filter. A half-wave retardation plate was also used in order to rotate the plane of polarisation of the pump beam to coincide with a desired axis of the polarisation-maintaining fibre, hence permitting launching into only one mode of the polarimetric interferometer.

The pump pulse intensity launched into the fibre was of the order of 1W and the pulse had a duration of 25 nanoseconds. This power was used in order to ensure low attenuation from stimulated Raman and Brillouin processes, yet be sufficient to cause a readily detectable phase change of the order of 10 milliradians between arms of the interferometer. Figure 5 shows the output of the polarimetric interferometer, as observed using a high speed APD photo-receiver. As can be seen, the pulse response, corresponding to a step phase change of approximately 10 milliradians, is flat topped, showing insignificant attenuation over the 80m length of fibre used. However, this result was taken with an intermediate 20m region of the fibre heated to 90°C, 70°C above the temperature of the remainder of the fibre, which was held at room temperature (20°C).

The flat topped pulse response is, therefore, rather disappointing since, from the 0.56% per degree centigrade Kerr dependence quoted in reference (8), a 70°C temperature difference might be expected to give rise to a 40% change in the Kerr phase shift. In the result shown in figure 5, we estimate that as little as 3% change in the Kerr phase shift would be discernable. Hence, we conclude, either that the temperature dependence of the optical Kerr coefficient in this configuration is at least an order of magnitude below the expected value or, alternatively, that the Kerr effect measured in the earlier publication had a component with a time constant which is too slow to respond under the high speed measurement conditions of our experiment.

Concluding remarks

Our early optimism for the sensing method was based upon the large temperature coefficient for the electrical Kerr effect reported in reference 8. This figure gave encouragement that there may be potential for good temperature resolution, even on a single-shot measurement, unlike other optical fibre distributed sensing methods which require averaging of multiple signal returns to achieve a satisfactory signal-to-noise ratio. Our somewhat disappointing measurements, however, indicate that for our particular configuration, the temperature dependence of the high-speed components of the optical Kerr effect is at least an order of magnitude smaller than that reported for direct electrical excitation at frequencies up to a few MHz.

The optical Kerr effect in fibres is reported in theoretical studies (9) to consist of two constituent parts, one due to purely electronic non-linearities and the other due to spatial redistribution of nuclear sites. Although the latter will also involve acoustic-wave excitation effects, our preliminary analysis suggests that these may be sufficiently rapid over the small dimensions of the fibre core to not seriously restrict the response to the optical pump pulse. Owyoung *et al* (9) concludes that the electronic contribution may dominate. As the electronic component is not expected to show a significant temperature coefficient, Owyoung's conclusion is supported by the measurements reported in this work. It appears from the high temperature variations observed in reference 8 that an additional low frequency component of the Kerr effect may have been responsible for the high temperature dependence.

With regard to the likelihood of future success with the new technique, it should be pointed out that a correspondence single-shot result for the successful Raman technique (4) is hardly discernable above the noise floor. Thus, with transient averaging facilities, it may in future prove to be a viable technique (albeit for more specialist applications), provided the optical Kerr effect proves to have a temperature dependent component with a sufficiently fast response time for the application.

Acknowledgements

The authors wish to thank Prof. A.J. Rogers and Dr. M.C. Farries for helpful discussions on the temperature dependence of the electrical Kerr effect.

References

1. A.J. Rogers, 'Polarization optical time domain reflectometry', *Electron Lett.*, Vol. 16, 1980, pp. 489–490.
2. A.H. Hartog, 'A distributed temperature sensor based on liquid-core optical fibres', *IEEE J Lightwave Technol*, LT-1, 1983, pp. 498–509.
3. E. Theocharous, 'Differential absorption distributed

thermometer' *Proc. 1st International Conf. on Optical Fibre Sensors*, 1983, pp. 10–12.
4. J.P. Dakin, D.J. Pratt, G.W. Bibby, and J.N. Ross, 'Temperature distribution measurement using Raman ratio thermometry', *SPIE Fiber optic and laser sensors III*, Vol 566, 1985 pp. 249–256.
J.P. Dakin, *UK Patent Application No. 8407945* Priority date 24/3/84.
6. J.P. Dakin and D.J. Pratt, 'Fibre-optic distributed temperature measurement—a comparative study of techniques', *IEE Coll Distributed optical fibre sensors, Digest No. 1986/74.*
7. J.P. Dakin, patent application.
8. M.C. Farries and A.J. Rogers, 'Temperature dependence of the Kerr effect in a silica optical fibre', *Electron Lett.*, Vol. 19, 1983, pp. 890–891.
9. A. Owyoung, R.W. Hellwarth and N. George, 'Intensity-induced changes in optical polarizations in glasses' *Phys. Rev. B.* Vol. 5, 1972, pp.628–633.
10. G.J. Reese, private communication.

Pressure/chemical sensing

THIS chapter covers methods of sensing pressure along the length of optical fibres and the relatively new area of distributed chemical sensing.

The first paper (G. Oscroft) describes a sensor based on microbending losses, in a special fibre cable, which has found extensive commercial application.

The second paper (Dunphy *et al*) describes a somewhat more complex concept, using mode coupling in a twin-cored optical fibre. The use of multiple-core fibres greatly enhances the number of 'degrees of freedom' for sensing, but it will remain to be seen whether they will find practical application in longer-length systems.

The third paper (Tsubokawa *et al*) describes an interesting sensor for location of pressure points on a continuous length of optical fibre, based on a coherence-multiplexing concept. The technique involves launching two different optical frequencies, derived by frequency-shifting of a short-coherence-length optical source into separate modes of a bi-modal fibre. The fibre must have two modes with significantly different velocity. When mode conversion occurs, due to external pressure, the beat signal at a detector, at the end of the fibre, will normally have a low amplitude, as a result of the differential path delay between the fibre modes. However, when a variable optical delay system is incorporated in the input signal to one of the fibre modes, to balance the delays between the two paths, the beat signal amplitude is maximised. Scanning of the input optical delay, therefore, allows interrogation of a length of fibre to determine the existence and position of mode-coupling points.

The fourth paper (Martinelli *et al*) describes a distributed strain sensor, based on an optical-fibre-hydrophone-type concept, where a fibre has a series of spheres attached to it, at periodic intervals, to form a 'necklace' type structure. The objective is to measure the resonant modes of a liquid reservoir. The eventual objective is for use in large water-storage reservoirs, where conventional detectors might have too poor a signal/noise ratio for practical application.

All distributed chemical sensing methods, so far, rely on evanescent mode-coupling to the detection region. Then the optical field can overlap with a region of the external medium (or the fibre cladding itself), which varies in its optical properties as a result of interaction with the chemical being monitored. The two main problems with current methods are:

- Undesirable thermal sensitivity due to refractive index variations of the polymer or liquid media which is in contact with the optical fibre. These variations then affect the evanescent mode-coupling.
- Imperfect chemical reversibility, either in terms of the extent of the reverse reaction, or in terms of too long a recovery time constant.

The papers by (Lieberman *et al*) and (Blyler *et al*) involve the use of polymer-clad silica fibres which, as the name implies, have an optical cladding of a plastic material. Reactive agents, in the form of organic dyes, can be incorporated into these polymer coatings and the chemicals to be sensed, in the external environment, can readily diffuse in, to react with the dye and change its absorption spectrum. The paper by Lieberman *et al* involves a quenching of a fluorescent dye by oxygen gas. The level of oxygen is then monitored by measuring the fluorescent intensity. The paper by Blyler *et al* describes a somewhat similar concept but, in this case, the *transmission* of the fibre is monitored and the dyed cladding detects ammonia due to the resulting pH change which affects the absorption of the dye.

None of the chmeical sensors described appears, as yet, to have been tested with OTDR-type systems, thereby enabling measurement of variations of chemical concentrations, as a function of length along the fibre. However, the future potential for such configurations is clearly very attractive.

Intrinsic fibre optic sensors

G. Oscroft
Herga Electric Ltd., UK

NUMEROUS articles have been written about the many applications of fibre optic sensors and their future potential. However, very few products are yet in production which incorporate the sensors.

The majority of applications to date have been at the most sophisticated level of sensing, i.e. fibre gyroscopes, ultra-sensitive hydrophones, etc.

The volume application of fibre optics to date has been in the telecommunications field where the objective is to use a glass fibre to transmit the maximum amount of information over the maximum distance, with the minimum of losses and as little distortion as possible, i.e. undesirable transmission losses have been reduced to a minimum.

With a fibre optic sensor, however, the converse is required. The light is changed, or distorted, as it passes through the optic system. This change is used to sense force, pressure, temperature, magnetic fields, etc. The light-loss mechanisms which are minimised in telecommunications are, therefore, maximised in the design of optical sensors.

All fibre optic sensors can be classified into two types according to their mode of operation.

Extrinsic sensors

In such sensors, the fibre itself is passive and is used only to transmit light from one location to another (see Fig. 1), e.g. from the light source to the sensing location and/or from the sensing location to a detector. The fibre plays no role in the sensing; sensing results from non-fibre-related effects, such as reflected or scattered light, interruption of a light beam, etc. Any influence of the fibre itself on such sensing is undesirable and will lead to inaccurate sensor response.

Extrinsic sensors are analogous to photo-electric controls with the same potential problems of being affected by dirt, dust, vibration, misalignment and possible stray light.

Intrinsic sensors

In this case, the fibre itself is the active element; propagation of light through the fibre is used not only for signal transmission, but for quantifying the parameter to be measured (see Fig. 2). The phenomenon to be measured or parameter to be quantified must influence fibre characteristics in some way so that the light flux transmitted by the fibre will be altered from the quiescent state. The received signal must be related to the measured quantity in some unambiguous way.

Many intrinsic fibre optic sensors use standard photo-electric heads connected by fibre optic cable to the sensing point. The major advantage is a great reduction in the size around the sensing area and the removal of all electrical components from a wet or hazardous measuring environment.

Although various classifications are possible, intrinsic sensors can be grouped into two main types:

Phase sensing These sensors utilise the phase difference caused by strain on an optical fibre. There are two distinct types of sensor in this category.

Phase modulation. Stress on a length of monomode or multimode fibre will cause a change in phase

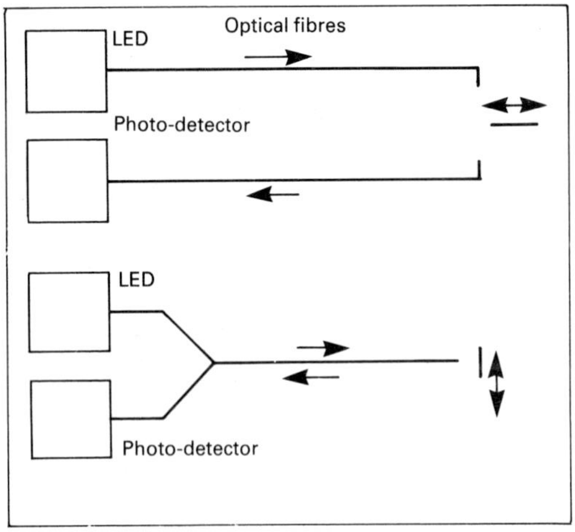

Fig. 1 Extrinsic sensors

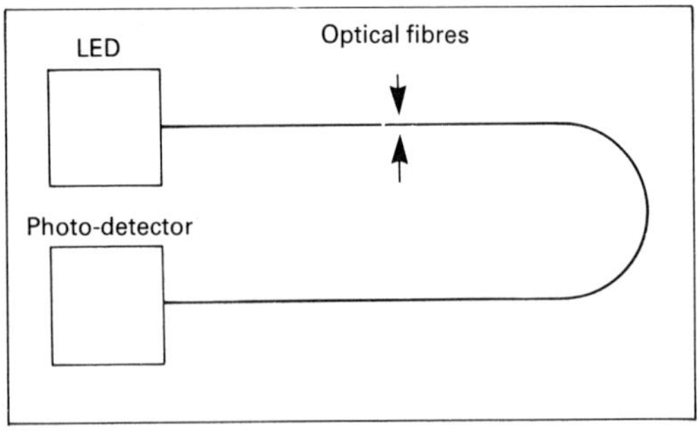

Fig. 2 Intrinsic sensors

due to length, diameter and refractive index effects (see Fig. 3). The dominant parameter is the length change; in a single mode fibre it is relatively easy to measure the phase change using interferometric techniques. With multimode fibre there can be difficulties in measuring the change in the speckle pattern produced by using a coherent source.

The major problem with this sensing technique is the sensitivity to other parameters, i.e. it is difficult to separate pressure effects from temperature effects. This cross-sensitivity can be reduced by fibre design or by applying specific coatings to the fibre to enhance the sensitivity to a particular parameter and minimise competing effects.

Polarization rotation. This technique is dependent on the polarisation of coherent radiation in a single mode fibre. A large birefringence can be introduced into a fibre by maintaining it under tension, i.e. coiling it around a cylinder under tension. The additional strain caused by pressure can then be detected because the resulting polarisation rotation causes a measurable amplitude variation in the output. An additional advantage of this method is that only one fibre is used rather than the two which are required with normal interferometric techniques. Considerable development work is taking place world-wide on polarisation-preserving optical fibres with the stress built into the fibre during manufacture, e.g. 'D' type fibres.

All these sensors require complex optical and electronic equipment. They are therefore expensive and, to my knowledge, are only used at present in sophisticated applications such as gyroscopes and hydrophones where the ultimate sensitivity is required. Integrated opto-electronic chips with built-in interferometers are being developed with potential cost reduction for volume production.

Amplitude sensors. These sensors use the optical fibre as a sensor, taking advantage of light loss through the cladding as a result of microbending, flexing or surface effects. An optical fibre consists of an inner silicone core and an outer cladding with a slightly different refractive index. Light guided down the optical fibre can be lost if the fibre is bent beyond the critical radius. The fibre can thus act as a strain or pressure sensor if it is fixed to a flexible element. Alternatively, the core and cladding refractive indexes can be carefully selected to produce fibres that are highly temperature sensitive as the refractive indexes may vary differentially with temperature. Relatively simple electronics can then be used to convert the light amplitude variation to variable signal outputs.

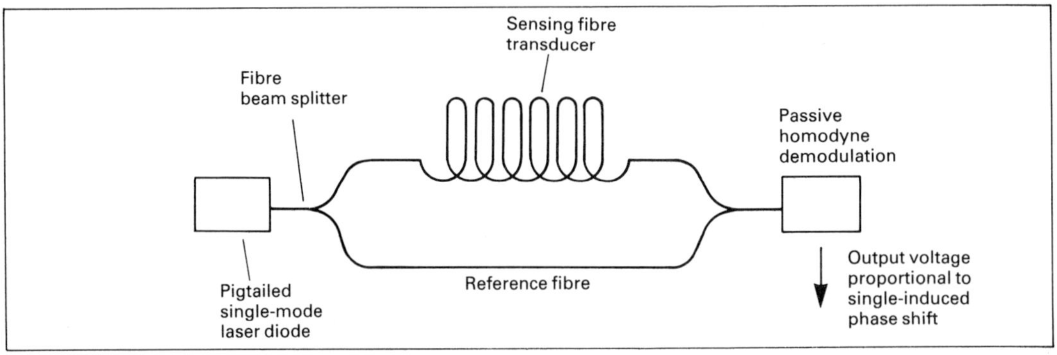

Fig. 3 Interferometer

The microbend sensor. The sensor that we have developed uses distributed microbending of the fibre to change the amplitude of the light signal being received. To explain the operation of the sensors, and some of the technical problems overcome during its development, a basic understanding of the different types of fibre is useful.

There are essentially two types of materials used for fibres: glass and plastic. Plastic fibres have a relatively high loss in light transmission and, consequently, are only used in short lengths. Termination of the fibre ends is relatively easy and carried out by a competent electrician. To my knowledge, plastic fibres have not been used for intrinsic fibre optic sensing.

In comparison, glass fibres have a relatively low loss and can be used over long distances. Their major disadvantage, however, is that termination of the fibre ends is much more difficult; this is because glass does not break in a regular and repeatable way. To pass light into or out of a fibre with minimal loss it is essential to polish the ends of the fibre or control the cutting method precisely.

There are two types of telecommunications glass fibre: monomode and multimode.

With monomode fibre the light goes directly down the middle of the fibre, bearing neither to left nor right—in other words, in one mode. However, it is also expensive to inject the light into the fibre, typically involving lasers and precision optical components.

With multimode fibre, the glass core has a high refractive index and the outer layer a lower refractive index. With step-index fibre the refractive index of the glass changes sharply between the inner and outer layer. With graded-index fibre the refractive index changes gradually (see Fig. 4). For our microbend sensor we use a graded-index fibre of 125μm diameter.

The purpose of the change in refractive indexes is so that the fibre acts as a waveguide. Any light which deviates from the centre core is bent back towards the centre by the glass on the outside of lower refractive index.

While some of the light will follow a path straight down the centre of the fibre, there are many other possible paths or modes of transmission, some spiral, some zig-zag—hence the term multimode. There are some 1500 different modes of transmission down a typical multimode fibre. If a fibre is bent at precise points along its length, some light is guided out of the core into the outer layer where it is dispersed. A simple analogy (see Fig. 5) is to consider a series of lenses in line with each other that allow the light to pass through with very little loss. As soon as the alignment of the lens is lost, light is thrown out at an angle and, hence, the light received is attenuated.

Similarly, when an optical fibre is bent in an alternating series of curves, or 'micro bends', the light travelling in the fibre is attenuated (see Fig. 6). To achieve this 'micro bending' we have would a spiral of flexible plastic around the outside of the fibre. When the outside of the spiral is compressed it applies pressure at regular spacing along the fibre. Possible microbending loss mechanisms for the light travelling in a bent fibre are:

- Radiative loss: light traversing a steep bend may be radiated sideways out of the fibre core into the cladding.
- Mode coupling loss: bending a fibre creates an interchange of energy between modes, normally into higher-order or more leaky modes. This may subsequently be attenuated by a lossy cladding.
- Mode conversion loss: this occurs in the transition region for a straight mode entering and leaving a curved region.

For the sensor, a regular steep bending radius achieves reasonable microbending effects in a small space. In this case, mode coupling losses normally dominate though other effects can, in some circumstances, play a significant role. It should be noted that the values of microbending loss vary constantly for different fibres. In general, the losses increase for lower numerical aperture (NA) values.

The sensor we have developed has, in common with most fibre optic sensors, a number of remarkable characteristics. The sensor is only 0.8mm high and can be between 20mm and 1km long. It is only necessary to

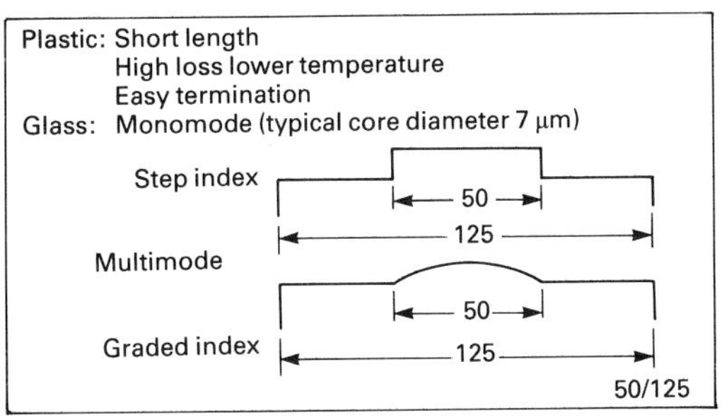

Fig. 4 Types of optical fibre

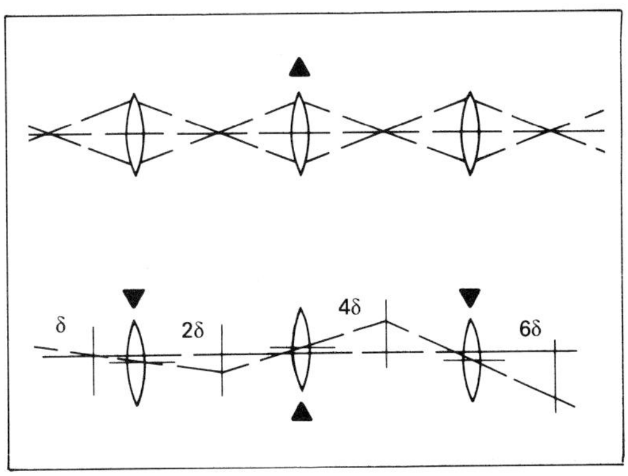

Fig. 5 A series of lenses in line with each other

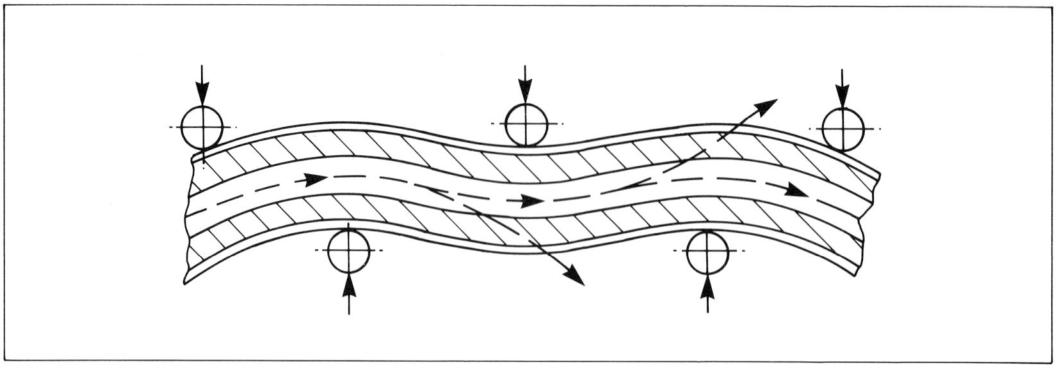

Fig. 6 Micro bends

displace the sensor 0.04mm over a 12mm length of the sensitised fibre.

After the displacement of 0.04mm the fibre can be compressed a further 0.2mm without damage to the sensor. Life cycle tests on the sensor have shown no detectable change in the attenuation level after 10 million operations. We have also tested the fibre over a temperature range of −40°C to +80°C. After 100,000 operations at these extreme temperatures, we could not detect more than a 3% change in the attenuation levels. The restriction on temperature range is totally dependent on the commercial availability of suitable buffer materials.

The relationship between applied force and attenuation is linear over 80% of the attenuation range. This in turn means that the relationship between applied force and the electrical output of the control connected to the sensor is also linear over this range. This shows a potential for use in transducers and other variable output devices.

The small size and minimal operating movement required by the sensor enable it to be built into very rugged designs.

The fundamental patents on this technology are held by the Battelle Institute of Columbus, Ohio. Herga funded development work there and now has the exclusive rights to the use of the patents in a number of fields.

Applications of sensor

Having dealt with the sensing fibre, we now have to make the sensor operate. Our initial concern was injecting light into the fibre and detecting the light output.

With our own devices we have avoided the use of in-line connectors in the fibre. These are both expensive to buy and to use. Instead, we take the fibre straight to the light emitters and detectors.

We simply use low cost emitters and detectors working in the infra-red spectrum. Significant advances have been made in recent years in the quality, performance and price of these devices. The latest devices enable the fibre end to be flooded with light, sufficient being captured by the core to provide an adequate working level for the latest high sensitivity detectors. With this knowledge we have worked out very simple techniques for connecting the fibre directly to these devices.

The initial attraction of using the fibre optic sensor was its inherent 'fail to safe' characteristic. This is an important feature of our electro-mechanical machine-guarding sensors and it was essential that any new sensor that we developed should have the same characteristic.

What does 'fail to safe' mean?

If the fibre, the sensor, the connecting cable or any other components in the control unit are broken, open circuited, short circuited or contaminated in any way, the system will shut down. The ability of a circuit to 'fail to safe' is determined by making a fault analysis in which every type of failure of every component is considered.

For our guidance in design we have taken the latest British Standard BS6491 Part 1, which is intended for photo-electric guards used on applications such as press-brakes. Because we are also dealing with light, the analogy is simple. However, our controls are only a fraction of the price of the photo-electric guard, bringing high integrity within the reach of most users.

The basic design philosophy is shown in figure 7. The main operating features of the control system are:

- The pulsed signal is used to create the 'fail to safe' design.
- The system will not run under any steady-state fault condition including short and open circuited components.
- The integral sensor amplifiers mounted in the sensor heads are designed to give steady-state output when any component fails.

- In addition the pulsed signal is coded. The code is designed to check the integrity of all system components each time it is injected at intervals of 5 milliseconds.

The simplest way to understand the system operation is to assume the system is operating and running.

The two sensors are driven in cascade. The control unit drives the LED in the first sensor head. Pulsed light from the LED is launched into the sensing fibre. The light intensity is sensed at the remote end using a photo-detector. The signal is amplified and injected into the next signal conditioner and so forth down the cascade.

The signal from the final processor is split for parallel processing. The processors recognise the pulsed signal and switch on the first transistor T1 (Relay Train). The code in the signal is then detected by each decoder which switches on the transistor T2: The output relays are maintained by contacts holding the circuits to their own coils.

The injection of the code signal effectively switches off the control, thus switching off transistor T1. The relays are then held in only by the capacitor C1. The decoder then detects the system has been switched off and uses a single pulse re-set to re-start the code generator.

When a sensor is operated the signal is attenuated until it drops below a pre-set level which in turn de-energises transistor T1 and trips out the control relays.

Before a re-start is possible, a number of conditions have to be validated to ensure that it is safe to re-start. These include, checking that the output relay contacts are open, the decoded transistors are switched off and the re-set button is open.

The output relays are of a special type called guided contacts security relays. With these, failure of one relay will either stop the system or will prevent it from starting again after a stoppage. In reality, for many applications two good quality relays in parallel are satisfactory where the level of risk is limited. However, it must be stressed that the system then would not be truly safe in that it depends on visual inspection during servicing to discover which one of the output contacts is faulty.

We have subsequently developed a range of controls, from the most simple form, using a single relay, to the most sophisticated type detailed above. We can now look at some of the applications to see why this, fail to safe, aspect is so important.

Pressure sensing mats for use around robots have, in the past, been of the electric contact type or air operated. All electric mats depend on normally open contacts which close when someone steps on the mat.

While it is possible to determine whether the

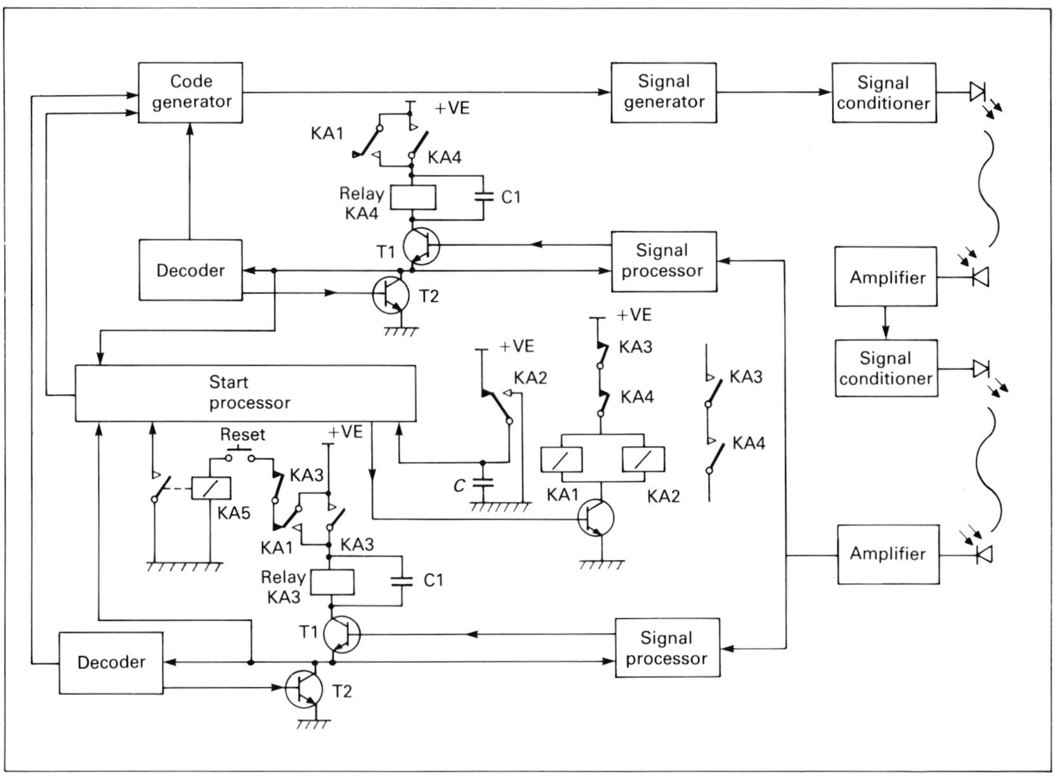

Fig. 7 Basic design philosophy

connections to the mat are in good condition, it is not possible to determine whether the contacts themselves will close when pressure is applied. So, if some contamination gets between the contacts the operator is unaware of it. It is not visible from the outside and no-one is aware of the failure until an accident has occurred.

This contrasts with a mechanical guard where it is normally easy to observe whether it is properly in use. Guarding system sare like brakes on a car: if they work reliably you depend on them and take greater risks than if they were not there. If they have broken down and you are not aware of the failure the results can be alarming.

With fibre optic sensors the condition of the sensor and all its associated components is monitored every 5 milliseconds. If anything fails the whole system 'fails to safety'.

Products application

The greatest amount of field experience has been in safety devices for machine guarding and bumpers for the guarding of Automated Guided Vehicles (AGV).

The construction of the safety mats (see Fig. 8) shows how the sensor can be built to very robust and simple designs. The sensor is sandwiched between a base plate and a hard rubber buffer layer. This is covered with a metal plate finished in a grit load epoxy paint. The metal plate folds over and is sealed to the base along the edge. The LED, photo-detector and amplifier are built into the base and the whole construction sealed.

Because of the small travel required to operate the sensor there is virtually no air gap inside the mat. The steel top will withstand both direct and shear loads as may be experienced by forklift trucks.

Another important feature of the design is that the mat is sensitive right to the edge; this allows several mats to be placed together (see Fig. 9), with no dead lines along the joints.

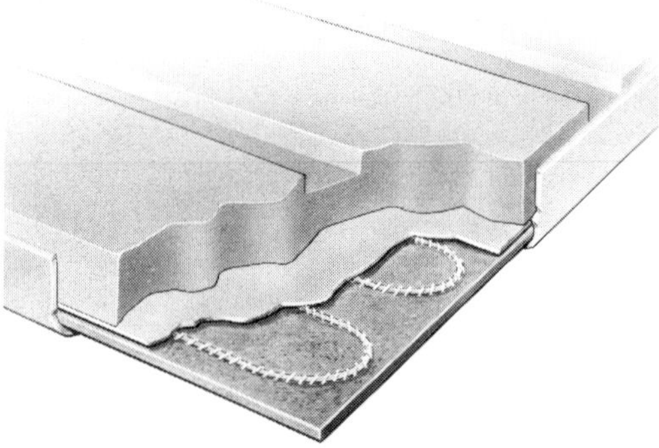

Fig. 8 Safety mat construction

Fig. 9 Safety mats in robotic work cell

Various special versions of mats for the food industry and pharmaceutical industry have been made.

This new mat system has now been tested by all the major safety authorities in Europen and, to date, has met and surpassed all their standards.

In a very different type of application, our fibre optic sensor is being used on pressure sensitive bumpers with AGVs, as shown in figure 10. The fibre optic sensor is fixed to a mounting plate built on to the front of the AGV and is itself covered by an actuating plate.

The foam core can be made to almost any shape and is coated with a polyurethane finish. The bumper can be designed to allow overtravel of up to 460mm after the sensor has been actuated. Because of the flexibility of the sensor the bumper unit can be made so that it wraps round the front of the vehicle to give additional protection when cornering (see Fig. 11).

Door sensing edges have been in use for some time at various military establishments in England. These have been fitted along the edge of hangar doors which are more than 20m long. They are to prevent personnel being hit by the door as well as guarding sophisticated nose and tail probes from impact, with a consequent $4 million bill. The construction is similar to the AGV bumper and it will sense movement normal or parallel to the plane of the door. Later designs incorporate replaceable foam sections and are now undergoing field trials.

Sensor mats with a total thickness less than 5mm are being developed for applications on mass transit systems. These are used in a number of specific applications. The most common is the monitoring of passengers entering and leaving buses, so helping to provide an automatic up-date of bus occupancy levels on any given route. From this information, route audits can be carried out and, where automatic monitoring facilities are avaialble, additional buses can be routed to relieve congestion.

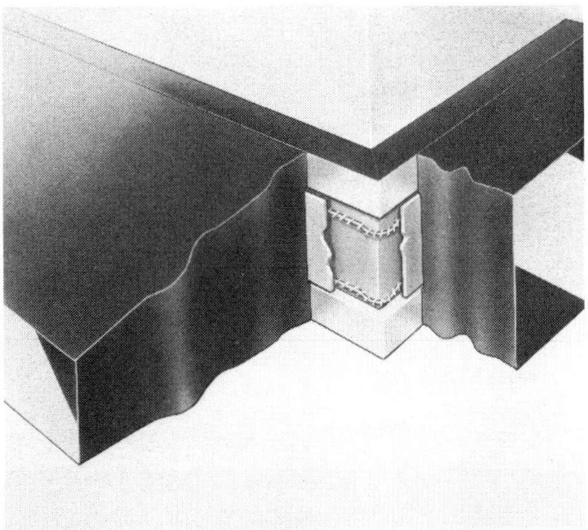

Fig. 10 Bumper sensor construction

The same sensing units can be used for operating doors on mass transit vehicles where vehicle movement past external objects tends to interfere with the more common radar and infra-red techniques currently used for door control in buildings.

Because of the very small physical dimensions of the sensor, it can be mounted in a very small space. This has enabled us to sensitise the door frame extrusions on car windows. This requirement has been brought about by the increasing incidence of injury to car occupants operating electric windows in car doors. A similar sensor can also be built into electrically operated sun-roofs in cars for the same purpose. In both these applications, the sensor and electronics will be packaged together, leaving all field connections in conventional wiring.

For the future, the linear relationship between force and signal will allow many applications in monitors for flow, pressure, temperature etc.

Fig. 11 Bumper guarding forward end of car assembly AGV

Distributed strain sensing with a twin core fibre optic sensor

J.R. Dunphy, G. Meltz and R.M. Elkow
United Technologies Research Center, USA

Theoretical development

AN optical waveguide formed from a pair of closely spaced, light-trapping cores in a common cladding (see Fig. 1) exhibits 'crosstalk' when the optical field in one core is coupled to the adjacent core. This occurs because light energy is not totally confined within a core, but extends to a small degree into the cladding and the adjacent core. When one core is illuminated, light couples back and forth between the cores as it propagates along the fibre. Complete energy exchange from the illuminated to the unilluminated core and back takes place in a beat length λ_b.

The variation in light intensity in each core at the end of a fibre of length L is a periodic function of the beat phase $\phi_b = \pi L / \lambda_b$. Both the beat length and the beat phase are sensitive functions of wavelength, varying nearly exponentially as the wavelength is changed. If the wavelength of the incident illumination is swept over a small range, then ϕ_b will vary linearly with wavelength, and the core contrast,

$$Q = (P_1 - P_2)/P_0,$$

will alternate between +1 and −1 with a dominant harmonic component at a frequency which is proportional to the fibre length. Thus, a 'crosstalk spectrum' will be generated by sweeping the wavelength. The harmonic content of this spectrum will change if the fibre is bent in a plane which contains both cores. If

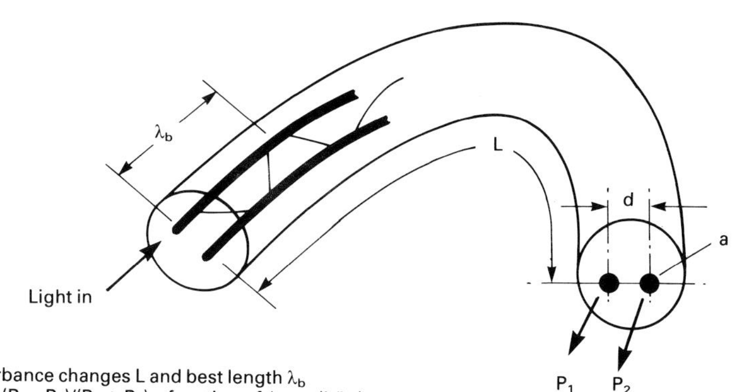

External disturbance changes L and best length λ_b
core contrasts $(P_1 - P_2)/(P_1 + P_2)$ a function of $\phi = n\,(L/\lambda_b)$

Fig. 1 Crosstalk twin-core fibre optic sensor

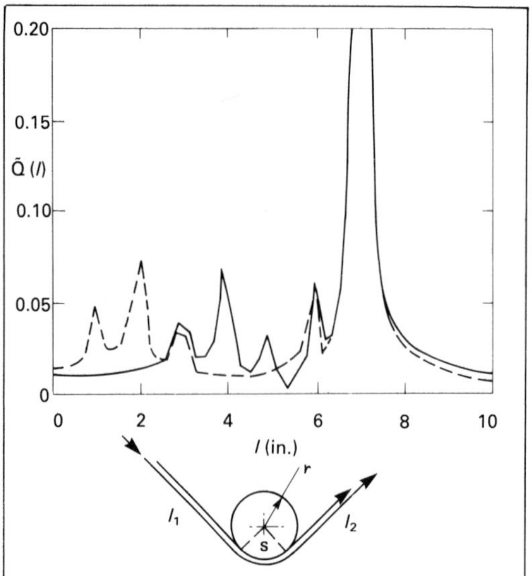

	r	s	l_1	l_2	L
-----	6.46 in	1 in.	4 in.	2 in.	7 in.
——	6.46 in	1 in.	5 in.	1 in.	7 in.

Fig. 2 Core contrast spectra for a twin-core sensor bent by a circular disc

the bend is gradual, as measured by the ratio of the beat length to the minimum radius of curvature, then the crosstalk spectrum can be related to the flexural strain distribution.

A simple illustration of the effect of bending on the spectrum can be derived from the change in crosstalk for a fibre that is formed into a circular arc of radius r (see Fig. 2). It can be shown[1,2] from coupled mode theory[1] when one core is illuminated that:

$$Q(K) = [1 - \tfrac{1}{2}(\delta/K)^2] \cos(2KL) + \tfrac{1}{2}(\delta/k)^2 \{ \cos[2K(l_1 - l_2)] + \cos[2K(L - s')] - \tfrac{1}{2}\cos[2k(l_1 - l_2 + s')]\} - \tfrac{1}{2}\cos[2k(l_1 - s')]\}$$

where: $K = \pi\lambda_b$, $\delta = (\pi/\lambda)(n_1 d/1.27\, r)$ and $s' = [1 + \tfrac{1}{2}(\gamma/K^2)]s$.

The parameter δ is proportional to the curvature of the bend. Experimentally, Q is a function of λ and not K. However, if the fractional variation in wavelength is small, K is nearly a linear function of λ. The two quantities, K and λ, are related by a crosstalk coupling factor which can be computed or measured. Thus, as the wavelength is swept, a measurable function $Q(K)$ is generated. It can be seen from the form of $Q(K)$ that its Fourier transform with respect to K, its 'spectrum', will contain a number of lines. The location of these spectral features will be determined primarily by the location and length of the bend, and their strength by the ratio λ_b/r or the sharpness of the bend. In the case of a single bend with unequal straight sections (see Fig. 2), the spectrum has five lines: a major component located at a point corresponding to the total length of the fibre (the first term in the preceding equation for Q); three small components grouped about $l = l_1 - l_2$ and a line at $l = L - s$. In the example of figure 2, the location of the bend is moved by 1 in. towards the end of the sensor. Note that the triplet of lines centred at $l_1 - l_2$ displaced by 2 in. but the other features of the spectrum remain nearly the same. The curvature or bending strain is approximately proportional to the square root of the amplitude ratio of the line at $l_1 - l_2$, or $L - s$, to the main component at $l = L$.

The locations of the spectral features are independent of the curvature when $\tfrac{1}{2}(\delta/K)^2 s \ll \pi$; i.e., for gradual bends. If the wavelength is held constant and the fibre is bent into a loop, Q will vary with radius at a rate which will increase as the loop becomes tighter. Thus, the core contrast will oscillate at a more rapid rate as the fibre loop is drawn in. These changes in the harmonic content of Q can be seen by inspection of the preceding equation for $Q(K)$. The spectral features for a sharp bend become more complicated because of these non-linear frequency contributions. In general, line shapes will be broadened and exhibit a ripple similar to the spectrum of a frequency chirped pulse.

Simple bends such as arcs or loops yield solutions to the coupled mode equations which can be expressed in the form of a transfer matrix that relates the core amplitudes to the incident illumination. Any disturbance composed of pure bends or arcs of constant curvature joined by straight sections of fibre can be analysed by multiplying the crosstalk transfer matrices of each section. With this analysis, the lowest-order solution to the coupled-mode equations takes the form of a weighted Fourier transform of the rate-of-change of bending strain. Therefore, an inverse Fourier transform operation on the output can be used to obtain the mechanical state of the structure. Computer models based on this formulation indicate the ability to diagnose both distributed bends and point-load concentrations.

Experimental demonstration

A demonstration apparatus (see Fig. 3) was configured around a pulsed dye laser as the variable wavelength source. In this arrangement, crosstalk measurements were conducted at each wavelength and then Fourier transformed in composite form to yield a signature corresponding to the simple bending strains imposed on the twin-core fibres. It is important to note that a light source as complex as the tunable dye laser is not required. For instance, miniature broadband sources,

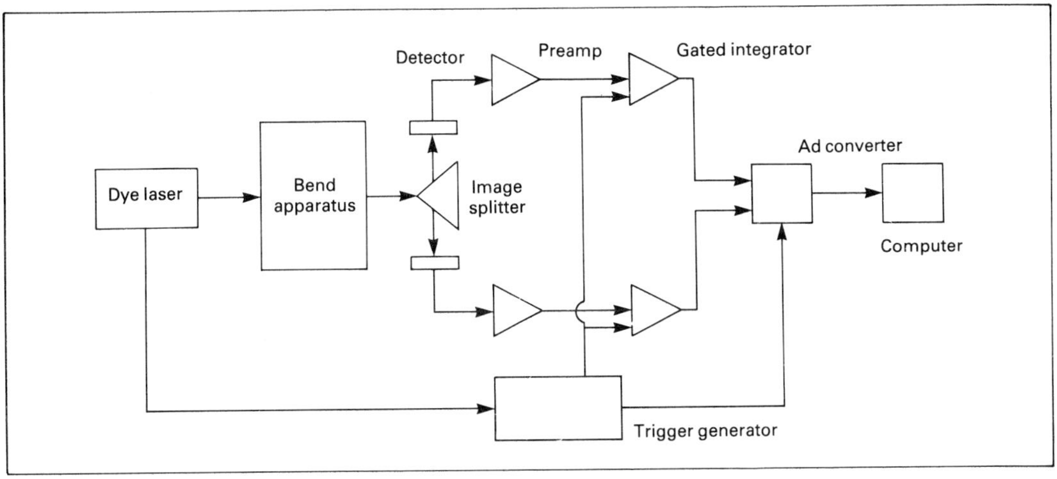

Fig. 3 Demonstration experiment

such as super-radiant laser diodes or even light-emitting diodes, could be used when appropriate wavelength demultiplexing of the crosstalk signals is available in the detector section of the instrumentation. In this form, a compact, practical, fieldable configuration could be realised.

The simplest demonstration of the sensor response is depicted in figure 4. This coarse sampling of the twin-core fibre output shows that bending strain switches light between the two cores at a rate which depends on the radius of the loop formed in the fibre in agreement with the theoretical model. In this example, a single wavelength was used to illustrate the basic effect of fibre curvature on sensor crosstalk. Other experiments were performed with simple bends and short arc lengths. During these, the swept wavelength data generation procedure outlined above was tested. Representative results are shown in figure 5 for a single right-angle bend. The data agree well with the model-predicted sensor output. The dominant signal component corresponds with a measurement of the entire sensor length. Lower frequency components provide the signature corresponding to the bending strain function imposed on the sensor. Frequencies higher than the dominant length reference are not important during interpretation of bend geometry. Further investigation revealed that cases with relatively long sensor sections exhibited spectral peak broadening due to dispersion in the relation between beat length and wavelength. This effect was also observable in predictions from the theory. Such correlation of the laboratory measurements with predictions from the theoretical model verifies the fundamental analysis introduced here.

Concluding remarks

An analytic proof has shown that the twin-core sensor output is the Fourier transform of the bending strain. A computer model based on the fundamental integral relations demonstrated the feasibility of measuring point loading simultaneously with distributed loading

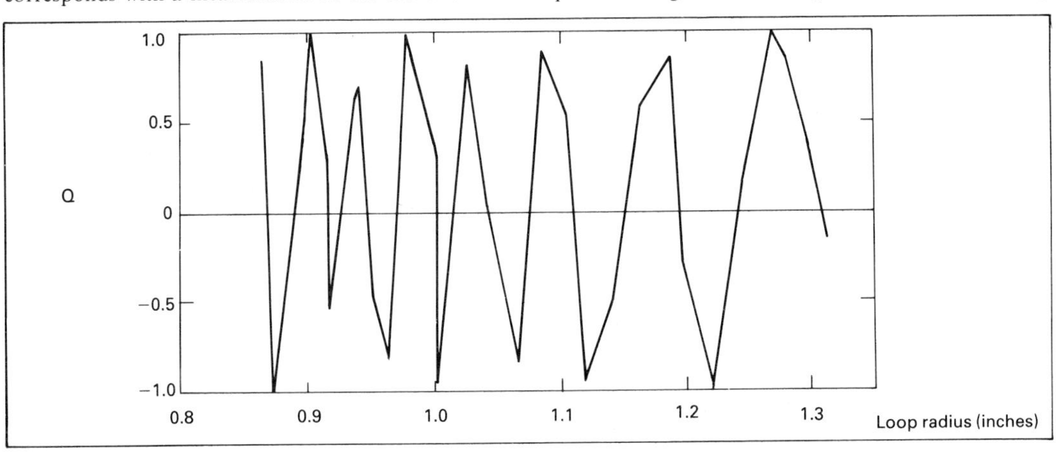

Fig. 4 Measured crosstalk vs *loop radius; fixed wavelength (633nm); sample AO4SQC/DSS7.00*

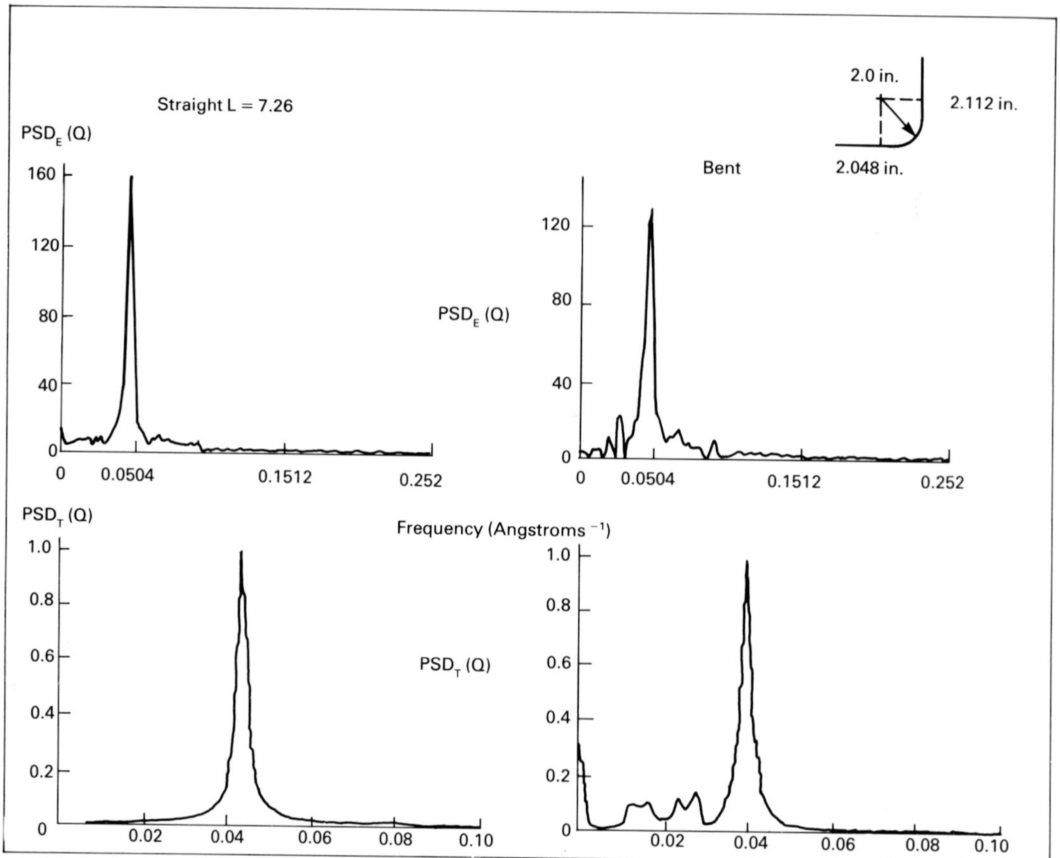

Fig. 5 Comparison of experiment and theory; round core; right-angle bend; sample AORDC/12.5.6

of a cantilevered beam. Further, experimental measurements with a device subjected to pure bends validated a second computer model based on the same physical formulation.

Acknowledgement

The work described in this paper was sponsored in part by the Department of the Air Force, Eglin Air Force Base, Florida.

References

1. Y. Murakami and S. Sudo, 'Coupling characteristics measurements between curved waveguides using a two-core fiber coupler', *Applied Optics*, 20, 1981, p. 417.
2. J.R. Dunphy, G. Meltz and R.M. Elkow, 'Composite integral response sensing', final technical report on Contract F08635-85-C-0287, USAF Armament Laboratory AFATL-TR-85-37 and United Technologies Research Center *Technical Report R85-926546-20*, April 1985.

Presented at the 32nd International Instrumentation Symposium/Aerospace Industries Division, Instrument Society of America.

Fibre optic sensor for measuring external forces distributed along a fibre

M. Tsubokawa and T. Higashi
NTT Transmission Systems Laboratories, Japan

S. Seikai
Ritsumeiken University, Japan

THE polarisation-maintaining ability of a birefringent fibre is closely related to anisotropic perturbation along a fibre due to inner waveguide structure and external force.[1] It is very important for fibre manufacturing to know the inner perturbation along a fibre. For fibre installing processes, and for fibre sensor design, it is also important to know how the modal conversion between two eigenmodes is affected by environmental changes in, for example, pressure, temperature, and electromagnetic field. Polarisation optical time domain reflectometry (POTDR) allows us to observe modal conversions distributed along a fibre.[2] With this method, however, length resolution and measurable dynamic range are usually restricted to a few hundred metres and 30dB due to input optical pulse width and intensity.

In this paper, a simple scheme for measuring the magnitude and location of modal conversion distributed along a fibre axis is presented. The method, which is based on two-beam heterodyne interferometry,[3-5] makes it possible to evaluate mode couplings with higher location-accuracy and a wider dynamic range of mode coupling magnitude. This method applied to the mutliplexing of various fibre optic snesors.[6-8]

A schematic diagram of the present fibre optic sensor is shown in figure 1. When distributed external forces, e.g. stress, bend, etc., are applied along a dual-mode fibre, mode conversions between two modes (1 and 2) occur at force applying positions. External force is applied at N points along the fibre with length L. Assuming a small power coupling coefficient, the electric fields on the detector are expressed as the sum of electric fields E_{22}, which transverses the fibre in mode 2, and E_{12n}, which is transferred from mode 1 to mode 2 at position z_n. The time-averaged interference signal $S(t)$ is represented as:

$$S(t)=r\,\mathrm{Real}\left[\left\langle \sum_{n=1}^{N} E^{*}_{12n}\cdot E_{22}\right\rangle\right] \tag{1}$$

with:

$$E_{12n}=\sqrt{h_n/(1-h_n)}\prod_{p=1}^{N}\sqrt{1-h_p}\int_0^{\infty}A(\omega) \times\exp i[(\beta_1-\beta_2)z_n+\beta_2 L-(\omega+\omega_b)t]d\omega \tag{2}$$

$$E_{22}=\prod_{p=1}^{N}\sqrt{1-h_p}\int_0^{\infty}A(\omega)\exp i[\beta_2 L +\omega(\Delta L/c-t)]d\omega \tag{3}$$

where ω is angular frequency, ω_b the angular frequency difference between two modes, r the constant, c light velocity in free space, ΔL the optical path difference in free space, β_j's (j = 1, 2) the propagation constants for both modes, and h_n the power coupling coefficient at z_n.

For simplicity, we assumed in equation (1) that fibre transmission loss is negligible and the coupling coefficient satisfies the symmetrical relation between the two modes.[9] Let us assume Gaussian power spectral distribution is $P(\omega)(=A(\omega)^{*}\cdot A(\omega))$ as $P(\omega)=\exp[-(\omega-\omega_0)^2/2\Delta\omega^2]$ where ω_0 and $\Delta\omega$ are centre angular frequency and spectral half-width, respectively. Beat amplitude is expressed from

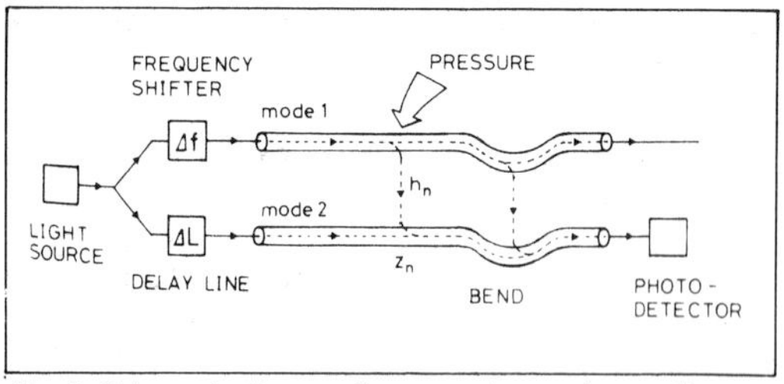

Fig. 1 Schematic diagram for measuring mode coupling distribution along a fibre

equations (1)–(3) as:

$$S=\sum_{n=1}^{N} r\sqrt{h_n/1-h_n}\prod_{p=1}^{N}(1-h_p) \times \exp\left[-(\Delta\omega/\omega_0)^2\phi_n^2/2\right] \quad (4)$$

with:

$$\phi_n=[(d\beta_1/d\omega)_0-(d\beta_2/d\omega)_0]\omega_0 z_n+\omega_0\Delta L/c \quad (5)$$

Here, $(d\beta_j/d\omega)_0$'s ($j=1, 2$) denotes the derivative of β_j at $\omega=\omega_0$. Normalising equation (4) with beat amplitude S_{ref} caused by beams traversing the fibre in the same mode 2, equation (4) is simplified as:

$$S/S_{ref}=\sum_{n=1}^{N}\sqrt{h_n/1-h_n}\exp\left[-(\Delta\omega/\omega_0)^2\phi_n^2/2\right] \quad (6)$$

with:

$$S_{ref}=r\prod_{p=1}^{N}(1-h_p) \quad (7)$$

If light source spectral width is broad, S/S_{ref} in equation (6) has discrete peaks at $\alpha_n=0$. Length resolution Δz is defined as fibre length-width for which the magnitude of S/S_{ref} becomes $1/e$ of the maximum:

$$\Delta z=\sqrt{2}/[\{(d\beta_1/d\beta)_0-(d\beta_2/d\omega)_0\}\,\Delta\omega] \quad (8)$$

Δz is realised as the ratio of coherence time T_c of the light to the group delay difference T_p per fibre length L. When external forces were applied to the fibre at positions separated by a minimum of Δz, beat amplitude due to mode conversion at z_n is expressed for $\phi_n=0$ as:

$$S_n/S_{ref}=\sqrt{h_n/1-h_n} \quad (9)$$

We then obtain information exclusively on h_n and z_n.

In this experiment, a birefringent fibre (PANDA fiber)[10] was used as a dual-mode fibre. Cut-off wavelength, fibre length, transmission loss and modal birefringence of the fibre were 1.29μm, 100m, 0.6dB/km, and 3.0×10^{-4}, respectively. A laser diode with an operation wavelength of 1.27μm and wavelength half-width of 4nm was used for this experiment.

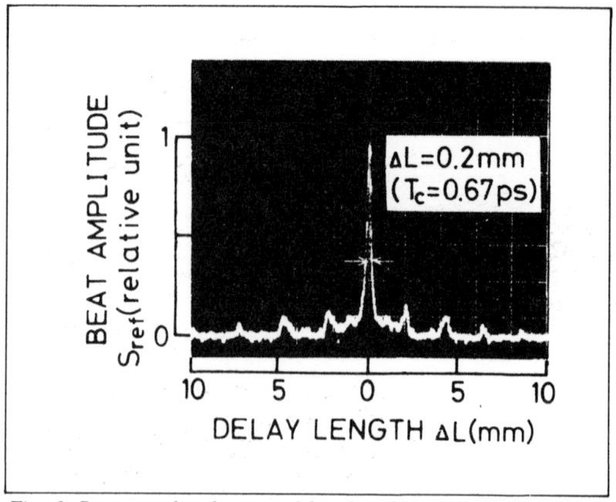

Fig. 2 Beat amplitude caused by the identically polarised HE_{11} mode beams versus delay length. Vertical scale shows relative units

Beat amplitude for the interference caused by the same HE_{11} mode beams traversing the fibre is shown in figure 2.

The power level for both beams was −22dBm at the detector, and the beam amplitude magnitude at $\Delta L=0$ was set at unity. Location of the peak at $\Delta L=0$ was measured with an accuracy of ±10μm in optical path difference, which was equivalent to ±0.03ps in group delay. From figure 2, coherence time T_c was 0.67 ps. The beat amplitude for interference between the beam traversing in the HE_{11}^y mode and the beam transferred from the HE_{11}^x mode to the HE_{11}^y mode in the no-disturbance scenario on the fibre is shown in figure 3(a). The beat amplitude with side-pressure on the fibre at $z=34$m and 35m is shown in figures 3(b) and 3(c). The horizontal scale in figure 3(c) is magnified to be ten times greater than that in figure 3(b). Here, the power of both the HE_{11}^x and HE_{11}^y mode beams was −22dBm. Vertical scales are magnified to a greater degree than in figure 2. Two peaks near the left and right margins of figure 3(a) represent interference signals due to modes leaked from polarisers at both ends of the fibre. Separation was 130ps in group delay difference T_p between the polarisation modes. This corresponds to the polarisation dispersion of a 100 metre-long fibre. Δz is then approximated to be 0.5m, from equation (8).

It can be seen that the two pressure-applying points separated by 1m are fully resolved in figure 3(c). If modal birefringence of the fibre is almost independent of fibre length, peak locations can be measured within an accuracy of ±0.03m in fibre length due to measurement error in the group delay. Magnitude of the coupling coefficient at $z=34$m is 1.8×10^{-3} per 1m($=2\Delta z$), as derived from the peak magnitude shown in figure 3(c) using equation (9). In this experiment, figure 3(a) shows coupling coefficient per $2\Delta z$ in the no-disturbance scenario to be less than 4×10^{-6}.

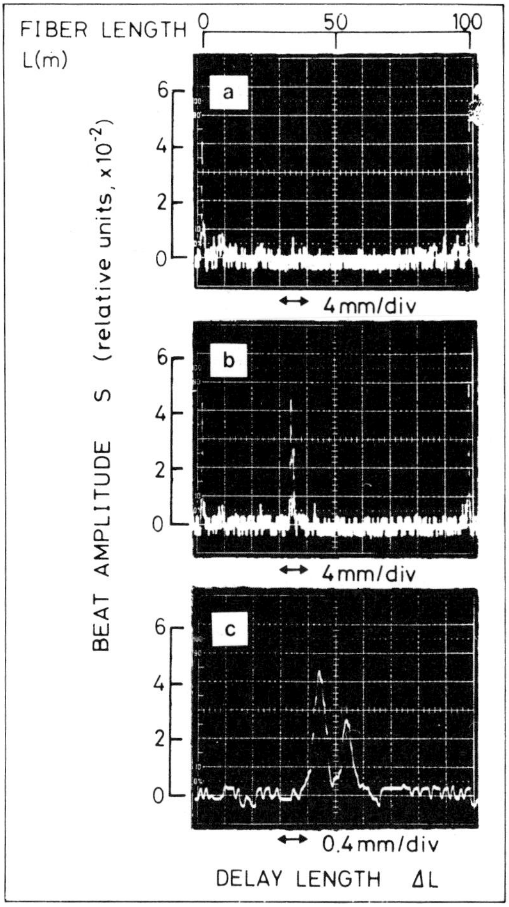

Fig. 3 Optical beat amplitudes for interference between beam propagating only in the HE^y_{11} mode and transferred from the HE^x_{11} to HE^y_{11} mode (a) without side-pressures and (b), (c) with side-pressures at z=34m and 35m. Horizontal scale in (c) magnified ten times over that in (b)

Where the total number of mode conversions for a pair of beams is three, e.g. when the mode conversions occur at $z=z_a$, z_b, and z_c along a fibre, and normalised peak magnitude of the interference signal can be represented as follows:

$$\tilde{S}_n/S_{ref} \leqq \sum_{a,b,c}^{N} [h_a h_b h_c/(1-h_a)(1-h_b)(1-h_c)]^{1/2} \quad (10)$$

The sum in equation (10) represents cases where interference signals overlap with the same ΔL. The equals sign in equation (10) represents a case where N coupling points are placed along the fibre with the same separation length between them. Calculating the sum in the simple case of $h_a=h_b=h_c=h_n$, the signal ratio $R(=\tilde{S}_n/S_n)$ is represented as:

$$R \leqq [N(N-1)/2+(N-1)^2/4][h_n/(1-h_n)] \quad (11)$$

This calculates the sum of the number of pairs of modes transferred once or twice, and the pairs of modes transferred zero and three times. From equation (11), we can calculate the maximum number N against the required values of R and h_n. For example, N must be less than 116 in the case of $R=0.1$ and $h_n=1\times10^{-5}\,m^{-1}$, and the dynamic range of a h_n retains 10dB for a general birefringent fibre with a coupling coefficient of 1×10^{-6}. It is emphasised that a much larger N number and wider dynamic range of h_n can be achieved for actual sensor systems, since the sensors can be arranged with random separation lengths.

Concluding Remarks

In conclusion, the interferometric scheme for measuring spatially distributed external forces is presented. This method provides much information concerning environmental changes related to model conversions in fibre.

Acknowledgements

We wish to thank N. Shibata, Y. Sasaki, Y. Negishi and N. Uchida for their useful suggestions and comments, and also N. Kojima for his continuous encouragement.

References

1. I.P. Kaminow IEEE *Quantum Electron* QE-17, 1981, p. 15.
2. M. Nakazawa, N. Shibata, M. Tokuda and Y. Negishi, *J. Opt. Soc. Am. A.* Vol. 1, 1984, p. 285.
3. N. Shibata, M. Tsubokawa and S. Seikai, *Electron. Lett.*, Vol. 20, 1984, p. 1055.
4. K. Takada, J. Noda and Y. Sasaki, *Electron. Lett.* Vol. 20, 1984, p. 119.
5. M. Tsubokawa, N. Shibata and S. Seikai, *J. Lightwave Technol.* LT-3, 1985, p. 850.
6. A.R. Nelson, D.H. McMahon and R.L. Gravel, *Appl. Opt.* Vol. 19, 1980, p. 2917.
7. I.P. Giles, D. Uttam, B. Culshaw and D.E.N. Davies *Electron. Lett.*, Vol. 19, 1983, p. 14.
8. J.L. Brooks, R.H. Wentworth, R.C. Youngquist, M. Tur, B.Y. Kim and H.J. Show, *J. Lightwave Technol.*, LT-3, 1985, p. 1062.
9. See e.g. D. Marcuse, *Theory of Dielectric Optical Waveguide*, Academic Press, New York, 1974, chapter 5.
10. J. Noda, K. Okamoto and Y. Sasaki, *J. Lightwave Technol.* LT-4, 1986, p. 1071.

Distributed fibre-optic interferometric sensor for the detection of pressure fluctuations in reservoirs

M. Martinelli, O. de Sanctis and P.G. Sona
CISE SpA, Italy

THE SENSITIVITY to pressure of the optical fibre has been the reason for its use as a hydrophone since the early development stages of fibre optic sensors.[1-4]

In the subsequent years, much effort was made to improve the sensitivity of such a sensor. After the development of interferometric techniques, this goal was attained by a proper choice of thicknesses and types of coating materials of the optical fibre. In such a way, the sensitivity was improved by roughly an order of magnitude.[5-7] More recently, attention has been focused on the development of a new sensing architecture[8,9] and of a sensor array, the basic sensing principle being the same.[10,11]. Notwithstanding this large effort in developing the fibre optic hydrophone, very little is known about the behaviour of a straight optical fibre sensor when immersed in a water volume with resonant modes. The question is far from being theoretical: tanks and reservoirs show a dynamic interaction between liquid motion and deformations of the boundary walls that can raise serious structural engineering problems. Moreover, because the reservoirs of most interest (such as the hydroelectric ones) are large, their resonant frequencies are usually very low, resulting in problems with noise when conventional detection techniques are used, and indicating the need to develop special fibre optic hydrophones.

The aim of this paper is to present the concept of a new fibre optic interferometric sensor (FOIS) designed to detect low-frequency resonant modes of reservoirs. The theoretical aspects of fibre optic detection are reviewed and a model of the new sensor is developed. An experiment sensor is described, and the results are discussed.

Theoretical aspects

Pressure versus elongation sensing

Let us consider the expressions of the radial pressure and elongation phase sensitivity for an intrinsic fibre optic interferometric sensor of length L:

$$\frac{\Delta\phi}{\Delta p} = C_p \beta \left\{ \frac{2\nu}{1-\nu} + \frac{n^2}{2}\left[p_{11} + p_{12} \left(1 - \frac{2\nu}{1-\nu}\right)\right]\right\} \cdot \frac{1-\nu}{E} L \tag{1}$$

$$\frac{\Delta\phi}{\Delta L} = C_E \beta \left\{ 1 - \frac{n^2}{2}\left[p_{12} - \nu\,(p_{11} + p_{12})\right]\right\} \tag{2}$$

where C_p and C_E are empirical factors as functions of the coating materials, β is the single-mode fibre propagation wavevector, n is the mean refractive index, p_{11} and p_{12} are the elasto-optic coefficients, and ν and E are the fibre Poisson ratio and Young modulus respectively. These expressions show the two basic mechanisms involved in acoustic detection. In the latter expression, the variation of the optical path length is proportional to the geometrical elongation of the fibre, with a constant of proportionality of the order of 0.8, and the fibre is acting as a strain gauge with sensitivity of the order of 10^7 rad/m. In the former expression, the mechanical characteristics of the fibre ν and E relate the fibre radial stress to an axial strain which is then converted to an equivalent variation of the optical path. In this case, the fibre is acting like a hydrophone with a typical sensitivity of the order of 10^{-4} to 10^{-3} rad/Pa·m.

The two expressions can be related as follows:

$$\text{expression (1)} = \frac{\Delta\phi}{\Delta p} = \frac{\Delta\phi}{\Delta L}\frac{\Delta L}{\Delta p} \tag{3}$$

$$= \text{expression (2)} \cdot \frac{\Delta L}{\Delta p}$$

That is, when FOIS is used as a hydrophone, the extremely high sensitivity to strain is multiplied by the factor $\Delta L/\Delta p$, and the fibre strain produced by a pressure change Δp. Equation (3) concisely points out that the efforts to enhance the FOIS sensitivity to water pressure must be directed towards any mechanism which can couple pressure fluctuations directly to fibre strain.

Resonant modes of reservoirs

Before going on, it may be useful to recall the typical parameters involved in reservoir resonances. Because of the relatively large dimensions of the reservoirs, they are expected to have low-frequency fundamental resonance,[12,13] which involve significant corresponding displacements of the water mass. In fact, from the Euler equation, the liquid displacements will be inversely proportional to the square of the resonant frequency. To illustrate the idea, let us consider the pressure distribution corresponding to a resonant mode of order m:[12]

$$p = p_0 \sin(m\pi z/L) \cos\omega t \tag{4}$$

where p is the pressure fluctuation, z is the spatial coordinate, L is the reservoir dimension along z, $\omega/2\pi$ is the resonant frequency. Associated with this pressure wave is a corresponding displacement wave, which has a peak value:

$$\Delta L_w = -\frac{1}{p\omega^2}[p'(L) - p'(o)] \tag{5}$$

where ρ is the liquid density. Thus, for m odd:

$$\Delta L_w = \frac{2m\pi p_0}{p\omega^2 L} \tag{6}$$

With typical parameters of reservoirs (L of the order of hundreds of metres and $\omega/2\pi$ below 10Hz) the $\Delta L_w/\Delta p$ value of the order of a few micrometers per Pascal is easily obtained.

The sensor concept

By analogy with the previous result, one may devise a mechanism by which the pressure fluctuation is coupled, more favourably, to the elongation of a sensing fibre immersed in the reservoir. By optimising $\Delta L/\Delta p$ in this way the sensitivity of FOIS might be improved. Such a mechanism can be realised by means of a fibre threaded with small spheres of volume u, density ρ_s (slightly greater than water density ρ), and equispaced with separation l. Under the mode pressure gradient, forces act on the spheres, which consequently strain the length of the fibre between them. The resulting necklace-type sensor is driven by forces which depend on the sphere parameters, as well as on the distance between them compared with the spatial distribution of the pressure.

In order to evaluate the behaviour of such a sensor, us introduce an equivalent cross section $\sigma = u/l$, relating to each periodic element of the sensor. The mass and volume of the fibre are neglected, along with the direct effect of the pressure on the bare optical fibre. In a first approximation, disregarding the hydrodynamic and viscosity forces, the differential equilibrium equation referred to the elementary length of the sensor will be:

$$dF_e + dF_i + dF_f = 0 \tag{7}$$

where:

$$dF_e = AE\frac{\partial^2 u}{\partial z^2}dz \quad \text{(elastic term)} \tag{8}$$

$$dF_i = -\rho_s\sigma\frac{\partial^2 u}{\partial t^2}dz \quad \text{(inertial term)} \tag{9}$$

$$dF_f = -\sigma\frac{\partial p}{\partial z}dz \quad \text{(forcing term)} \tag{10}$$

where A is the fibre cross section, E is the fibre Young modulus, and $u(z,t)$ is the axial displacement. Substituting equations (8)–(10) in the equilibrium equation (7) we obtain:

$$AE\frac{\partial^2 u}{\partial z^2} - \rho_s\sigma\frac{\partial^2 u}{\partial t^2} = \sigma\frac{\partial p}{\partial z}. \tag{11}$$

One particular integral of this differential equation can be evaluated in correspondence of the odd modes reported in the previous section. Let us assume free boundary conditions. Then, from equations (4) and (11), the peak value of displacement $U(z)$ becomes:

$$U(z) = \frac{\sigma p_0 \frac{m\pi}{L} \cdot \cos\frac{m\pi}{L}z}{\rho_s\sigma\omega^2 - AE\left(\frac{m\pi}{L}\right)^2} \tag{12}$$

As a consequence, the maximum fibre elongation will

be:

$$\Delta L = \left|U(L) - U(o)\right| = \frac{2m\pi p_0}{\rho_s \omega^2 L} \cdot \frac{1}{\left|1 - \frac{AE(m\pi/L)^2}{\rho_s \sigma \omega^2}\right|} \quad (13)$$

Equation (13) describes the behaviour of the translational symmetrical fibre sensor when subjected to a pressure gradient $(m\pi p_0/L) \cdot \cos(m\pi z/L) \cdot \cos \omega t$.

Let us write equation (13) in terms of the equivalent sound velocity for the fibre sensor:

$$C_c = \left(\frac{AE}{\rho_s \sigma}\right)^{1/2} \quad (14)$$

and of the equivalent propagation velocity of the particular resonant mode with axial order m and angle frequency ω:

$$C_R = \frac{\omega L}{\pi m} \quad (15)$$

With the approximation $\rho/\rho_s \simeq 1$ and introducing equation (6), we obtain:

$$\Delta L = \Delta L_w \cdot \frac{1}{\left|1 - (C_c/C_R)^2\right|} \quad (16)$$

In order to discuss the physical behaviour of this necklace sensor, let us consider the trend of equation (16) as a function of the characteristic ratio σ/A, angular frequency ω being kept constant. Four situations can be analysed.

a) $\sigma/A \rightarrow 0$

The sensor reduces to the simple fibre. Because of the initial approximations made in developing the model, equation (16) becomes equal to zero. Actually, ΔL takes the value corresponding to the bare fibre subjected to the same pressure field. This ΔL value is the source of the fibre sensitivity expressed by equation (1).

b) $\frac{\sigma}{A} = \frac{E}{\rho_s} \cdot \frac{1}{2C_R^2}$ or $\frac{C_R}{C_c} = \frac{1}{\sqrt{2}}$

The whole fibre elongation equals the water column elongation. The sensor behaves as a theoretical strain sensor that directly transduces the water displacement. While σ/A is approaching this value, this fibre sensor experiences the pressure field with increasing sensitivity in comparison with previous steps (a). For higher values, the sensor gains in sensitivity also in comparison with the theoretical strain sensor.

c) $\frac{\sigma}{A} = \frac{E}{\rho_s} \cdot \frac{1}{C_R^2}$ or $\frac{C_R}{C_c} = 1$

The sensor is in resonant condition. The fibre elongation is limited by the damping factors (not considered herein). The resonant condition means that the elastic wave velocity in the sensor exactly equals the equivalent propagation velocity of the particular resonant mode of the reservoir.

d) $\frac{\sigma}{A} > \frac{E}{\rho_s} \cdot \frac{1}{C_R^2}$ or $\frac{C_R}{C_c} > 1$

The case deals with the region on the other side of the resonance and ratio $\Delta L/\Delta L_w$ decreases with increasing σ/A, nevertheless still being > 1. The whole fibre elongation remains larger than the water column elongation.

In summary, we have shown a new linear distributed sensor that seems promising to detect the low-frequency resonant modes of reservoirs. Since it depends on the value of the characteristic parameter σ/A, the sensitivity can be much larger than that of a hydrophone made with a bare fibre of the same length. In fact, in the fibre pressure sensor model mentioned above, the limiting factor $\Delta L/\Delta p$ in equation (3) can be greatly increased by satisfying the resonant condition. Moreover, the distributed characteristics of such a sensor make it particularly suitable for the detection of long wavelength pressure distributions, such as those present in reservoirs.

Experimental set up

Hydraulic set up

In order to investigate the previous statements, a model of a resonant tank was performed. It consists of a water-filled cylindrical container, 1.8m high and 0.6m in diameter. The steel cylinder wall is a very thin 2mm in order to give low first resonant frequency. The tank is excited by means of an outer shaker set at mid-height of the cylinder. Then, only the resonant modes with odd axial number m can be excited. Inside the tank, a conventional pressure transducer, Brüel & Kjaer model 8101, is used to detect the presence of the mode.

Fibre sensor and optical set up

A straight fibre can be considered unidimensional and consequently suitable to comply with a resonant mode in one of its principal axes. In the present experiment, the fibre sensor is set vertically (see Fig. 1) so that only the axial mode distribution is tested (m number). The optical fibre is threaded with small spheres having density just higher than the water density. The spheres

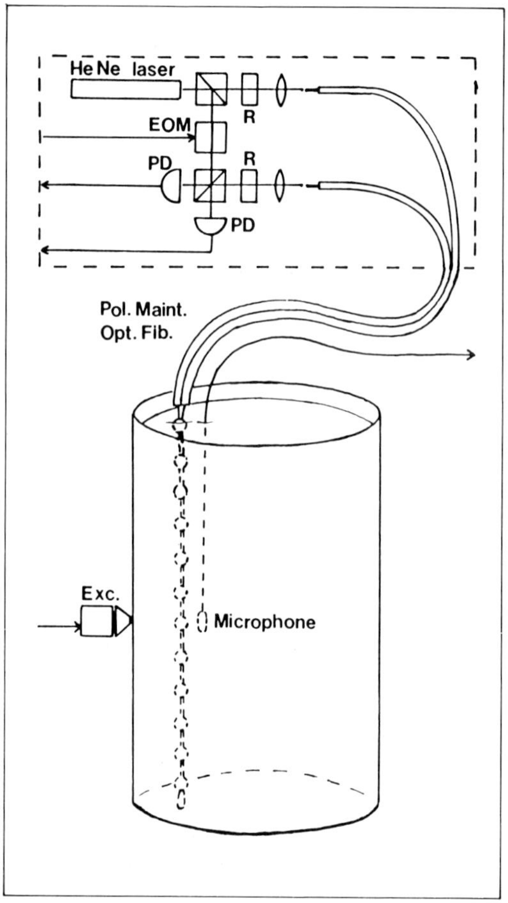

Fig. 1 Experimental set up

are 10cm spaced and cemented on the fibre itself. Care was taken to ensure that neither vibrations nor any other unwanted environmental perturbations affect the fibre sensor. The optical processing of the fibre sensor is a classical FOIS scheme. In order to collect the launched light without any loss, the output end of the sensing fibre is returned to the optical processing unit. Hence, the fibre passes twice through each sphere (see Fig. 1). The fibre is of the polarisation-maintaining kind and a couple of retarder plates, $R1$ and $R2$, match the light polarisation with the fibre birefringence principal axis. An electro-optic crystal modulator (EOM) gives a frequency shift to the interferometric signal. FOIS utilises a pseudoheterodyne scheme, and the signal processing follows that reported in reference (14). The signals from the fibre sensor, from the acoustic transducer and from the shaker are real-time processed by a four-channel FFT, Nicolet Mod.804 A.

Table 1. York high-birefringence fibre

Single-mode silica made
Cladding diameter: 125μm
Operating wavelength: 633nm
Polarisation maintaining
Beat length: 5mm
Primary coating: Acrylate
Secondary coating: None
Average pressure sensitivity: 1.10^{-3} rad/Pa·m
Elongation sensitivity: $1.16 \cdot 10^{7}$ rad/m

Measurements and results

Before going on with the experiments, preliminary tests were made to measure the pressure and elongation sensitivity of the fibre employed. The elongation sensitivity was evaluated by means of the methods reported in reference (14) and was 1.16×10^{7} rad/m, in excellent agreement with the theoretical predictions. The pressure sensitivity was measured by immersing a conventional 10m-long fibre coil in water and exciting it with an acoustic wave generated by a loudspeaker. The average value turned out to be 1×10^{-3} rad/Pa · m in the frequency range 10 = 500Hz, in agreement with the conventional pressure sensitivity for a simple

Table 2

m		*ω/2n*	*Po*	*Δφ*	*ΔLw*	*Cc*	*CR*	*Δφ*	*Δφ*
Mode number	*Theoretical axial mode shape*	*Experimental resonant frequency*	*Experimental pressure peak*	*Expected Δφ simple fibre sensor*	*Water column elongation*	*Distributed sensor elastic wave velocity*	*Tank resonance equivalent velocity*	*Expected Δφ distributed sensor*	*Experimental Δφ distributed sensor*
		Hz	*Pa*	*rad*	*m*	*m/sec*	*m/sec*	*rad*	*rad*
1	$\sin \frac{\pi}{L} z$	*32.25*	*10*	$11.5 \cdot 10^{-3}$	$8.63 \cdot 10^{-7}$	*386*	*116.2*	*1*	*2.32*
3	$\sin \frac{3\pi}{L} z$	*154.00*	*10*	$3.8 \cdot 10^{-3}$	$1.12 \cdot 10^{-7}$	*386*	*184.9*	*0.39*	*0.80*
5	$\sin \frac{5\pi}{L} z$	*263.75*	*10*	$2.3 \cdot 10^{-3}$	$6.34 \cdot 10^{-8}$	*386*	*189.9*	*0.23*	*0.60*

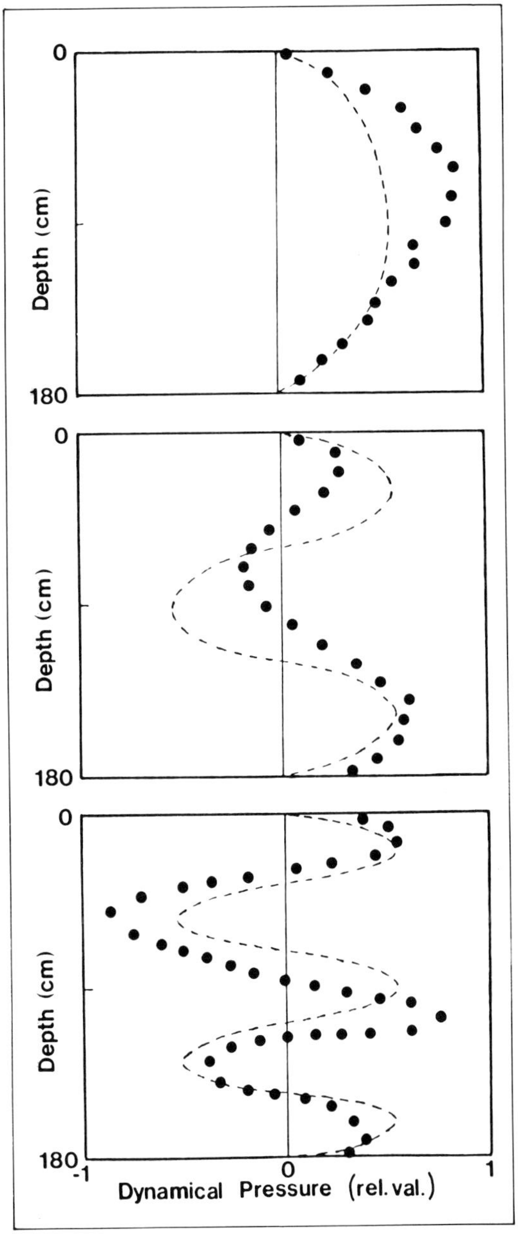

Fig. 2 Experimental (dots) and theoretical (dashed line) pressure behaviours of the first three odd axial modes.

primary-coated fibre. Both the data are reported in table 1. Another set of preliminary measurements was made in order to study the resonant behaviour of the cylindrical tank. Attention was focused to identify three modes with different m number at least. Each mode was defined in terms of the axial pressure behaviour $p(z)$ and by the circumferential pressure behaviour $p(\theta)$. Figure 2 reports the experimental data compared with theoretical curves $p(z) = p_0 \cdot \sin(m\pi z/L)$ for modes $m = 1$, $m = 3$ and $m = 5$. The experimental curves demonstrate the presence of the wanted mode, but with significant differences in comparison with the theoretical shapes. The disagreement is mainly due to differences between the boundary conditions of the laboratory tank and those assumed in the mathematical model. These differences will introduce an important approximation factor into the theoretical evaluation of the sensor behaviour.

For a preliminary evaluation of the developed theory and of the behaviour of the sensor, a set of measurements was made. The mechanical properties of prototype sensor and the values of the tank resonant frequency led to an experimental situation where C_c is always greater than $\sqrt{2 \cdot C_R}$. Hence, case (b) of the theoretical discussion applies. All the results are summarised in table 2 together with the theoretical predictions. The theoretical mode shapes and the experimental resonant frequencies are reported in columns 2 and 3. Each mode was excited to a pressure amplitude of 10Pa (column 4). By means of the experimental pressur esensitivity reported in table 1 and the theoretical mode shapes in column 2, the expected phase shifts for the simple fibre are given in column 5. In other words, column 5 reports the expected response of a hydrophone made with a simple straight length of fibre and sensitive to the whole axial mode pressure. Column 6 shows the values of ΔL_w, calculated by means of equation (6) with the data from columns 1, 3, and 4. The equivalent sound velocity C_c of the described sensor (column 7) is 386m/s. C_R (column 8) is evaluated for each mode with equation (15). The expected phase shifts for the prototype (column 9) are obtained by using data ΔL_w, C_c, C_R, equation (16) and the experimental elongation sensitivity (table 1). The last column in table 2 gives the measured phase shifts. Each resonant mode is excited at the proper frequency and the phase shift datum is obtained from the power spectrum of the sensor signal.

Concluding remarks

Remarks arise from table 2:

- The marked difference between the expected phase shift values for a simple fibre pressure sensor (column 5) and the experimental data reported in column 10, indicate the sensitivity enhancement obtained with the necklace sensor.
- The good agreement between columns 9 and 10 supports the proposed theoretical analysis of the sensor behaviour. The residual difference between the two sets of data is attributed to the approximate nature of some of the assumptions (e.g. column 2 as the mode shape).
- The ratio between C_c and C_R values reported in

columns 7 and 8 points out that the sensor almost follows the water displacement. Then, by using other types of sensor, a further increase in sensitivity can be obtained.

In summary, the behaviour of a new intrinsic-type fibre optic interferometric sensor was investigated. The sensor is configured as a necklace, where spaced spheres strain the fibre when they are subjected to a pressure gradient. A simple theoretical analysis was proposed in order to explain the sensor behaviour. Preliminary results show a rather good agreement between the theoretical predictions and the experimental data and demonstrate a sensitivity two orders of magnitude larger than the simple fibre pressure sensitivity. The developed theory shows that our distributed fibre optic sensor is a concept open to different implementations. The particular realisation of the sensor proposed herein addresses only a fraction of the sensing possibilities. Other regions, e.g. the resonance one, can be explored and make our fibre optic distributed sensor concept worth further work.

Acknowledgements

The authors wish to thank the directors of the CRIS for encouragement and authorisation to publish the material. They would also like to thank their co-workers in the Coherent Optics Section for continuous support.
This work was supported by the ENEL-Hydraulic and Structural Research Center (CRIS) under a 'PRESS-OPT' contract.

References

1. B. Culshaw, D.E.N. Davies, and S.A. Kingsley, 'Acoustic sensitivity of optical-fibre waveguides, *Electron. Lett.*, Vol. 13, 1977, p.760.
2. J.A. Bucaro and T.R. Hickman, 'Measurement of sensitivity of optical fibers for acoustic detection', *Appl. Opt.*, Vol. 18, 1979, p. 938.
3. G.B. Hocker, 'Fiber-optic sensing of pressure and temperature', *Appl. Opt.*, Vol. 18, 1979, p. 1445.
4. J.H. Cole, R.L. Johnson and P.G. Bhuta, 'Fiber-optic detection of sound', *J. Acoust. Soc. Amer.*, Vol. 62, 1977, p. 1136.
5. R. Hughes and J. Jarzynski, 'Static pressure sensitivity amplication in interferometric fiber-optic hydrophones', *Appl. Opt.*, Vol. 19, 1980, p. 98.
6. J. Jarzynski, R. Hughes, T.R. Hickman and J.A. Bucaro, 'Frequency response of interferometric fiber-optic coil hydrophones', *J. Acoust. Soc. Amer.*, Vol. 69, 1981, p. 1799.
7. T.G. Giallorenzi *et al*, 'Optical fiber sensor technology', *IEEE J. Quantum Electron.*, Vol. 18, 1982, p. 626.
8. M. Matsumoto *et al*, 'Fiber-optic acoustic sensor based on Fabry-Perot interferometer', *Proc. 4th Int. Conf. on Optical Fiber Sensors*, Tokyo, Japan, October 1986, pap. 2.8.
9. S.W. Thornton and M.L. Henning, 'Experimental performance of an optical hydrophone with downlead insensitivity', *Proc. 2nd Int. Conf. on Optical Fiber Sensors*, Stuttgart, FRG, September 1984, p. 369.
10. J.L. Brooks *et al*, 'Coherence-multiplexing of fiber-optic interferometric sensors', *Proc. OFC/OFS '85*, San Diego, CA, pap. ThBB4.
11. C.A. Wade and J.P. Dakin, 'Optical fiber hydrophone array recent progress', *Proc 2nd Int. Conf. on Optical Fiber Sensors*, Stuttgart, FRG, September 1984, p. 375.
12. U.S. Lindholm, D.D. Kana and H.N. Abrahamson, 'Breathing vibrations of a circular cylindrical shell with an internal liquid', *J. Aerospace Sci.*, Vol. 29, 1962, p. 1052.
13. A.A. Lakis and M.P. Paidoussis, 'Free vibration of cylindrical shells partially filled with liquid', *J. Sound and Vibration*, Vol. 19, 1971, p. 1.
14. M. Martinelli and A. Barberis, 'Interferometric Michelson-type optical-fiber sensor: Comparison between phase-modulation and frequency-modulation detection', *J. Opt. Soc. Amer. A*, Vol. 2, 1985, p. 603.

A distributed fibre optic sensor based on cladding fluorescence

R.A. Lieberman, L.L. Blyler and L.G. Cohen
AT&T Bell Laboratories, USA

ENVIRONMENTALLY induced changes in the properties of the cladding of an optical waveguide can cause changes in the loss characteristics of the waveguide, due to the interaction of the evanescent fields of the core modes with the optical properties of the cladding material. In many cases this sensitivity to the environment is unwanted, but recently such effects have been exploited to create sensors.[1,2,3] Recent work[4] has shown the feasibility of creating distributed sensors in which chemical dopants in polymer-based optical cladding materials enhance the dependence of fibre transmission loss on the chemical composition of the surrounding environment. A new approach to distributed sensing, embodied by the oxygen sensor described in this paper, uses the inverse process; optical energy generated by fluorescent dye molecules in the cladding is coupled into the core by means of the evanescent fields of the guided modes. These sensors differ from fibre optic sensors based on direct evanescent-wave spectroscopy[5,6,7] in that active substances having optical properties that depend on their chemical environment are made an integral part of the waveguide structure.

Fluorescence has been used in miniature transducers, bonded to the tip of optical fibres, to monitor oxygen concentration.[8,9] These transducers, called 'optodes' or 'optrodes'[10] are structures in which a fluorescent organic dye is isolated behind an oxygen-permeable membrane. Many optrodes based on evanescent-wave fluorescence of short coated sections have also been reported.[11,12,13] Light is transmitted to the optrode by an optical fibre, and fluorescence emission is collected by another fibre and returned to a photo-detector. The fluorescence emission intensity depends on the concentration of oxygen in the optrode due to the Stern-Volmer ('quenching') effect.[14] The membrane allows this concentration to rapidly equilibrate with the local environment, and may also serve to screen the dye from the presence of other quenching agents. The optrode approach, however, is not suitable for the distributed sensing of oxygen.

Materials and methods

Our early calculations indicated that the coupling of fluorescent energy generated in the fibre cladding to guided modes in the core can be appreciable, especially for fibres with large V numbers, and recent work by others[15] has shown that oxygen sensitive dyes may be dissolved in polydimethyl siloxane, (a material which has a high gas permeability) acrylate used in the early Bell Laboratories experiments. The fibre used in these experiments was made of fused silica, drawn to 125μm and coated to a final diameter of 230μm with polydimethyl siloxane (PDMS) doped with a fluorescent dye, 9,10-diphenylanthracene (9,10-D). The dye-doped silicone material was prepared by mixing a freshly-made solution of 9,10-D in an organic solvent with PDMS 'base' prior to drawing, and mixing in PDMS 'hardener' immediately prior to the draw. The silicone prepared in this way was added to a coating cup and heat-cured in place on the fused silica fibre

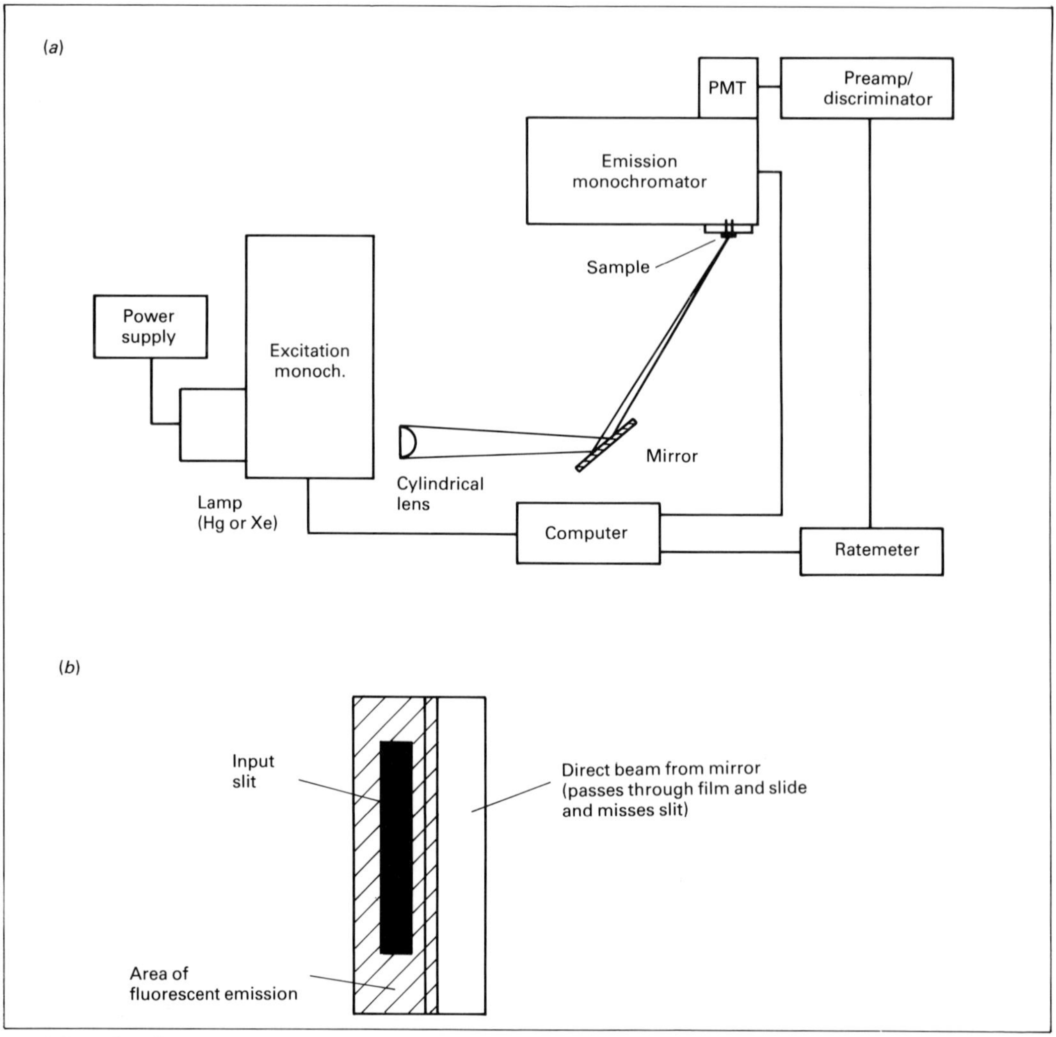

Fig. 1 Recording fluorescence from a thin layer of dye-doped silicone.
(a) Overall set-up
(b) Detail: light input to emission monochromator

during draw by means of a 'curing tube' placed downstream from the coating cup. Reference samples were created by spreading thin (25μm) layers of coating material on Pyrex microscope slides a few hours after the fibre was drawn.

To record the fluorescence spectra, and to collect data for fibre attenuation measurements, a two-monochromator photon-counting system was used (see Fig. 1(a)). The monochromators were 0.25m focal length, f/3.4 units (Jarrell-Ash model 82–410), controlled by an Apple IIe computer.[16] Photon counting was accomplished by using an SSR model 1120 amplifier/discriminator and an EG&G-Ortec model 449-2 ratemeter. Ratemeter output was recorded by the computer during scanning. To record total fluorescence output the monochromator/photon counting system at the output end of the fibre was replaced with a silicon photo-detector coupled through a current preamplifier (Ithaco model 1211) to a lock-in amplifier (Ithaco model 391), and input to the excitation monochromator was modulated by a mechanical chopper.

Results

Fluorescence spectra

Figure 2 shows the fluorescence spectra (excitation and emission) of one of the 0.25μm reference films made from the fibre coating material. The experimental arrangement used to record the data is shown in figure 1. Input and output slits on the excitation monochromator, which used a grating ruled at 1180 grooves/mm were 500μm; for the emission mono-

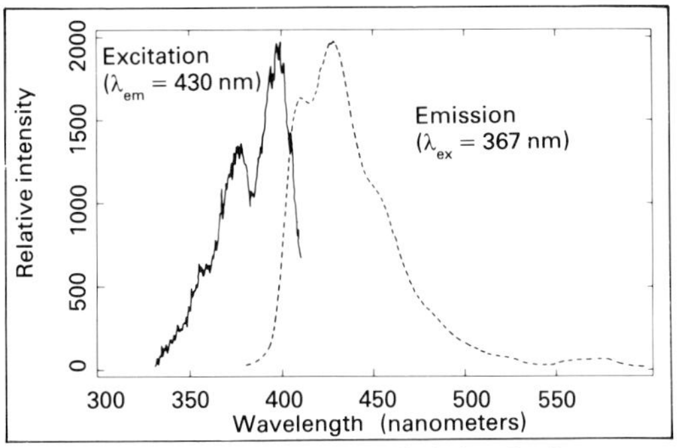

Fig. 2 Fluorescence spectrum of 9,10-diphenylanthracene in polydimethyl siloxane

chromators. The cylindrical lens pictured in figure widths were 250μm. This arrangement resulted in bandwidths of approximately 0.3nm for both monochromators. The cylindrical lens pictured in figure 1(a) was arranged so that the image of the output slit of the excitation monochromator was focused directly in front of, but was not incident on, the input slit of the emission monochromator. This arrangement assured that, when the reference sample was placed in front of the emission monochromator, a maximum amount of fluorescence emission and a minimum amount of scattered excitation light were collected. To record the emission spectrum the excitation monochromator was set at a bright line (367nm) so that a bright line in the spectrum of a 75W mercury arc lamp could be used as the excitation source, and the emission monochromator was scanned. The 367nm line is within the excitation band of the fluorescent-dye/silicone combination with which the silica fibre is clad (see Fig. 2). For the excitation spectrum, the emission monochromator was set at 430nm, near the fluorescence peak, and the excitation monochromator was scanned; to minimise the dependence of source intensity on wavelength, a 75W Xe lamp was used during the excitation scan.

The fluorescence excitation peak near 390nm and the fluorescence emission peak near 430nm are consistent with fluorescence parameters reported in other literature.[17] The slight shifts in fluorescence wavelength reported for similar systems,[18] and the 'self absorption' effect[17] arising from the choice of excitation wavelength are also observable in figure 2.

Propagation of cladding fluorescence emission by the fibre core

Figure 3 shows one arrangement used to measure the coupling of fluorescence from the cladding to guided modes in the fibre core. Cladding fluorescence was excited in a short (one metre) section of fibre illuminated from the side by arranging it into a flattened coil of roughly 20 turns taped to the front surface of a mirror. The coil was arranged perpendicular to the light exiting from the monochromator in such a way that it just filled the output light field. The mirror reflected light which had traversed the fibre without being absorbed, thus effectively doubling the optical path length of the excitation light in the fluorescent cladding. For these experiments, the input monochromator was set at 367nm. Input and output slits of the monochromator were 4mm and 5mm, respectively. Matching slits were unavailable, but taking the larger slit dimension as the determining factor in monochromator output resolution, this combination resulted in an excitation bandwidth of about 8nm, comparable to the width of the mercury emission peak.

Fluorescent light coupled into propagating modes in the fibre core was guided in an 11-metre path to the emission monochromator. A 250μm entrance slit assured that the fibre was properly aligned with the input end of the second monochromator. Opening the output slit to its widest possible setting assured that the photo-multiplier tube collected the maximum possible amount of fluorescent output light.

Using this arrangement, a fluorescence emission spectrum was collected in the vicinity of the fluorescence peak (430nm) of 9,10-diphenylanthracene. After data collection, ten metres were removed from the section of fibre between the illuminated region and the detector, and the experiment was repeated. The transmission loss was calculated in the standard way, using the equation:

$$\alpha = \frac{10}{L(\mathrm{km})} \log \left(\frac{P_{short}}{P_{long}} \right)$$

where α is the transmission loss in dBkm, L is the length of fibre removed by cleaving (0.01kM in this case), P_{short} is the power detected at the end of the fibre after cleaving, and P_{long} is the power detected at the end of the fibre before cleaving. Applying the above formula to the average of the relative intensities at 430±2nm results in a mean loss of 0.52dB/m at the peak fluorescence output wavelength. A similar experiment (see Fig. 3(b)), in which lock-in detection was used to monitor the signal from a silicon detector with no intervening monochromator, measured the average loss for the entire spectral range of the fluorescence light. This measurement resulted in a computed loss of 0.57dB/m. This slightly higher figure may be due to the broad band nature of the detection scheme; essentially all of the fluorescence output light is collected by the detector, including wavelengths below 410nm, for which the loss is considerably higher than at 430nm.

Lock-in detection was also used to perform a more conventional 'spectral loss' measurement in which the attenuation of light launched into the fibre core by a monochromatic source placed at the end was measured as a function of wavelength. The experimental

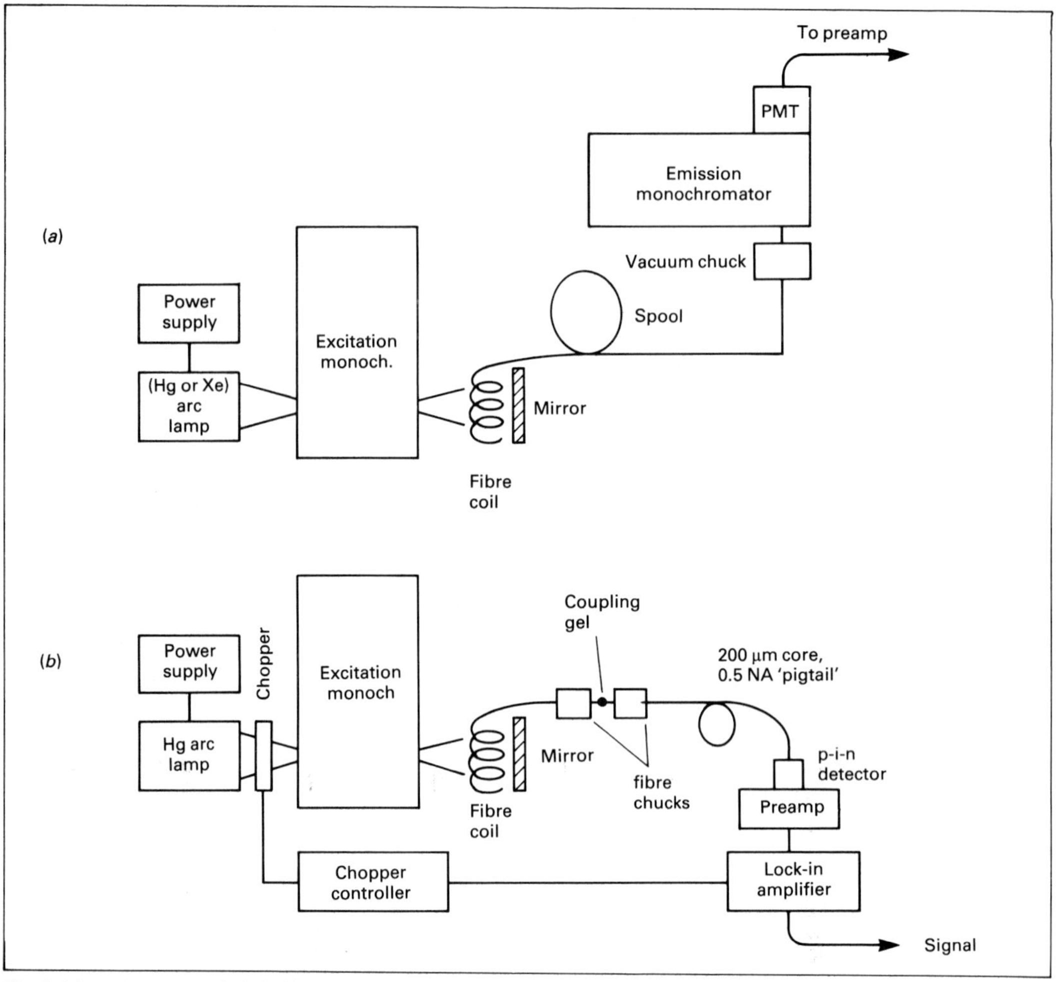

Fig. 3 Measuring core-guided cladding fluorescence. (a) Double monochromator photon counting detection system (b) Broad band (lock-in detection) system

configuration used was similar to the one shown in figure 3(b), with the mirror-mounted coil of fibre replaced by a 5mm long piece of fibre with its cladding removed. The stripped section was held by a ferrule in a way that allowed only the end face of the core to be illuminated, to ensure that the attenuation of end-coupled light, and not of cladding fluorescence generated by side illumination, was measured. Input monochromator slits were set at 150μm, and a silicon detector was used on the output end of the fibre. Since the light intensity increased rapidly for wavelengths above 450nm the system gain was changed at this point to preserve dynamic range in the loss measurement. A plot of the resulting loss curve, shown in figure 4, shows that the loss at 430nm is 0.29dB/m, 0.27dB/m less than the loss measured for light coupled into the fibre from the fluorescent cladding (taken to be 0.56dB/m, the average of the losses computed by the two methods described in the preceding paragraph). One possible source of this lower loss is the fact that the fluorescence emission guided by the fibre core is carried predominantly by the higher order bound modes, which may be more susceptible to loss mechanisms such as macrobending, cladding absorbance and scattering.

Figure 4 also shows three peaks at 390, 720 and

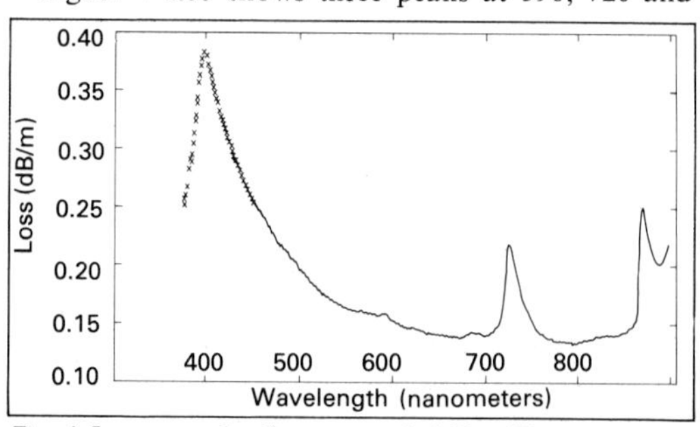

Fig. 4 Loss curve for fluorescent-clad fibre.(Transmission loss of end-coupled light–gratings changed at 450nm)

875nm. The strongest peak, at 390nm, corresponds to the fluorescence excitation maximum (see Fig. 2) of the 9,10-D/PDMS system. The high-absorbance regions at 720nm and 875nm are due to C-H bonds in the PDMS.[19]

Dependence of fluorescence output on illuminated length

For weakly guiding fibres with negligible loss at the fluorescence wavelength, the amount of cladding-generated fluorescent light coupled to guided modes in the fibre core is directly proportional to the length of the fibre, and the coupled fluorescent power measured at any point along the fibre outside the fluorescent region will be equal to the fluorescent power transmitted across the plane defining the end of the illuminated region. All optical fibres, however, exhibit transmission loss over the entire spectral range; in the case of 9,10-D/PDMS clad silica this loss is quite appreciable in the fluorescence output range. When loss is taken into account, the dependence of coupled fluorescence intensity on fibre length is non-linear; there is an upper limit on the amount of fluorescent power which can be coupled to guided modes in the core. The length at which the coupled fluorescence reaches 90% of this maximum value is:

$$x_{90} = \frac{10}{\alpha_1}$$

where α_1 is the attenuation coefficient of the fibre at the fluorescence emission wavelength. For coupled cladding fluorescence propagating in 9,10-D/PDMS clad silica fibres, the measurements of the preceding section imply $\alpha_1 = 0.56$dB/m, and the 90% length is $x_{90} = 18$ metres. For short lengths ($x \leq 1.2/\alpha_1$) the dependence of fluorescence output upon illuminated length will be linear to within 20%.

To investigate the length dependence of coupled fluorescence, an illuminated section of 9,10-D/PDMS clad silica fibre was gradually shortened by removing individual coils from an arrangement similar to that shown in figure 3(b). To ensure uniform illumination the fibre was arranged in a spiral approximately 3cm in diameter on a quartz plate, held in place with tape, and illuminated from the back (through the plate). The fibre tip at the end of the spiral was masked from the source to avoid end-coupling of excitation light, and was coated with index-matching gel to minimise the effect of end-reflections on the measurement of coupled fluorescence. Figure 5 shows the results of this experiment, and confirms that the dependence of output intensity on illuminated length is linear to within 20% for lengths in the 0–2 metre range.

Oxygen response

Figure 6 illustrates the oxygen-sensing capabilities of

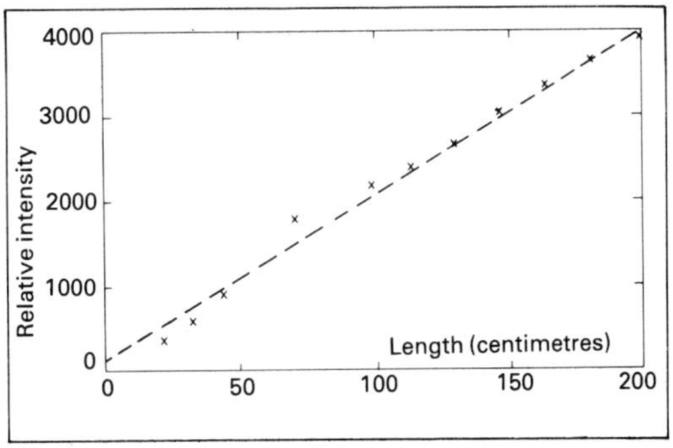

Fig. 5 Dependence of fluorescent output intensity on illuminated length

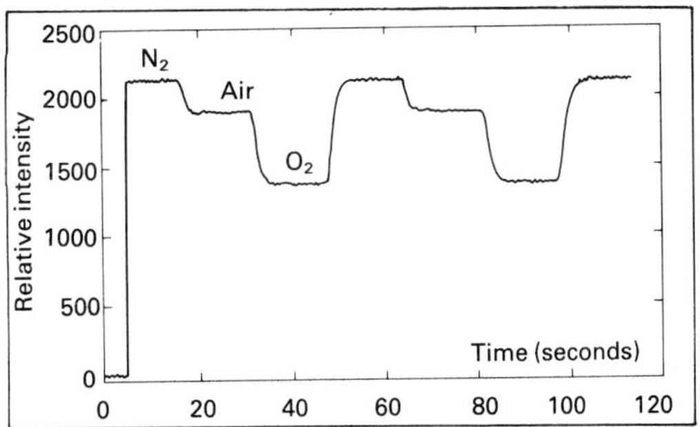

Fig. 6 Oxygen response of fluorescent-clad fibre

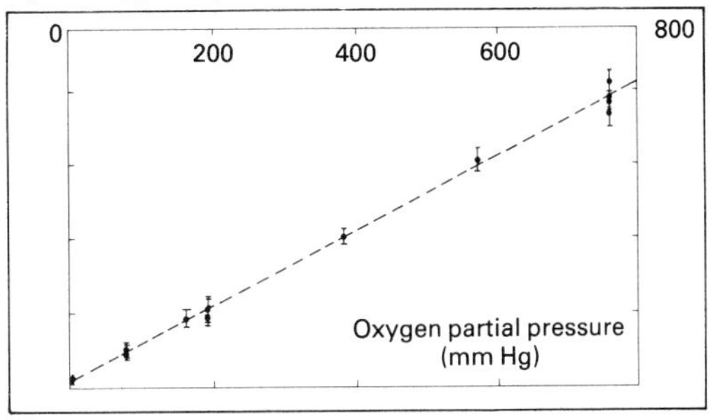

Fig. 7 Stern-volmer plot

9,10-D/PDMS clad silica fibre. A 130cm segment of sensor fibre was coiled in an arrangement similar to that shown in figure 3, and placed in a test chamber having a volume of approximately 25cc. The test chamber was continuously flushed with predetermined concentrations of oxygen in nitrogen, namely pure oxygen, pure nitrogen, air (21% O_2:N_2), and four commercially available mixtures (10%, 25%, 50% and 75% O_2:N_2). The environment was changed by manually opening and closing valves leading to a manifold connected to the test chamber. In this way, output intensities at seven different oxygen concentrations were obtained; figure 7 shows a plot of the ratio (I_0/I) of the intensity measured in an oxygen-free (pure nitrogen) atmosphere to the intensities measured when oxygen was present in the test chamber. The error bars on this plot correspond to the standard deviations of the segments of the experimental curves (e.g. the 'flat' sections in figure 6) which were averaged to obtain the values plotted in figure 7. Figure 7 also shows a linear least-squares fit of the data to the Stern-Volmer equation.[25]

$$\frac{I_0}{I} = 1 + \frac{P_{O_2}}{P'}$$

where I_0 is the fluorescence intensity in the absence of oxygen, I is the fluorescence intensity obtained when the oxygen partial pressure is P_{O2}, and P' is the 'half-quench pressure' characterising the oxygen response of the system.

The half-quench pressure of 9,10-D in the PDMS cladding of the fibre, determined from figure 7, is $P' = 1588$mmHg. In analysing this data, it has been assumed that all of the light intensity at the silicon detector of figure 3(b) is due to fluorescence emission from the 9,10-D/PDMS cladding. The excitation of fluorescence in the core, and coupling of excitation light into guided modes by scattering in the cladding, could add a small oxygen-independent component to the intensity measured at the detector. Analysis of the data presented in figure 7 shows that this added background intensity, whose presence would have the effect of artificially raising the apparent value of P', is less than 10% of the unquenched intensity.

The sensor response time, τ_r, defined as the time it takes the sensor output to drop by *1-1/e* of the difference between equilibrium values when the oxygen concentration is increased, was calculated using the data in figure 6. The results show that the sensor responds with a time constant less than 5 seconds when the source of the gas flow into the test chamber is changed. This upper limit on the response time is considerably longer than the time constant (0.4s) used in the ratemeter during data collection, the fluorescence emission time constant of 9,10-D (of the order of 10ns) and the time (146ms) for oxygen to diffuse through the cladding to the core of the fibre (estimated using the diffusion coefficient, 3.56×10^{-5} cm^3/s, for oxygen in PDMS[4,15]). From these considerations, it seems probable that the determining factor in the system response is the time associated with changing the gas environment during the experiment. The system response time is shorter than 13–17s. '1/e' response times derived from reports on fluorescence-quench based optrode oxygen sensors.[9] This may be due to differences in sensor layer thickness and composition, or to differences in the time required to change atmospheres in the test chambers used.

Discussion

Fluorescence coupling to optical fibres

Although it is not obvious, from the 'ray-approximation' point of view, that optical energy originating sources in a fibre cladding can be truly guided by the core, this phenomenon does indeed occur, and forms the basis for the operation of the distributed fluorescence sensor. In the ray-approximation, light entering a fibre core from the side will always enter at an angle greater than the critical angle for total internal reflection, and will thus enter as 'leaky rays'; energy is coupled back out to the cladding at each reflection of such rays as they propagate down the fibre. A more complete treatment of the process, however, using the weak guidance approximation to describe the guided modes[20] shows that light originating in the cladding energy can indeed by coupled to guided modes in the fibre core and transmitted over long distances without appreciable loss. The evanescent fields of the guided modes provide a path for this coupling, but it is difficult to form a simple geometrical picture which accounts for this coupling. In a sense, the process may be thought of as the inverse of the well-known cladding loss phenomenon, whereby optical absorbance centres in the cladding cause attenuation of light guided in the fibre core. Cladding loss can be understood by the simple ray-approximation, if the evanescent electromagnetic field of a bound ray is expressed as a simple decaying exponential function in the cladding, matched to an oscillating exponential in the core at the instant of total internal reflection. In the absence of absorbence centres in the evanescent 'tail', all of the energy in the cladding comes back into the core; if an absorbence centre is present, energy is lost in the cladding. One might stretch the analogy to the cladding source case, imagining that light emitted by a source in the cladding can momentarily have a spatial dependence that resembles the decaying exponential associated with a ray bound in the core. This would result in the excitation of an oscillating exponential in the core, giving rise to a bound ray. This analogy is somewhat strained, and a better means of understanding the coupling phenomenon is through careful consideration of the more mathematically correct mode-theoretical approach.

Fluorescence coupling efficiency

Theoretical calculations related to this series of experiments have been performed.[20] The results show that the efficiency of the coupling of cladding fluorescence to the guided modes depends strongly upon the 'V number' of the fibre:

$$V = \frac{2\pi}{\lambda} a(n_1^2 - n_0^2)^{1/2}$$

where λ is the vacuum wavelength of the guided light, a is the radius of the fibre core, and n_1 and n_0 are the indices of refraction of the core and cladding, respectively.

Fibres with higher V numbers exhibit stronger coupling coefficients since these fibres have a larger number of modes near cut-off, and the evanescent fields of such modes penetrate more deeply into the cladding. For the 9,10-D/PDMS clad silica fibre used for the experiments reported here, with a 125μm diameter core, $V = 365$ at 430nm; for this fibre, the numerical model predicts that the fraction of the total fluorescence output which is coupled into guided modes is:

$$P_{core}/P_{tot} > 1.65 \times 10^{-4}$$

While it is difficult to determine the collection efficiency experimentally, it may be estimated by comparing the fluorescence intensities recorded as in figure 1 with a similar experiment in which sections of fibre, rather than a thin film, are illuminated (see Fig. 3). The data in figure 2 represents the fluorescence output from a 25μm thick sheet of material on a slide placed 1mm away from the 250μm input slit of an f/3.4 monochromator. The illuminated area completely covered the 'acceptance wedge' of the monochromator, and so may be approximated as a Lambertian source. Using this approximation the total output power measured from the slide is related to the power density generated in the film by:

$$I_s^F = R\left(\frac{\theta_{crit}}{\pi/2}\right)\frac{lwt}{2}$$

where θ_{crit} is the half angle of the acceptance wedge of the monochromator ($\theta_{crit} = Tan^{-1}\ (1/2f)$), f is the 'f-number' of the monochromator, I_s^F=measured intensity of slide fluorescence, R is the power density (energy per unit volume) of light emitted by the thin film, w is the width of the fluorescing stripe, l is the length of the stripe and t is the thickness of the stripe.

For f=3.4, the monochromator collects only 4.5% of the light generated (isotropically) in the stripe. To compare the collected fluorescence with the total cladding fluorescence output, the effect of transmission loss on the total power transmitted to the output end of the fibre must be considered. For the arrangement shown in figure 3, but with the fibre spool arranged so that illuminated sections are separated by unilluminated sections, and with an appreciable length of fibre connecting the spooled section to the detection system (output monochromator and photo-multiplier), the total coupled power is:

$$I_f^F = KRA\delta x\left(\frac{P_{core}}{P_{tot}}\right)$$

where:

$$K = \left(10^{-\alpha(2\pi rN + l_0)}\right)\left(\sum_{n=1}^{n=N} 10^{2\pi r\alpha n/10}\right)$$

α =fibre loss at the emission wavelength, I_F^f=measured intensity fluorescence, r =spool radius, n =number of windings on spool, l_0 =length of fibre from spool to detector, RA =amount of power generated per unit length of illuminated fibre, A =cross sectional area of fluorescent layer and δx =length of illuminated segments.

Combining the above equations and accounting for differences in the optical arrangement used to illuminate the two types of sample:

$$\frac{P_{core}}{P_{tot}} = \left(\frac{C}{K\delta x}\right)\left(\frac{I_s^0}{I_f^0}\right)\left(\frac{I_f^F}{I_s^F}\right)$$

where:

$$C = \frac{2\theta_c w f t}{\pi^2(b^2-a^2)}$$

is a geometrical factor, a =fibre core radius, b =fibre cladding radius, I_s^0=intensity of excitation light for the slide and I_f^0=intensity of excitation light for the fibre.

Using the values obtained for measurements on the spooled fibre and illuminated film, the calculated coupling efficiency ratio is:

$$P_{core}/P_{tot} = 2.04 \times 10^{-4}$$

where account has been taken of the fact that the incident power densities and ratemeter gain settings were different for the two experiments being compared. Due to uncertainties in the numerous experimentally determined constants which were used to derive the above ratio, it should be considered an 'order of magnitude' estimate. The surprisingly good agreement of this number with the predicted value verifies the general validity of the model. The fact that the experimental results indicate higher coupling efficiencies than the theoretical model, even though the test fibre had a thinner coating (b/a=2) than assumed in the model (b/a=5), may be due to light carried by 'leaky modes' (low-order unbound modes) which are not treated in the model.

Oxygen sensitivity

Although our experimental equipment does not allow direct determination of the minimum detectable oxygen partial pressure, this quantity may be estimated by finding the oxygen partial pressure which causes a change in fluorescence intensity comparable to the measurement 'noise' (the standard deviation of the intensity measured in an oxygen-free environment). By this criterion, the minimum oxygen concentration which can be detected with the manifold system used in these experiments is roughly 0.01atm (8mmHg) for the 130cm length of fibre used. Assuming that the standard deviation in the signal is independent of signal intensity, the ultimate sensitivity of the system, reached by illuminating a section of fibre longer than the power saturation length (x_{90} = 18 metres) would be Δ_{min} (P_{o_2}) = 1.2mmHg. Illuminated lengths of fibre short compared to the saturation length would have a minimum detectable partial pressure inversely proportional to the length of the fibre, with the constant of proportionality being 10mmHg–m (e.g. a one metre length of fibre would have a minimum detectable P_{o_2} of 10mmHg). This is to be compared with minimum detectable P_{o_2} values of 1–2mmHg reported for optrodes using similar dye/silicone systems.[9]

The minimum detectable oxygen partial pressure could be improved by improving the signal-to-noise ratio of the detection arrangement. For example, since an arc lamp is used as the excitation source, source flicker is a likely cause of noise in the overall measurement. The effect of source fluctuations could be reduced by stabilising the source through a feedback control mechanism, or by monitoring source intensity and using a ratioing technique.

Microbending, induced when the flowing gas stream used to control the environment shakes the fibre, is another possible noise source. This effect could be reduced by the use of a less turbulent means of controlling the gas environment, and by implementing mounting procedures which would fix the fibre more firmly in the test chamber. Microbending noise could also be minimised by coupling light into the fibre at a wavelength near to, but not within, the fluorescence output wavelength range, and detecting the output light intensity at two wavelengths. Output intensity at the non-fluorescent wavelength would be affected by microbending, but would be independent of oxygen concentration, and so could be used as a reference intensity for signal processing techniques such as ratioing. If the reference intensity were derived from the same source as the excitation light, this scheme would also minimise source-fluctuation effects.

The range of oxygen concentrations for which a fluorescent dye is most sensitive to changes in oxygen concentration is determined by the half-quench pressure (P') of the dye in the medium in which it is held. Although a suitable response was obtained for large excursions in oxygen partial pressures in the range 0–1atm with 9,10-D (P' = 1588mmHg), many other fluorescent dyes exhibit oxygen quenching; dyes with different values of P' might be more suitable for other measurements. Specifically, dyes with lower values of P' (for example, diphenyl methane, acenapthene, or napthalene) could be used to detect very low oxygen concentrations (of use in sensors to be used as oxygen 'alarms' in anaerobic environments); dyes with high values of P' (such as 2-phenyl indole) are suitable for measurements of oxygen concentration in high-pressure environments.

Use of fluroescent cladding fibre sensors

The distributed nature of the fluorescent-cladding sensor technique means that it is suitable for situations where point measurements are impossible or unsuitable. For example, a distributed oxygen sensor would be superior to point sensors in ensuring the integrity of anaerobic environments such as those used in many biotechnological and chemical processes. In such an application, a fluorescent-clad fibre could be placed around the seal of an anaerobic chamber, and any leak across the seal would be detected as a change in fluorescence yield. This would assure an immediate response to a leak anywhere on the perimeter of the seal, with much less complexity than any arrangement using point sensors such as Clark electrodes or fluorescent optrodes.

Fluorescent dyes which are sensitive to compounds other than oxygen are also available. Such compounds could form the basis of other cladding fluorescence based sensors, and could be used in conjunction with oxygen sensors in a number of industrial and environmental applications. For example, pH changes affect the fluorescence output of 8-hydroxypyrene-1,3,6 trisulfonic acid ('HPTS' or 'pyranine'),[21] calcium causes changes in the excitation spectrum of fura-2 and 11-(9-carbazole) undecanoic acid is sensitive to quenching by chlorinated hydrocarbon pesticides.[22] In addition, a host of fluorescent-labelled antibodies and antigens may be purchased commercially; such substances could be used in the distributed sensing of biomolecules in research, in biotechnological instrumentation and in clinical and environmental monitoring applications.

Concluding remarks

We have shown that light generated in an optical fibre cladding applied during draw may be coupled to guided modes in the fibre core. Using this phenomenon, we have demonstrated the feasibility of using a fluorescent-clad fibre as a distributed chemical sensor.

Acknowledgements

We thank A.C. Hart, R.G. Huff and F.V. DiMarcello for drawing the fluorescent-clad fibre used in these experiments. We would also like to thank D. Marcuse for discussions leading to better understanding of evanescent-field coupling of optical energy, and

S. Siddiqui for his advice on the addition of fluorescent dyes to polymer matrices.

References

1. E.E. Hardy, D.J. David, N.S. Kapany and F.C. Unterleitner, *Nature*, 257, 1975, p. 666.
2. J.T. Giuliani, H. Wohltjen and N.L. Jarvis, *Optics Lett*, 8, 1983, p. 54.
3. A. Arie, R. Karoubi, Y.S. Gur and M. Tur, *Appl. Opt.*, 25(11), 1986, p. 1754.
4. L.L. Blyler, J.A. Ferrara and J.B. MacChesney, *A Plastic-Clad Silica Fiber Chemical Sensor*, AT&T Bell Labs. Technical Memorandum, 1986.
5. A. Saafai-Jazi, C.K. Jen and G.W. Farnell, *Appl. Opt.*, 24(15), 1985, pp. 2341–2345.
6. M.T. Wlodarczyk and D.J. Vickers, *Fiber Optic and Laser Sensors IV*, SPIE Vol. 718, 1986, pp. 192–196.
7. H. Yoshikawa, M. Watanabe and Y. Ohno, *OFS '86 Fourth International Conference on Optical Fiber Sensors,* OITDA, Tokyo, Japan, 1986, pp. 105–108.
8. J.I. Peterson, S.R. Goldstein, R.V. Fitzggerald and D.K. Buckhold, *Anal. Chem.*, 52(6), 1980, p. 864.
9. O.S. Wolfbeis, L.J. Weis, M.J.P. Liener and W.E. Ziegler, *Anal. Chem. Acta*, 205, 1988, p. 1.
10. D.W. Lubbers and N. Optiz, *Sensors and Actuators*, 4, 1983, p. 641.
11. R.M. Sutherland, C. Dahne, J.F. Place and A.S. Ringrose, *Clin. Chem., 30(9), 1984, pp. 1533*–1538.
12. J.D. Andrade, R.A. Vanwagenen, D.E. Gregonis, K. Mewby and J.-N. Lin, *IEEE Trans El. Dev*, ED-32(7), 1985, pp. 1175–1179.
13. W.F. Love and R.E. Slovacek, *OFS '86 Fourth International Conference on Optical Fiber Sensors*, OITDA, Tokyo, Japan, 1986, pp. 143–146.
14. O. Stern and M. Volmer, *Physik Z*, 20, 1919, p. 183.
15. J.W. Parker, M.E. Cox and B.S. Dunn, *Fiber Optic Sensors*, SPIE Vol. 586, 1985, p. 156.
17. I.B. Berlman, *Handbook of Flourescence Spectra of Aromatic Molecules*, Second Edition, Academic Press, NY, 1971.
18. M.E. Cox and B. Dunn, *Appl. Opt.*, 24(14), 1985, p. 2114.
19. S. Tanaka, K. Inada, T. Akimoto and M. Kozima, *Electron. Lett.*, 11, 1975, p. 153.
20. D. Marcuse, *IEEE J. Lightw. Tech.*, 6(8), 1988, p. 1273.
21. O. Wolfbeis, E. Furlinger, M. Kroneis and H. Marsoner, *Fresenius' Z. Anal. Chem.*, 314(2), 1983, p. 119.
22. J.R. Lakowicz, D. Hagen and G. Omann, *Bioch. Bioph. Acta*, 471, 1977, p. 401.

A plastic clad silica fibre chemical sensor for ammonia

L.L. Blyler and J.B. MacChesney
AT&T Bell Laboratories, USA

J.A. Ferrara
University of Washington, USA

PLASTIC-CLAD silica (PCS) fibres, having a rubbery silicone cladding, offer a distinct advantage for chemical sensing applications in that small molecule chemical species will diffuse through the cladding very rapidly. If a dye is incorporated in the silicone which is altered in colour by a chemical of interest, the dye will be affected throughout the silicone volume. Further, because the evanescent tail of the light energy propagating in the fibre core extends into the cladding where it interacts with the dye, the absorption loss spectrum of the fibre is altered by the presence of the dye and by any colour changes which it undergoes. The fibre thus represents a distributed chemical sensor in that absorption loss changes which occur anywhere along its length may be detected.

Oxazine perchlorate has been used to detect ammonia by earlier workers.[1] It is blue in acidic and neutral media with a peak absorption at a wavelength of 665nm, and red in basic environment with a peak absorption at 534nm.

In this work a dye-doped PCS fibre was fabricated for the detection of ammonia. The dye was dissolved into a photo-curable silicone acrylate formulation ($n_d = 1.42$) used to clad the silica core.

Several hundred metres of fibre were drawn from a high purity synthetic silica (Suprasil 2) preform rod and clad with the dye-doped, UV-curable silicone acrylate. A test length of this fibre on a plastic spool was placed in a 1-gallon container which could be purged with anhydrous ammonia, and/or air, in any desired mix ratio. The ends of the fibre were led out of the container to a spectral loss test set.

Figure 1 shows the spectral loss characteristics of the dye-doped silicone acrylate-clad fibre in air and in 100% ammonia. The principal difference in the loss curves is the dye-related absorption peak at 665nm which disappears in the ammonia atmosphere. The fibre loss at 665nm changes from a value of 450dB/km in air to 90dB/km in 100% ammonia.

With the assumption that the normalised transmission loss change at 665nm is linearly related to normalised ammonia concentration, the time-dependent response of the PCS ammonia sensor was analysed assuming molecular (Fickian) diffusion

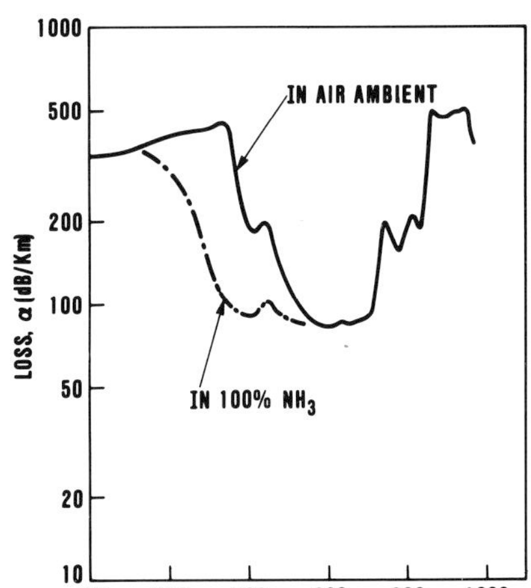

Fig. 1 Spectral losses of PCS fibre ammonia sensor in air and in 100% ammonia

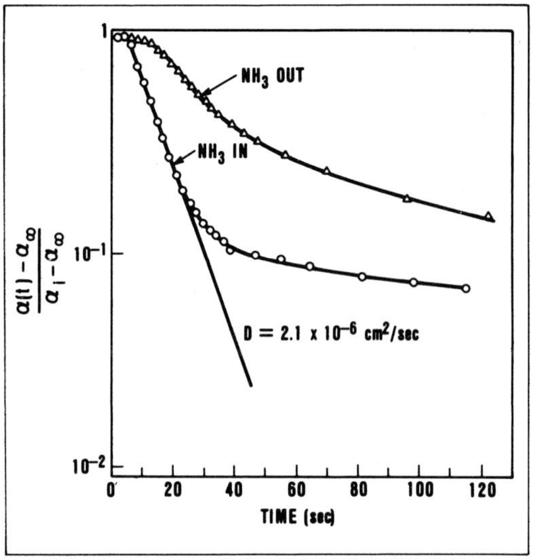

Fig. 2 Time-dependent loss changes of 665nm for PCS ammonia sensor

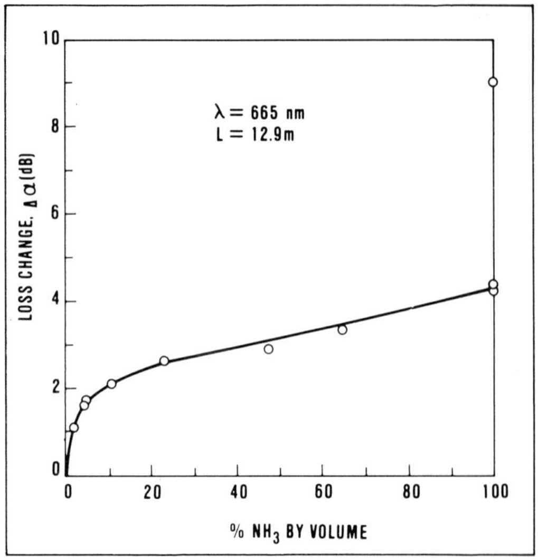

Fig. 3 Concentration-dependent loss changes for PCS ammonia sensor

mechanism.[2] The logarithm of the normalised loss change is plotted against linear time in figure 2 for both sudden ammonia exposure after equilibration in air and sudden air exposure after equilibration in 100% ammonia. For the former case there is an initial time delay of 7s, principally related to container fill time, followed by linear region from 7–25s. From the slope of this curve the diffusivity of ammonia in the rubbery polymer cladding is determined to be 2×10^{-6} cm^2/s. The sudden air exposure response of the sensor is considerably slower than the ammonia exposure response, and reversibility is incomplete, implying an interfering chemical interaction between ammonia and a component of the silicone acrylate cladding. The equilibrium response of the sensor to changes in ammonia concentration is depicted in through the rubbery silicone to be the controlling figure 3. The concentration dependence is non-linear with highest sensitivity in the low concentration range.

The dye-doped PCS fibre has demonstrated several advantages for chemical detection, which include high sensitivity, rapid response and distributed sensing. Deficiencies observed in this work, such as incomplete reversibility, non-linear concentration response and non-specificity, may be addressed through appropriate dye and cladding polymer formulation.

References

1. J.F. Guiliani, H. Wohltjen and N.L. Jarvis, *Optics Lett.*, 8, 54, 1983.
2. J. Crank, *The Mathematics of Diffusion*, Oxford University Press, London, UK, 1975, chapter 5.

Authors' Organisations and Addresses

H Arditty
Photonetics SA
BP 39
52 Avenue de L'Europe
78160 Marly Le Roi
France

I Bennion
Plessey Research Caswell
Towcester
Northants NN12 8EQ
UK

J J Bernard
Photonics & Electronics Dept
Laboratoires de Marcoussis
CGE Research Centre
Route de Nozay
91460 Marcoussis
France

G W Bibby
Central Electricity Generating Board
Central Electricity Research Laboratories
Kelvin Avenue
Leatherhead
Surrey KT22 7SE
UK

L L Blyler
AT&T Bell Laboratories
600 Mountain Avenue
Murray Hill
NJ 07974-2070
USA

A C Boucouvalas
Hewlett-Packard Laboratories
Filton Road
Stoke Gifford
Bristol BS12 6QZ
UK

L G Cohen
AT&T Bell Laboratories
600 Mountain Avenue
Murray Hill
NJ 07974-2070
USA

J P Dakin
York Ltd
York House
School Lane
Chandler's Ford
Hampshire SO5 3DG
UK

E Depresler
Photonics & Electronics Dept
Laboratoires de Marcoussis
CGE Research Centre
Route de Nozay
91460 Marcoussis
France

F X Desforges
Photonetics SA
BP 39
52 Avenue de L'Europe
78160 Marly Le Roi
France

J R Dunphy
United Technologies Research Center
Mailstop 92
Silver Lane
East Hartford
CT 06108
USA

C Edge
Plessey Research Caswell
Towcester
Northants NN12 8EQ
UK

R M Elkow
United Technologies Research Center
Mailstop 92
Silver Lane
East Hartford
CT 06108
USA

R E Epworth
STC Technology Ltd
London Road
Harlow
Essex CM17 9NA
UK

J K A Everard
Dept of Electronic & Electrical Engineering
King's College London
Strand
London WC2R 2LS
UK

C Farries
Plessey Research Caswell
Towcester
Northants NN12 8EQ
UK

M E Fermann
Dept of Electronics & Computer Science
University of Southampton
Southampton SO9 5NH
UK

J A Ferrara
Dept of Chemical Engineering
University of Washington
Seattle
WA 98195
USA

R B Franks
Dept Electronic & Electrical Engineers
University of Sheffield
Applied Science Building
Mappin Street
Sheffield S1 3JD
UK

M J Goodwin
Plessey Research Caswell
Towcester
Northants NN12 8EQ
UK

P Graindorge
Photonetics SA
BP 39
52 Avenue de L'Europe
78160 Marly Le Roi
France

M Greaves
GEC Research Ltd
Hirst Research Centre
East Lane
Wembley
Middx HA9 7PP
UK

M Groos
Laboratoire d'Electronique
Physique Appliquée & Productique
Ecôle Centrale de Paris
Grande Voie des Vignes
92295 Chatenay-Melabry
France

L Guenadez
Laboratoire d'Electronique
Physique Appliquée & Productique
Ecôle Centrale de Paris
Grande Voie des Vignes
92295 Chatenay-Melabry
France

A H Hartog
York Ltd
York House
School Lane
Chandler's Ford
Hampshire SO5 3DG
UK

S Hattori
Tokyo Electric Power Co
Telecommunication Engineering Dept
Computer & Communication Research Centre
Irifune 1-4-10
Chuo-Ku
Tokyo 104
Japan

T Higashi
NTT Transmission Systems Laboratories
Tokai
Ibaraki-ken 319-11
Japan

L Jeunhomme
Photonetics SA
BP 39
52 Avenue de L'Europe
78160 Marly Le Roi
France

N E Jolley
STC Technology Ltd
London Road
Harlow
Essex CM17 9NA
UK

S A Kingsley
Fiberdyne Optoelectronics
545 Northview Drive
Columbus
OH 43209
USA

K Kurosawa
Tokyo Electric Power Co
Engineering Research Centre
Nishi-Tsutsuji – Gaoka 2-4-1
Chofu-shi
Tokyo 182
Japan

P Lecoy
Laboratoire d'Electronique
Physique Appliquée & Productique
Ecôle Centrale de Paris
Grande Voie des Vignes
92295 Chatenay-Melabry
France

R A Lieberman
AT&T Bell Laboratories
600 Mountain Avenue
Murray Hill
NJ 07974-2070
USA

J B MacChesney
AT&T Bell Laboratories
600 Mountain Avenue
Murray Hill
NJ 07974-2070
USA

M Martinelli
CISE SpA
Via Reggio Emilia 39
20134 Segrate (Milano)
Italy

V D McGinniss
Battelle Columbus Division
505 King Avenue
Columbus
OH 43201-2693
USA

R S Medlock
ABB Kent plc
Lea Road
Luton
Beds LU1 3AE
UK

G Meltz
United Technologies Research Center
Mailstop 92
Silver Lane
East Hartford
CT 06108
USA

G Oscroft
Herga Electric Ltd
Northern Way
Bury St Edmunds
Suffolk IP32 6NN
UK

D A J Pearce
Madge Networks
100 Lodge Lane
Chalfont St Giles
Bucks HP8 4AH
UK

D J Pratt
Plessey Research Roke Manor Ltd
Roke Manor
Romsey
Hampshire SO51 0ZN
UK

N S Rayit
GEC Research Ltd
Hirst Research Centre
East Lane
Wembley
Middx HA9 7PP
UK

A J Rogers
Dept of Electronic & Electrical Engineering
King's College London
Strand
London WC2R 2LS
UK

J N Ross
Dept of Electronics & Computer Science
University of Southampton
Southampton SO9 5NH
UK

O de Sanctis
CISE SpA
Via Reggio Emilia 39
20134 Segrate (Milano)
Italy

S Seikai
Faculty of Science & Engineering
Ritsumeikan University
Kyoto-shi 603
Japan

P G Sona
CISE SpA
Via Reggio Emilia 39
20134 Segrate (Milano)
Italy

A P Strong
Plessey Research Roke Manor Ltd
Roke Manor
Romsey
Hampshire SO51 0ZN
UK

E Theocharous
BP Research Centre
Chertsey Road
Sunbury-on-Thames
Middlesex TW 16 7LN
UK

R Thomas
Dept of Electronic & Electrical Engineering
King's College London
Strand
London WC2R 2LS
UK

W Torruellas
81 Avenue Cartemany
Les Escalpes
Principality of Andorra
France

M Tsubokawa
NTT Transmission Systems Laboratories
Tokai
Ibaraki-ken 319-11
Japan

C A Wade
Plessey Research Caswell
Towcester
Northants NN12 8EQ
UK

S Wright
Hewlett-Packard Laboratories
Filton Road
Stoke Gifford
Bristol BS12 6QZ
UK

T Yoshino
Dept of Electrical Engineering
Faculty of Engineering
Gunma University
Tenjin-cho
Kiryu 376 Gunma
Japan

R C Youngquist
Edward L Ginzton Laboratory
W W Hanser Laboratory of Physics
Stanford University
Stanford
CT 96305
USA

Source of material

Multiplexed and distributed optical fibre sensor systems
First published in Journal of Physics E: Scientific Instruments, Vol. 20, No. 8, August 1987, pp. 954–967.
Reprinted courtesy of the author and Institute of Physics.

Review of modulating techniques for fibre optic sensors
First published in the Journal of Optical Sensors, Vol. 1, No. 1, 1986, pp. 43–68.
Reprinted courtesy of the author and Taylor & Francis Ltd.

Distributed optical-fibre sensors for the measurement of pressure, strain and temperature
First published in the Journal of the Institution of Electronic and Radio Engineers, Vol. 58, No. 5, July/August 1988, pp. S113–S122.
Reprinted courtesy of the author and The Institution of Electrical Engineers.

Advances in distributed FODAR (Fiber Optic Detection and Ranging)
First presented at the Fibre Optic and Laser Sensors IV Conference, 22–24 September 1986, Cambridge, MA, USA. Proceedings of SPIE, Vol. 718, pp. 66–77
Reprinted courtesy of the author and Society of Photo-Optical Instrumentation Engineers (SPIE)

Point and distributed polarimetric optical-fibre sensors
First presented at the Fibre Optic and Laser Sensors IV Conference, 22–24 September 1986, Cambridge, MA. USA. Proceedings of SPIE, Vol. 718, pp. 28–38
Reprinted courtesy of the author and Society of Photo-Optical Instrumentation Engineers (SPIE)

A coherent time multiplexed distributed sensing system
First presented at the IEE Colloquium on Distributed Optical Fibre Sensing, 12 May 1986, London, UK, pp. 1/1–4
Reprinted courtesy of the authors and The Institution of Electrical Engineers

A novel distributed optical fibre sensing system enabling location of disturbances in a Sagnac Loop Interferometer
First published in the Journal of Optical Sensors, Vol. 2, No. 6, 1987, pp. 403–406
Reprinted courtesy of the authors and Taylor & Francis Ltd.

Distributed sensors based on differential absorption
First presented at the IEE Colloquium on Distributed Optical Fibre Sensing, 12 May 1989, London, UK, pp. 13/1–4
Reprinted courtesy of the authors and The Institution of Electrical Engineers

Correlation-optical-time-domain-reflectometry for high resolution distributed fiber-optic sensing
First presented at the Fibre Optic and Laser Sensors V Conference, 17–19 August 1987, San Diego, CA, USA. Proceedings of SPIE, Vol. 838, pp. 206–209
Reprinted courtesy of the authors and Society of Photo-Optical Instrumentation Engineers (SPIE). Updated August 1989

Distributed optical fibre temperature sensor using spread-spectrum techniques
First published in the Electronics Letters, Vol. 25, No. 2, 19 January 1989, pp. 140–142
Reprinted courtesy of the authors and The Institution of Electrical Engineers

Progress in OTDR optical fibre sensor networks
First presented at the Fibre Optic and Laser Sensors IV Conference, 22–24 September 1986, Cambridge. MA, USA. Proceedings of SPIE, Vol. 718, pp. 225–236
Reprinted courtesy of the authors and Society of Photo-Optical Instrumentation Engineers (SPIE)

Birefringent stress location sensor
First presented at the Fibre Optic Sensors Conference, 26–27 November 1985, Cannes, France. Proceedings of SPIE, Vol. 586, pp. 84–89
Reprinted courtesy of the authors and Society of Photo-Optical Instrumentation Engineers (SPIE). Updated August 1989

Distributed fibre-optic sensor using forward travelling light in polarisation maintaining fiber
First presented at the Fibre Optic Sensors II Conference, 31 March–3 April 1987, The Hague, The Netherlands. Proceedings of SPIE, Vol. 798, pp. 36–41
Reprinted courtesy of the authors and Society of Photo-Optical Instrumentation Engineers (SPIE)

Special shaped fibres for distributed sensors
First presented at the IEE Colloquium on Distributed Optical Fibre Sensing, 12 May 1986, London, UK, pp. 8/1–4.
Reprinted courtesy of the authors and The Institution of Electrical Engineers

Distributed temperature sensor based on liquid-core optical fibres
First published in the Journal of Lightwave Technology Vol. 1, No. 3, September 1983, pp. 498–509
Reprinted courtesy of the author and The Institution of Electrical and Electronic Engineers

Temperature measurement using intrinsic optical fibre sensors
First published in the Journal of Optical Sensors, Vol. 1, No. 2, 1986, pp. 101–114
Reprinted courtesy of the author and Taylor & Francis Ltd.

Distributed fiber-optic hot-spot sensors
First presented at the Fiber Optic and Laser Sensors IV Conference, 22–24 September 1986, Cambridge. MA, USA. Proceedings of SPIE, Vol. 718, pp. 218–224.
Reprinted courtesy of the authors and Society of Photo-Optical Instrumentation Engineers (SPIE)

Distributed optical fibre Raman temperature sensor using a semiconductor light source and detector
First published in the Electronics Letters, Vol. 21, No. 13, 20 June 1985, pp. 569–570

Reprinted courtesy of the authors and The Institution of Electrical Engineers

New fibre optic distributed temperature sensor
First presented at the Fibre Optic Sensors II Conference, 31 March–3 April 1987, The Hague, The Netherlands. Proceedings of SPIE, Vol. 798, pp. 131–136
Reprinted courtesy of the authors and Society of Photo-Optical Instrumentation Engineers (SPIE)

Temperature sensing by thermally-induced absorption in a neodymium-doped optical fibre
First presented at the Fibre Optic Sensors II Conference, 31 March–3 April 1987, The Hague, The Netherlands. Proceedings of SPIE, Vol. 798, pp. 46–48
Reprinted courtesy of the authors and Society of Photo-Optical Instrumentation Engineers (SPIE)

A distributed fibre temperature sensor using the optical Kerr effect
First presented at the Fiber Optic Sensors II Conference, 31 March–3 April 1987, The Hague, The Netherlands. Proceedings SPIE, Vol. 798, pp. 149–155
Reprinted courtesy of the authors and Society of Photo-Optical Instrumentation Engineers (SPIE)

Intrinsic fibre optic sensors
First published in the Journal of Optical Sensors, Vol. 2, No. 4, August 1987, pp. 269–279
Reprinted courtesy of the author and Taylor & Francis Ltd. Updated August 1989

Distributed strain sensing with a twin-core fibre optic sensor
First published in the ISA Transactions. Vol. 26, No. 1, 1987, pp. 7–10
Reprinted courtesy of the authors and The Instrument Society of America

Fiber-optic sensor for measuring external forces distributed along a fibre
First published in the Japanese Journal of Applied Physics, Part 2, Letters, Vol. 26, No. 5, May 1987, pp. L587–L589
Reprinted courtesy of the authors and the Japanese Journal of Applied Physics

Distributed fibre-optic interferometric sensor for the detection of pressure fluctuations in reservoirs
First published in the Journal of Lightwave Technology, Vol. 6, No. 9, September 1988, pp. 1435–1439
Reprinted courtesy of the authors and The Institute of Electrical and Electronics Engineers

A distributed fibre optic sensor based on cladding flourescence
First published in the Journal of Lightwave Technology, Vol. 7, 1989
Reprinted courtesy of the authors and The Institute of Electrical and Electronics Engineers

A plastic-clad silica fibre sensor for ammonia
First presented at the Optical Fibre Sensors Conference, 27–29 January 1988, New Orleans, LA, USA, pp. 369–372 (Technical Digest Series Vol. 2)
Reprinted courtesy of the authors and the Optical Society of America